Yong Sang
System and Measurements

Also of interest

Yong Sang

System and
Measurements

—

DE GRUYTER   Science Press
Beijing

Authors
Prof. Yong Sang
School of Mechanical Engineering
Dalian University of Technology
Dalian, Liaoning
China
Sang110@163.com

ISBN 978-3-11-062437-3
e-ISBN (PDF) 978-3-11-062439-7
e-ISBN (EPUB) 978-3-11-062479-3

Library of Congress Control Number: 2019951941

Bibliographic information published by the Deutsche Nationalbibliothek
The Deutsche Nationalbibliothek lists this publication in the Deutsche Nationalbibliografie;
detailed bibliographic data are available on the Internet at http://dnb.dnb.de.

Preface

This textbook is mainly applicable to bilingual or full English teaching of measurement technology in mechanical engineering. Measurement technology in mechanical engineering is usually a compulsory professional course for mechanical students. At present, there are only Chinese textbooks in China and no relevant textbooks in English or bilingual. I have been teaching measurement technology since 2007. In my teaching work, I pursue a teaching study with a rigorous attitude and to get a good teaching effect. Since 2012, I have offered a full English course in measurement technology. This course is mainly for international mechanical classes (all English) and regular classes (bilingual). In the training mode, I refer to the experience of well-known foreign universities and use a small class to teach. I have received three teaching-reform funding projects and won the Dalian University of Technology Award for Teaching Quality in 2015 and 2016, respectively. During teaching, I carefully prepared handouts, PPTs, problem sets and experimental tutorials. The self-edited lecture has been used for 6 years, and the teaching effect is good, which is unanimously recognized by the students. The publication of this textbook is based on the pre-edited English handouts. This textbook is conducive to expanding international cooperation and exchanges, introducing advanced foreign education, teaching concepts and models, deepening education and teaching reform, innovating training models and training a group of high-level international innovation talents with solid disciplines. The content of the textbook also has guiding significance for the design of mechanical equipment, especially the measurement and control system of the major equipment.

I have been engaged in teaching for 11 years and have adopted the Chinese textbook *Measurement Technology of Mechanical Engineering* (3rd Edition, only available in Chinese, Xiong Shibo, Huang Changyi) in Chinese teaching. My Chinese teaching experience has laid a solid foundation for the publication of the full English textbook. In writing this textbook, I have comprehensively considered the current situation of teaching in China and the characteristics of internationally renowned textbooks, and have compiled various chapters. It is expected to be compatible with Chinese textbooks and with the textbooks of similar universities in other countries. I thank the education-reform fund of Dalian University of Technology (no. JC2018030) and National Natural Science Foundation of China (no. 51975082). I also thank Prof. Wang Dianlong, Prof. Duan Fuhai, Prof. Zhang Jun and others who presented valuable suggestions. My graduate students Sun Weiqi, Sun Peng and Zhao Jianlong have made important contributions in checking grammatical errors and drawing high-resolution schematics. Editors from Science Press in China and De Gruyter Press in Germany offered a lot of help. Without their help, this textbook cannot be published smoothly.

Finally, I express my gratitude to my family. This work could not have been finished without their support, encouragement and love.

https://doi.org/10.1515/9783110624397-202

Contents

Preface —— V

1	**Introduction to measurement technology** —— **1**	
1.1	History of measurement technology —— 1	
1.2	Basic concepts of measurement technology —— 4	
1.3	Applications of measurement technology —— 7	
1.3.1	Industrial automation —— 7	
1.3.2	Equipment monitoring —— 11	
1.3.3	Quality performance inspection —— 12	
1.3.4	Building monitoring and security system —— 13	
1.3.5	Home and office automation —— 14	
1.3.6	Other field applications —— 16	
1.4	New trends in measurement technology —— 17	
1.4.1	New sensors —— 17	
1.4.2	Intelligent sensors —— 19	
1.4.3	Virtual instrument —— 20	
1.5	Manufacturers —— 21	

2   **General description of signal** —— **27**
2.1   Introduction —— 27
2.2   Classification of signal —— 27
2.3   Signal synthesis —— 33
2.4   Fourier decomposition of a sawtooth wave —— 35
2.5   Spectrum of periodic signal —— 37
2.5.1   Trigonometric Fourier series representation —— 37
2.5.2   Exponential Fourier series representation —— 44
2.6   Characteristics of periodic signal —— 48
2.7   Aperiodic transient signal —— 50
2.7.1   Definition of the Fourier transform —— 50
2.7.2   Properties of Fourier transform —— 55
2.7.3   Spectrums of typical aperiodic transient signals —— 61
2.8   Random signals —— 66

3   **Characteristics of measuring systems** —— **71**
3.1   Introduction —— 71
3.1.1   Classification of measuring instruments —— 71
3.1.2   Description of the characteristics —— 75
3.1.3   Linear system and its main properties —— 76
3.2   Static characteristics —— 79
3.2.1   Accuracy —— 79

3.2.2	Precision —— 81	
3.2.3	Range or span —— 81	
3.2.4	Linearity —— 82	
3.2.5	Sensitivity —— 82	
3.2.6	Resolution —— 84	
3.2.7	Bias —— 85	
3.2.8	Dead zone —— 85	
3.2.9	Threshold —— 85	
3.2.10	Hysteresis —— 86	
3.2.11	Drift —— 87	
3.2.12	Repeatability —— 87	
3.2.13	Stability —— 88	
3.2.14	Tolerance —— 88	
3.2.15	Error —— 88	
3.3	Dynamic characteristics —— 88	
3.3.1	Types of dynamic inputs —— 88	
3.3.2	Physical meaning of convolution —— 89	
3.3.3	Mathematical description —— 90	
3.3.4	Series connection and parallel connection —— 102	
3.3.5	Mathematical models of the measuring systems —— 103	
3.4	Step responses of measuring system —— 114	
3.4.1	The response of the first-order system —— 114	
3.4.2	The response of the second-order system —— 115	
3.4.3	Transient response specifications —— 117	
3.5	Nondistortion measurement —— 120	
3.6	Identification of dynamic characteristics —— 123	
3.7	Loading effect —— 127	
3.8	Anti-interference technology —— 131	
4	Signal conditioning, processing and recording —— 135	
4.1	Introduction —— 135	
4.2	Bridge circuit —— 135	
4.2.1	Classification —— 136	
4.2.2	Stress and strain —— 136	
4.2.3	DC (Wheatstone) bridge circuit —— 138	
4.2.4	AC bridge circuit —— 145	
4.3	Modulation and demodulation —— 153	
4.3.1	Definitions —— 153	
4.3.2	Amplitude modulation process —— 154	
4.3.3	Frequency modulation —— 162	
4.3.4	Typical applications —— 165	
4.4	Filter —— 169	

4.4.1   Classification ▬ 169
4.4.2   Real filter ▬ 170
4.4.3   Low-pass filter ▬ 173
4.4.4   High-pass filter ▬ 175
4.4.5   Band-pass filter ▬ 177
4.4.6   Filter bank ▬ 178
4.4.7   Typical applications ▬ 181
4.5   Sampling and aliasing ▬ 183
4.5.1   Sampling ▬ 183
4.5.2   Nyquist–Shannon sampling theorem ▬ 184
4.5.3   Aliasing ▬ 185
4.6   Displaying and recording signal ▬ 188
4.6.1   Analog and digital display ▬ 188
4.6.2   Pen strip chart recorder ▬ 189
4.6.3   Circular chart recorder ▬ 189
4.6.4   Paperless chart recorder ▬ 190

5   **Correlation measurement** ▬ 193
5.1   Introduction ▬ 193
5.2   Correlation analysis ▬ 193
5.2.1   Correlation coefficient ▬ 195
5.2.2   Autocorrelation function ▬ 197
5.2.3   Cross-correlation function ▬ 200
5.3   Spectral analysis ▬ 203
5.4   Digital image correlation measurement ▬ 204
5.4.1   Introduction ▬ 204
5.4.2   Basic principle of two-dimensional DIC ▬ 205
5.4.3   Applications of two-dimensional DIC ▬ 209

6   **Sensors** ▬ 211
6.1   Introduction ▬ 211
6.2   Temperature sensors ▬ 211
6.2.1   Thermocouple ▬ 212
6.2.2   Resistance temperature detector ▬ 213
6.2.3   Thermistors ▬ 216
6.2.4   Integrated circuit temperature sensors ▬ 218
6.2.5   Infrared temperature sensors ▬ 219
6.2.6   Bimetallic temperature devices ▬ 220
6.3   Pressure sensors ▬ 221
6.3.1   Strain gauge pressure sensors ▬ 221
6.3.2   Piezoelectric pressure sensors ▬ 222
6.3.3   Piezoresistive pressure sensors ▬ 223

6.3.4   Capacitive pressure sensors —— 225
6.3.5   Variable reluctance pressure sensors —— 226
6.3.6   Potentiometric pressure sensors —— 227
6.3.7   Bourdon tube pressure sensors —— 228
6.4   Flow sensors —— 230
6.4.1   Differential pressure flowmeter —— 230
6.4.2   Turbine flowmeter —— 233
6.4.3   Vortex flowmeter —— 234
6.4.4   Thermal flowmeter —— 235
6.4.5   Doppler ultrasonic flowmeter —— 237
6.4.6   Transit-time ultrasonic flowmeter —— 238
6.4.7   Laser Doppler anemometer —— 239
6.5   Displacement sensors —— 240
6.5.1   LVDT displacement sensor —— 240
6.5.2   Capacitive displacement sensor —— 242
6.5.3   Inductive displacement sensor —— 243
6.5.4   Laser displacement sensor —— 244
6.5.5   Resistive displacement sensor —— 246
6.5.6   Ultrasonic displacement sensor —— 247
6.5.7   Grating displacement sensor —— 248
6.6   Velocity sensor —— 250
6.6.1   Piezoelectric velocity sensor —— 250
6.6.2   Tachometer —— 251
6.6.3   Electrodynamic velocity sensor —— 252
6.7   Acceleration sensors —— 253
6.7.1   Piezoelectric accelerometer —— 254
6.7.2   Strain gauge accelerometer —— 255
6.7.3   Variable reluctance accelerometer —— 256
6.7.4   MEMS accelerometer —— 257
6.7.5   Angular accelerometer —— 258
6.8   Force/torque sensors —— 259
6.8.1   Strain gauge force/torque sensor —— 259
6.8.2   Piezoelectric force sensor —— 262
6.9   Level sensors —— 263
6.9.1   Ultrasonic level sensor —— 263
6.9.2   Radar level sensor —— 264
6.9.3   Capacitance level sensor —— 266
6.9.4   Resistance level sensor —— 267
6.9.5   Float level switch —— 268
6.9.6   Rotary paddle level switch —— 269
6.10   Image sensors —— 270
6.10.1   CCD sensor —— 271

6.10.2	CMOS sensor — **272**	
6.11	Proximity or presence sensors — **273**	
6.11.1	Photoelectric proximity sensor — **273**	
6.11.2	Ultrasonic proximity sensor — **274**	
6.11.3	Inductive proximity sensor — **275**	
6.11.4	Capacitive proximity sensor — **276**	
6.12	Color, contrast and luminescence sensors — **278**	
6.12.1	Color sensor — **278**	
6.12.2	Contrast sensor — **279**	
6.12.3	Luminescence sensor — **280**	
6.13	Environmental sensors — **281**	
6.13.1	Sound level sensor — **281**	
6.13.2	Humidity sensor — **282**	
6.13.3	CO_2 sensor — **283**	
6.13.4	Particulate matter sensor — **285**	
6.13.5	Volatile organic compound sensor — **286**	

References — **289**

Index — **297**

1 Introduction to measurement technology

System and measurements is an important technological course in mechanical engineering education, which provides students with basic concepts and the fundamental principles of measurement such as experimental methods, calibration techniques, signal conditioning methods, real-time acquisition and processing of field data, selected sensors for mechanical engineering and display of data results. After learning this course, students will acquire the following abilities:
(1) An ability to grasp the methods of data acquisition and data processing
(2) An ability to analyze instruments with different input conditions (zeroth-order system, first-order system and second-order system)
(3) An ability to apply the sampling and signal conditioning principles learned in the classroom to measuring instruments
(4) An ability to use software and hardware for data acquisition
(5) An ability to synthesize an individual data acquisition project
(6) An ability to select the suitable sensors for temperature, displacement, force, acceleration, pressure and flow and the corresponding instrumentations
(7) An ability to describe experimental facilities and experimental procedures, collect data, write reports, analyze data and represent field data
(8) An ability to complete group work and share results with others

In this chapter, students will know the history of measurement technology, basic concepts of measurement technology, applications of measurement technology, new trends of measurement technology and world's famous manufacturers in the measurement field.

1.1 History of measurement technology

The emergence of measurement technology can be traced back to the distant ancient times, and the development of barter trade promoted the measurement of weight and the emergence of scale. Original measurement technologies are also used in other areas, such as measuring buildings and land, measuring calendar time and measuring astronomical geography. In ancient times, people can accurately measure the angle of refraction of light and determine the meridian arc of the earth. In the sixteenth and eighteenth centuries, the rapid development of physics greatly facilitated the advancement of measurement technology. Physics based on experiments relies entirely on measurement technologies, and has invented some advanced measuring devices such as microscopes, barometers and thermometers. At the turn of the seventeenth century, measurement accuracy was further improved, using an accurate astronomical measurement device that enabled Johannes

https://doi.org/10.1515/9783110624397-001

Kepler to determine the rotation of the planet in an elliptical orbit. Great scientists such as Galileo Galilei and Isaac Newton pioneered measurement theory and developed many instruments based on physical phenomena.

With the development of steam engines, machining accuracy requirements of mechanical parts are getting higher and higher, which greatly increases the level of measurement technology. In the field of machining, Vernier caliper was invented by French mathematician Pierre Vernier in 1631, which greatly improved the measurement accuracy of the parts. At the same time, the emergence of other high-precision dimensional measuring devices has been promoted. In the nineteenth century, people established the principles of measurement technology and metrology theory, and used the metric system to unify the measurement standards in machinery manufacturing. Johann Carl Friedrich Gauss made significant contributions to measurement technology and developed the least squares method, random error theory and unit absolute system. Measurement technologies are not only applied in scientific research, but also used industrially for measurement of heat and electricity. At the beginning of the twentieth century, the development of measurement technology entered a new stage: advanced technologies such as electronics, optics and computers have also been introduced into the field of mechanical measurement achieving high-precision measurement of many physical quantities, and the instrument manufacturing industry has been further improved. After the Second World War, the development of measurement technology has made a leap forward. With the advent of automation and computer technology, measurement technology has become the core of control measurement and data acquisition.

Measurement technology is very important for all scientific and technical fields (mechanical, electrical, optical and acoustic signals, etc.). It is the most important part of mechanical equipment in every manufacturing process. Modern measuring equipment is more powerful. The measurement results can be automatically recorded and mathematically processed. Remote automatic control can also be implemented using the network. The process of making measurements with modern measuring devices involves purposefully converting the measurand into the most convenient form for human or machine perception. The electrical measuring instrument such as ammeter or voltmeter can be used to convert the electrical quantity into a mechanical displacement indicating the magnitude of the current or voltage. Some mechanical measuring devices such as calipers, micrometers, spring scales, mercury thermometers and spring pressure gauges work in this way. Further developments in measurement technologies can convert measurements into electrical quantities (such as current, voltage and frequency) rather than mechanical displacement. It is more convenient for all subsequent operations, including the transmission of measurement data, recording and mathematical processing. The main advantages of electrical measurement technology are high sensitivity, low inertia of electrical equipment, the ability to measure many different properties simultaneously and the ability to measure slow and fast changes. Electrical measuring devices can also easily transport data over long distance wires and convert the measurement results into signals during the monitoring process.

Modern measurement technologies have a lot to offer in the measurement of mechanics, optics, acoustics, thermal and physical chemistry. Electromagnetic measurement, radio measurement, frequency measurement and radiation measurement can be realized. For different measurement scales, the measurements are used in a completely different way. Measurement technologies can be used to measure the mechanical stress on the surface of the part, the vibration of the machine and the random nature of the machining accuracy (correlation function, power spectrum). In recent years, with the development of computers, digital measurement technology has emerged, which converts analog quantities into digital quantities, and it is very easy to transfer measurement data into computers through the form of field buses. A number of new measuring devices with high integration, intelligence, information and modernization have been developed. These devices or technologies include fiber-optic grating measuring device based on optical fiber material, low-light image measuring device that breaks the application of human visual limit, intelligent measuring device that can take full advantage of big data, femtosecond laser detection technology based on femtosecond technology, quantum measurement technology widely used in medical and health fields, and seismic transverse wave measurement technology applied in geological exploration, petroleum and coal field exploration.

The current development trends of measurement technology are very clear and the most important features are as follows: (1) measurement accuracy, reliability and measurement speed are significantly improved, and measurement error is reduced to less than 0.01%; (2) measuring instruments toward miniaturization and low power consumption; (3) expanding the measurement range of measuring equipment, including measuring the quantity that cannot be measured before; (4) the digital instrumentation equipment has achieved unprecedented development, and the development of embedded single-chip microcomputer has enhanced the measurement function of instrumentation; (5) the development of measurement technology is moving toward network and big data; and (6) the latest scientific results have facilitated measurement devices based on new working principles.

Many measuring instruments are currently used in various fields, and measurement technologies are part of the curriculum of technical higher education institutions in the world for students to learn. Many scientists and researchers are engaged in the research of measurement technology, and their latest research results are published in the professional journals of measurement technology for peer learning. Worldwide, the following journals devoted to the problems of measurement technology are published regularly: *IEEE Transactions on Instrumentation and Measurement* (published by the Institute of Electrical and Electronics Engineers Inc., USA), *IEEE Instrumentation and Measurement Magazine* (published by the Institute of Electrical and Electronics Engineers Inc., USA), *IET Science, Measurement and Technology* (published by the Institution of Engineering and Technology, UK), *Measurement and Control* (UK) (published by SAGE Publications Ltd, UK), Measurement Science and Technology (published by the Institute of Physics Publishing, UK), *Measurement, Journal of Dynamic Systems, Measurement and Control, Transactions of the ASME* (published by the American Society of Mechanical Engineers, USA), *Instrumentation Science and Technology* (published by Taylor and Francis Inc., USA), *Journal of Instrumentation* (published by the Institute of Physics Publishing, UK), *Optoelectronics, Instrumentation and Data Processing* (published by Allerton Press Incorporation, USA), *Instrumentation Mesure Metrologie* (published by Lavoisier, France), *Instruments and Experimental Techniques* (published by Maik Nauka Publishing/Springer SBM, Russia), *Modelling, Measurement and Control A/B/C* (published by AMSE Press, France), *Mediterranean Journal of Measurement and Control* (published by Softmotor Ltd., UK), *Journal of Applied Measurement* (published by JAM Press, USA), *Measurement Science Review* (published by Slovak Academy of Sciences – Inst. Measurement Science, Germany), *Measurement Techniques* (published by Springer New York LLC, USA), *Metrology and Measurement Systems* (published by Polish ACAD Sciences Committee Metrology and Res Equipment, Poland), *Transactions of the Institute of Measurement and Control* (published by SAGE Publications Ltd, UK), *Flow Measurement and Instrumentation* (published by Elsevier Ltd, UK), *Measurement: Journal of the International Measurement Confederation* (published by Elsevier B.V., Netherlands), *Radiation Measurements* (published by Elsevier Ltd, UK) and others. Monographs, handbooks and booklets both in specific fields and on general problems of measurement technology and instrument-making are also published.

1.2 Basic concepts of measurement technology

Measurement technology is indispensable in the production process, experimentation and scientific research, and it plays the role of human senses. It is an important part of automated machining. The state and data of the production process are obtained through measurement technology. In many developed countries, despite the high level of instrument manufacturing technology, measurement technology is still highly valued.

Figure 1.1 shows an alcohol-in-glass thermometer. This alcohol-in-glass thermometer consists of a glass capillary labeled with Celsius or Fahrenheit. Alcohol expands or contracts with temperature to increase or decrease the level of liquid, thereby indicating the temperature. This thermometer is a simple system with only one module and no electronic conversion and analysis circuitry in the system. The change in temperature will be translated directly into a change in liquid volume. Similarly, in a mercury-in-glass thermometer, the mercury liquid responds to its expansion and contraction, and the temperature change is also marked on the scale in the form of a liquid level.

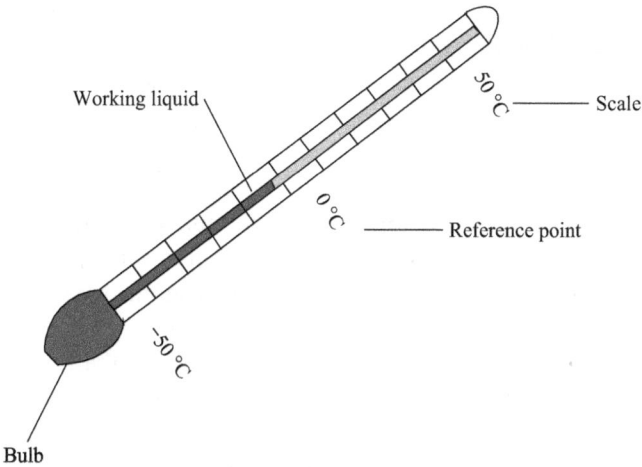

Figure 1.1: Alcohol-in-glass thermometer.

Figure 1.2 shows two sound level meters (UNI-T Group Co., Ltd. China). They measure the sound intensity through airborne sounds. The sound level meter in the figure is usually a hand-held instrument with a wide bandwidth microphone. The diaphragm of a high-sensitivity microphone responds to rapid changes in air pressure caused by high-speed sound waves. The left sound level meter (in Figure 1.2) features a compact, lightweight design. Its design is perfect for the palm of the

Figure 1.2: Sound level meter (UNI-T Group Co., Ltd. China).

hand. The meter is red with a gray frame screen and black buttons. The right sound level meter (shown in Figure 1.2) has a larger LCD. The large LCD provides large, clear digital readings and a fast-responding analog output that displays sound level readings. Such devices typically have multiple signal outputs and can record maximum/minimum values over time.

By comparing Figures 1.1 and 1.2, you can see that the alcohol-in-glass thermometer is simple with low precision and cannot be used for automatic measurement. However, in order to improve measurement accuracy and automation, sound level meter working together with other automatic device can convert the measured physical quantity into electricity and then output the electrical signals.

Typically, the measurement system is divided into three sections: the sensor (detector/ transducer), the signal conditioner, and the indicator/recording device, as shown in Figure 1.3. A detector is a device for detecting and responding to the physical measurand. The transducer is actually a smart device that can receive a signal of certain energy and change it into another, more convenient signal form, which is used to convert the measurand into a noted electrical signal that is easier to measure. The electrical signal output obtained from the transducer can be easily read or can be used as an input for the controller. The input form of the electric transducer can be strain, displacement, temperature, speed, flow and so on, and the output form can be voltage, current or changes of inductance, resistance and capacitance. The output signal allows for easy calibration to accurately measure the actual input value. For example, here is a system for measuring force. First, the detector/sensor converts the change in force into a change in displacement. Then, it converts the variation of the displacement into a variation of the resistance, which makes it easier to measure. Finally, the resistance change can be converted to a voltage change by applying a bias current across the resistor using a current source to produce an output voltage.

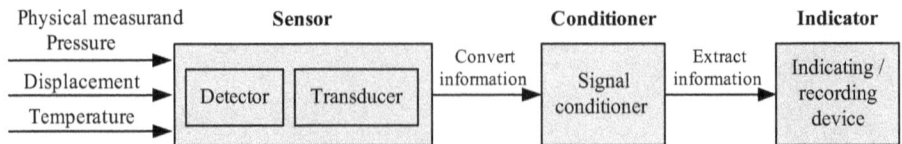

Figure 1.3: A measurement system.

A signal conditioner is an important special device that converts an electronic signal into another form of signal. Its main purpose is to change the signal that is difficult to obtain using traditional instruments into a more readable form. Many functional components can be used when performing this conversion, including amplification, filtering, electrical isolation, A/D converters, integrators and differentiators. Amplification: When the signal under test is used for amplification, the total amplitude of the signal increases. For example, a −2 to 2 mV signal can be converted into a −5 to 5 V signal during strain gauge measurement. Filter: In signal processing, a filter is a device that removes unwanted components or features from the signal. Electrical isolation: Electrical isolation typically destroys the current path from the input signal to the output signal, and it can be assumed that there is no actual physical connection between the input signal and the output signal. By disconnecting the current path between the input signal and the output signal, you can prevent unwanted signals on the input from entering the output. Especially when measuring under high pressure, isolation is required. In fact, isolation can also be used to prevent the generation of ground loops. A/D converter: In electronic devices, analog-to-digital converters (ADC or A/D) are systems that convert traditional analog signals into modern digital signals. Using A/D converter, the light entering the digital camera is converted into a digital signal. Integrator: The integrator is an important component in measurement and control applications, and its output signal is the integral of its input signal over time. It implements cumulative input over a specified time to produce an output. Differentiator: In electronics, a differentiator is a circuit designed. The output value of the differential circuit is roughly proportional to the magnitude of the rate of change of the simultaneous input signal. Active differentiator uses some amplifiers and the passive differential circuit consists only of resistors and capacitors.

1.3 Applications of measurement technology

Measurement technology products based on various mechanical and physical phenomena play key roles in industrial automation, aerospace engineering, mechanical engineering, electrical engineering, petrochemical engineering, scientific experiments, product development, production monitoring, quality control and others.

1.3.1 Industrial automation

In the field of industrial automation, typical applications of measurement technology including robot, automated guided vehicle (AGV) and manufacturing process monitoring are described in this book.

1.3.1.1 Robots and sensors

In the industrial and automation fields, robots require sensors to provide all the information they need to perform their operations. Adding sensors to different robots greatly increases the adaptability of the robot. Sensors are tools for robots to know the state of the world around them. Figure 1.4 shows some current well-known robots

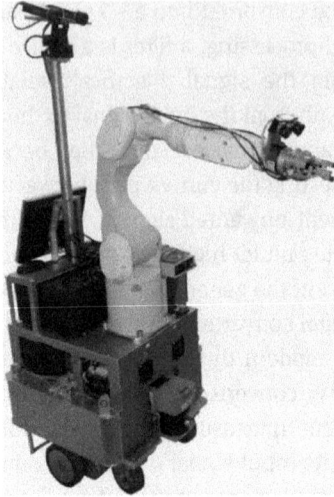

Jido MP-L655 (LAAS-CNRS, France)

Robot Maggie (The Robotics Lab, Spain)

SONY AIBO ERS-7 (SONY, Japan)

X-80H Mobile Robot (Dr Robot,Canada)

Figure 1.4: Well-known robots in the world.

in the world, including Jido MP-L655 (France), Robot Maggie (Spain), SONY AIBO ERS-7 (Japan) and X-80H Mobile Robot (Canada). Many special robot sensors can be found on these robots, including light sensor (light sensor is used to detect light and generates voltage differences), sound sensor (this sensor detects sound and returns voltage proportional to the sound level), temperature sensor (the sensor detects the temperature around the robot), contact sensor (the sensor requires physical contact with other objects to sense the contact force), proximity sensor (the sensor detects the presence of nearby objects, without any physical contact including infrared transceiver, ultrasonic sensors and photoresistors), distance sensors (ultrasonic distance sensor, infrared distance sensor, laser range sensor, encoder and stereo camera are best suited for distance measurement), pressure sensor (pressure sensor is used to measure pressure), tilt sensor (the sensor measures the tilt of the object), the navigation/positioning sensor (the navigation/positioning sensor is used to approximate the position of the robot including GPS and digital magnetic compass), acceleration sensor (accelerometer is a measuring device to measure acceleration) and gyroscope (a gyroscope is used to measure the rate of rotation around a particular axis). It is not possible to list all available sensors in this book. Apart from those sensors mentioned earlier, there are many other sensors used for specific applications.

1.3.1.2 Automatic guided vehicle

The AGV is an efficient mobile robot that follows signs or special wires on the ground. AGVs designed by HI-TECH RS (India) and SSI Schaefer Weasel (Germany) are shown in Figure 1.5. These robots are widely used in the manufacturing industry to move parts and materials around production lines, manufacturing facilities or warehouses. The first AGV was introduced to the market by Barrett Electronics Corporation in Northbrook (USA) in the 1950s, and it was just a tow truck, following the wires on the floor instead of the rails. A new type of AGV emerged on the market that followed the invisible UV markings on the floor rather than being pulled by the chain. The first such system was deployed at Willis Tower (Chicago, USA) to deliver mail throughout its offices. The technology has become more complex, and today's automated vehicles are primarily laser navigation, called laser-guided vehicles (LGV). During the automation process, the programmed LGV can communicate with other special robots to ensure that the product passes through the warehouse. Today, AGV plays a key role in the design of new plants and warehouses, safely transporting goods to legitimate destinations.

AGV (HI-TECH RS, India)   AGV (SSI Schaefer Weasel, Germany)

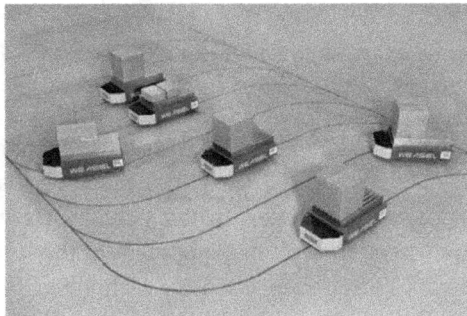

Figure 1.5: Automatic guided vehicles.

1.3.1.3 Manufacturing process monitoring

A lot of research has been done to monitor the manufacturing process. The performance innovation of CNC machine tools is attributed to the development of advanced feed drives and high-precision spindle motion controllers as well as machine and high-precision coolant temperature control capabilities. Advances in CNC machine tools require professional machine tool operators and part programmers to use measurement techniques to determine actual machining and proper operating parameters, such as how to select the feed rate, spindle optimum speed, tool and the best depth of cut. Measurement of state information such as force, noise, distance and proximity during manufacturing becomes very important. Cutting force measurements can be used to monitor the condition of the tool and avoid breakage during machining. The dynamic signal of the cutting force contains information about the chip formation mechanism and is a very important monitoring parameter. Figure 1.6 shows the measurement principle of the cutting force during the manufacturing process. In the manufacturing process monitoring, vibration information can be measured with an accelerometer and laser interferometers can be used to detect tool wear, breakage and chattering. Sound sensors can be used to monitor vibration of the motor, machine tool spindles, gearboxes, pumps and so on.

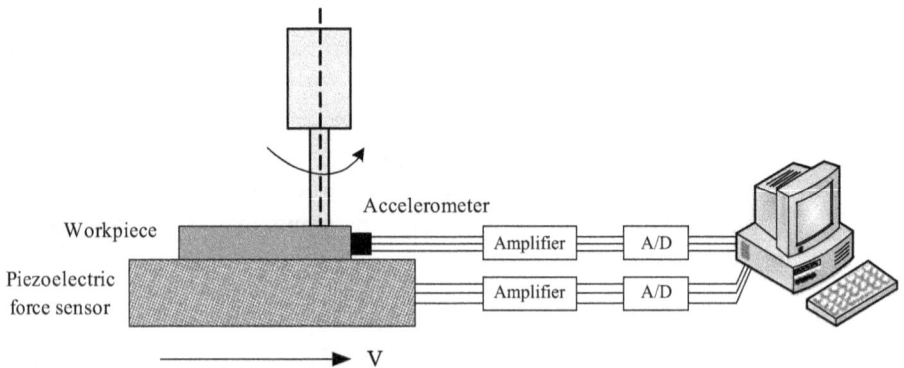

Figure 1.6: The setup of force and vibration measurements during face milling.

A digital factory is a visible virtual copy of a current or future plant, including all production lines, drive parts, machines, auxiliary equipment, installation and maintenance services. In the virtual factory, all the production lines, production equipment and testing equipment of the factory can be fully 3D visualized. Digital factories offer comprehensive technology-based service information of seamlessly integrated software and hardware to support manufacturing factories to increase the efficiency and flexibility of manufacturing parts and shorten the time to market for production. Figure 1.7 provides an efficient digital factory solution for smart

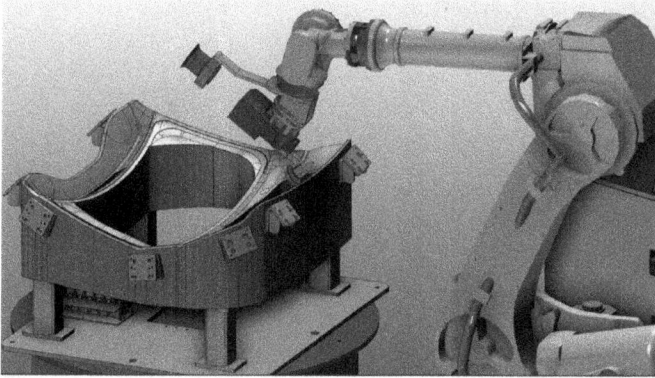

Figure 1.7: Digital factory (CENIT AG, Germany).

manufacturing (CENIT AG, Germany). The construction of the digital factory is inseparable from the development of measurement technology. The platform is based on mechatronics simulation of resources and components, enabling system integrators and manufacturers to simulate real-world production environments. The fourth industrial revolution can be achieved through digital factories. It is reported that the survival of the automotive industry depends on digital factory that can save 10–20% of the cost of production. The current well-known vendors that offer digital factory integration systems are CENIT AG (Germany), Siemens AG (Germany), SAP (Germany), ABB (Germany), Festo (Germany), IBM (USA) and GE (USA).

1.3.2 Equipment monitoring

In most nonmanufacturing control processes, critical equipment needs to be monitored (24 h) and controlled or protected in real time. A sudden failure of the equipment can cause serious process disturbances, downtime and severe damage to the entire process. To ensure the reliability of the entire control process, real-time online monitoring becomes especially important. Real-time monitoring can use various cost-effective sensors to achieve an automatic diagnosis and early warning. In addition, remote process monitoring can be achieved by means of network technology. Process monitoring can increase equipment availability, reduce unplanned downtime and minimize unplanned maintenance activities. The process monitoring operation status involves the entire production process of many industries such as power, metallurgy and petrochemical/chemical industry. Figure 1.8 shows an electric power monitoring (Sensor Synergy, USA), and an unattended real-time and automatic monitoring system is established in this system to obtain 24-h continuous online monitoring data in time.

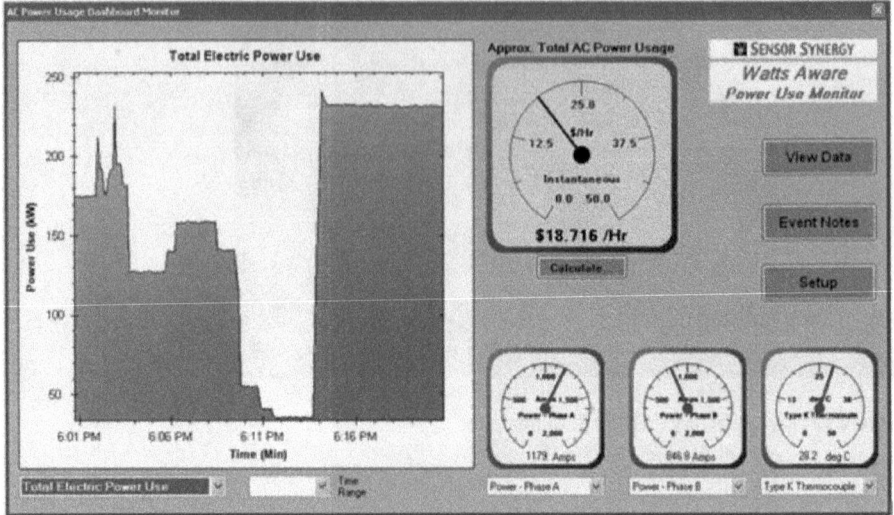

Figure 1.8: Electric power monitoring (Sensor Synergy, USA).

1.3.3 Quality performance inspection

Product quality is a topic of greatest concern to consumers, and measurement of product quality needs to be achieved through measurement technologies. For example, when equipment (automobiles, machine tools, etc.) and parts (motors, engines, etc.) are ready for sale, factory inspections must be performed to measure their performance and quality. There are often many safety hazards in vehicles that do not use measurement technology correctly or do not have the relevant measurement technology. Once many automobiles are found being ignored, the manufacturer needs to recall all the automobiles and re-examine them, greatly increasing production costs. Therefore, it is very important to continuously monitor the quality of the product during the manufacturing process. In the field of automobile manufacturing, online quality inspection is often used, and some laser radars are installed on six-axis robot arms located on both sides of the production line. This type of robot is very common in automotive production facilities and is very sturdy. Lidar is a noncontact measurement system with an accuracy of <0.1 mm for online quality inspection. Figure 1.9 shows the schematic of a car manufacturing factory (Hitachi, Japan). An automobile factory inspection may require more than 100 sensors, each of which is individually targeted to different functions on the vehicle. These features include seat height and comfort, backrest angle, oil and gas usage, air conditioning, heating system, ventilation system, noise level, vehicle speed, glass transmittance, acceleration, steering, engine characteristics, brakes, locks and tires.

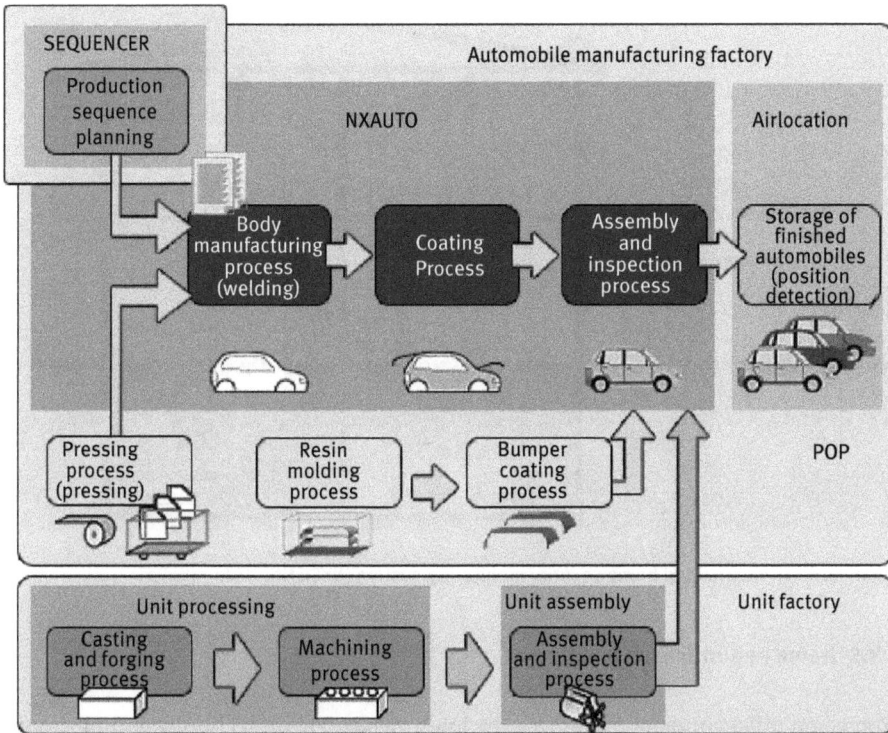

Figure 1.9: Automobile manufacturing factory (Hitachi, Japan).

1.3.4 Building monitoring and security system

Measurement technology is used in many buildings to ensure good operation and intelligent management. Building monitoring and security systems provide 24-h home security monitoring to prevent burglary, fire and carbon monoxide hazards while monitoring other critical conditions such as indoor and outdoor temperatures and ventilation. Figure 1.10 shows the fire alarm and home security monitoring system (NYCONN, USA). The door sensor is used to detect the opening of the door. The water sensor is used to trigger an alarm when detecting water. The smoke sensor is used to detect smoke and the heat sensor is used to detect rapid temperature rise, and the motion sensor is used to detect a person moving in a special area and the camera is used to monitor live video for activity. Building monitoring and security systems typically include access control subsystems, intercom systems, CCTV (closed-circuit television camera) subsystems and security alarm subsystems. These subsystems are connected to the coordination system by means of software, computer networks and structured cables.

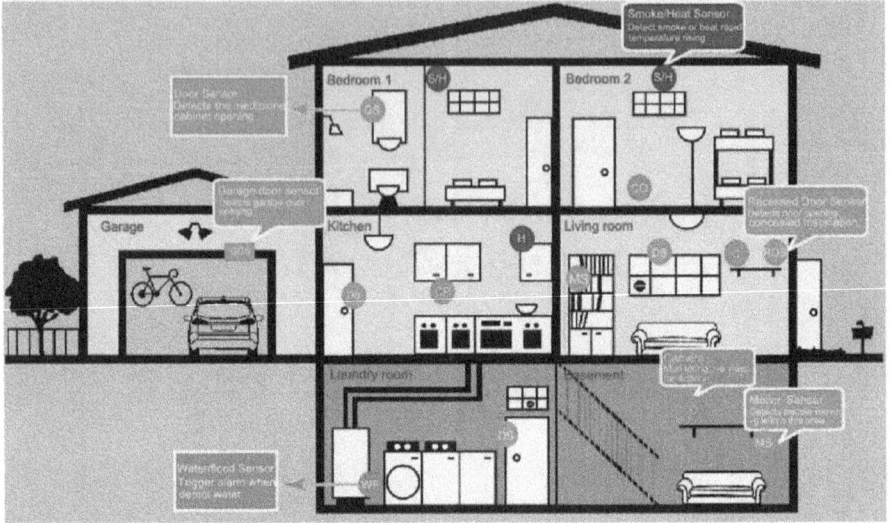

Figure 1.10: Fire alarm and home security monitoring system (NYCONN, USA).

1.3.5 Home and office automation

Home and office automation utilizes the latest measurement technology to fully control electric curtains, lighting, blinds, air conditioners and appliances. Figure 1.11 shows the home and office automation system (AVCS, Kenya). This automation system also integrates many other systems to provide enhanced comfort and security,

Figure 1.11: Home/office automation system (AVCS, Kenya).

such as card access systems, fire alarms, CCTV, audio and video systems and garden watering systems. Curtain control is an ideal solution for blinds and curtains in the bedrooms, living rooms and office. Climate control is a good system for improving energy efficiency, and multifunction control can control all lighting. TV/DVD, ventilation and others can be achieved with an intelligent controller. The office automation system is a combination of hardware, software and information systems to handle office activities at all levels of the office, including word processing, electronic documents, email, information exchange, data storage, data and voice communications. Some office automation devices with fingerprint and face recognition technology are shown in Figure 1.12.

Fingerprint ID access
(Apple iphone, USA)

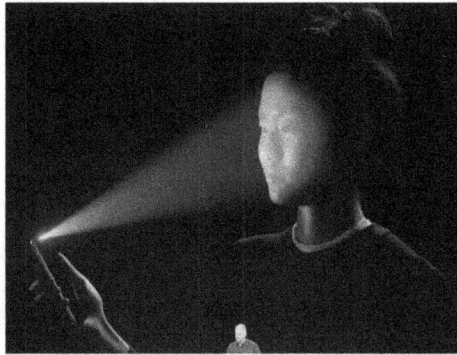

Face ID access
(Apple iphone X, USA)

Fingerprint door lock
(SekureID, USA)

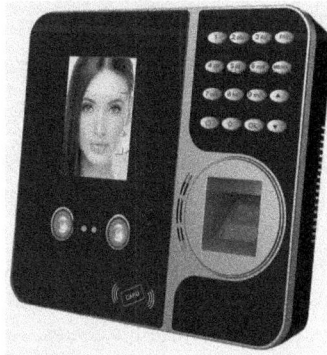

Face recognition time Att.
(KO Tech Co., China)

Figure 1.12: Office automation devices.

1.3.6 Other field applications

Measurement technologies are also widely used in other fields, such as agriculture, aerospace, transportation and medical treatment.

1.3.6.1 Agriculture

Measurement technology provides long-term, independent monitoring of meteorological parameters for almost all agricultural production applications. The monitoring process will use a number of meteorological sensors to monitor the following parameters: wind direction, wind speed, wind vector, solar radiation, atmospheric temperature, soil temperature, water temperature, relative humidity, precipitation, dew point, blade humidity, atmospheric pressure and steam pressure. For example, in Washington, it was reported that stink pest (in Figure 1.13) became an agricultural concern for farmers, and agricultural experts had attempted to develop a monitoring system to know when stink pests were coming in.

Agriculture   Aerospace   Transportation   Medical treatment

Figure 1.13: Other applications.

1.3.6.2 Aerospace

Measurement technology is used in aerospace applications, from the development of individual components to the use of aircraft or rockets. High reliability and high quality put forward higher requirements for measurement technology. Modern aircraft manufacturers face challenges such as unstable weather conditions, temperature, atmospheric pressure and heavy structural loads, which require rigorous testing with high-precision sensors to meet environmental challenges, weight specifications and flight launch issues. Advanced measurement technology guarantees safety, durability, robustness and economy in aerospace equipment.

1.3.6.3 Transportation

Transportation is the movement of humans, animals and goods from one place to another. Transportation modes include air, rail, waterway, road and pipeline. The continuous innovation of test technology, coupled with microchips, radiofrequency identification technology and low-cost smart sensing technology, has promoted the emergence of the global intelligent transportation system. The sensors used in transportation are as follows: infrastructure sensors are indestructible equipment for installing or embedding roads. Vehicle sensing systems can use video vehicle detection technology, inductive loop detection technology, audio detection technology or Bluetooth detection technology to increase continuous monitoring of the vehicle. Truck transportation systems need to measure average traffic speed, traffic volume, congestion delays, vehicle operating costs, parking supply and so on.

1.3.6.4 Medical treatment

Measurement technology is the foundation for more reliable and effective health diagnosis and treatment. Doctors and medical scientists rely on medical equipment to get the job done right. Measurement technologies are not only important in diagnosis but also important in the treatment itself. A variety of advanced diagnostic equipment include human B-type ultrasonic diagnostic equipment, CT or CAT scanner, X-ray scanner, pacemakers, insulin pumps, operating room monitors, defibrillators and brain stimulators.

1.4 New trends in measurement technology

The new trends in measurement technology are mainly manifested in three aspects: new sensors, intelligent sensors and virtual instrument.

1.4.1 New sensors

New sensors have been developed based on newly discovered reactions in biology, physics and chemistry. Biosensors are primarily analytical devices for detecting analytes. They typically consist of bioconductor components, biometric sites and electronic systems, including signal amplifiers, displays and processors. The general purpose of designing biosensors is to perform fast and convenient inspections. There are many potential applications for various types of biosensors. For example, a blood glucose biosensor is designed to check the concentration of glucose. Figure 1.14 shows the detection mechanism of the thermoelectric enzyme sensor (Chuhong Lin, Enno Kätelhön, etc. 2017). The enzyme moves freely in the liquid while continuously catalyzing the entry of substrate molecule A into product molecule B. If the enzyme is located near the surface of the charged sensor, electrons can jump from product molecule B to the sensor surface, converting B to another product C. The sensor uses a thin-film thermopile associated with an immobilized enzyme. The thermopile detects an increase in temperature that occurs when the enzyme catalyzes a particular chemical substrate. In thermoelectric enzyme

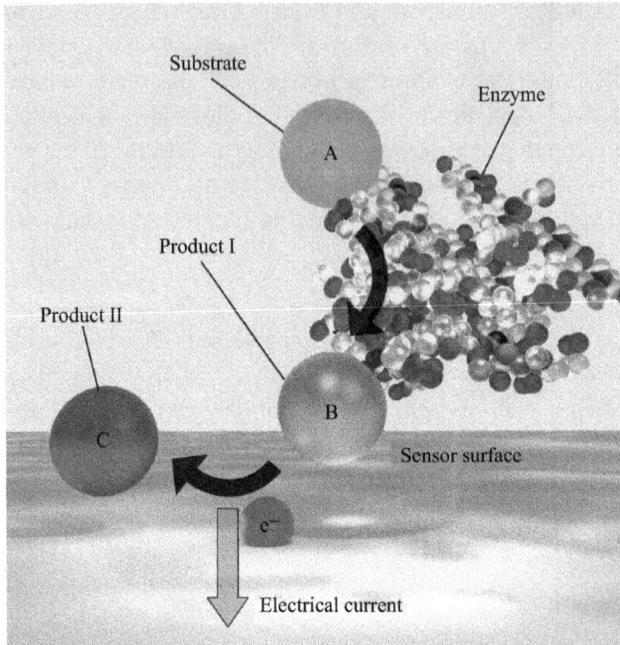

Figure 1.14: Detection mechanism of thermoelectric enzyme sensor (Chuhong Lin, etc.).

sensors, the enzyme is typically immobilized on the surface of a current measuring electrode that reacts with the substrate and produces a current that is dependent on the analyte concentration.

The development of nanotechnology has promoted the development of biosensors reported by the Centre for NanoHealth, Swansea University. Biosensor diagnostics are based on biofunctionalized semiconductor devices and are a milestone in the development of ultrasensitive sensors for early detection of disease biomarkers. Nanochannel devices and nanowires will play a vital role in the future of biosensor applications. There is also a need to fabricate silicon nanowires, graphene channels and metal oxide nanowire sensors. These nanoscale materials have excellent electronic properties, and sensors are expected to be the next generation of next-generation electronics and medical diagnostics. These sensors are capable of detecting ultra low concentration disease biomarkers. In this way, doctors can not only detect disease biomarkers in the blood, but also detect them in saliva or urine, providing minimal invasive testing.

Distributed fiber-optic sensors have been further developed in recent years. The new sensor is based on fiber optics, which quickly detects the structural health of bridges, buildings and dams. In fact, scientists are very interested in using distributed sensors to continuously monitor the structural health of various important building structures. The newly developed fiber-optic distributed sensor has 1 million sensing points, providing a faster structural health check than it currently does.

Fiber-optic distributed sensor is an excellent choice for monitoring infrastructure because it can be used in very harsh working environment and in area where there is a lack of nearby power. If you place a single fiber along the length of the road, structural changes along any sensing point of the fiber will result in a detectable change in light propagating along the fiber. According to a report by Dominguez-Lopez of the University of Alcala and the Swiss Federal Institute of Technology and his colleagues, the first fiber-optic distributed sensor was introduced, which can sense 1 million transmissions (strain and temperature) on a 10 km fiber within 20 min.

Novel printed/flexible (P/F) sensors were invented by Imec and Holst Centre. Imec (Belgium) is a world's leading innovation and research center in the field of digital technology and nanoelectronics and the Holst Center (established by Imec and TNO of the Netherlands) is a completely independent research and development center dedicated to the development of flexible electronic technology and wireless autonomous sensor technology. These P/F sensors include a solid-state ion-selective electrode for monitoring pH, Cl, Na and K and sensor labels based on ultra-thin (<150 μm) polyester film for humidity, temperature, chemicals and gases.

Quantum entanglement is a special physical phenomenon that was first recognized by Einstein, Rosen, Podolski and Schrödinger. It has been found that the quantum states of multiparticle systems cannot be decomposed into the product of single-particle wave functions, even if the particles are separated by a large separation distance. The entangled state has been generated in the laboratory and uses the Bell inequality to test the contradiction between classical local hidden variable theory and quantum mechanics. A research group from China demonstrates the satellite-based distribution of entangled photon pairs to two locations separated by 1,203 km on the Earth in 2017, through two satellite-to-ground downlinks with a summed length varying from 1,600 to 2,400 km. The results illustrate the possibility of a future global quantum communication network.

1.4.2 Intelligent sensors

As the name implies, an intelligent sensor is a sensor device that performs many intelligent functions. Intelligent sensors have the abilities to self-test, self-verify and self-adapt, as well as self-identify. Intelligent sensors can manage their functions through the stimulation of external functions. This shows that the intelligent sensor has an advanced learning, adaptive and signal processing architecture, all in one integrated circuit. Intelligent sensors require dedicated hardware called signal conditioning circuitry to monitor and control themselves and other devices. Figure 1.15 is an intelligent sensor based on STM32. The sensor must perform the following tasks: provide digital signals, be able to transmit signals and can perform logic functions and instructions. The components of the intelligent sensor include the main sensitive components, excitation generation, amplitude amplification,

Figure 1.15: Intelligent sensor.

high-speed analog filtering, signal compensation, data format conversion, digital processing module and digital communication module. To be a smart sensor, the sensor and the corresponding processor must be in exactly the same physical unit. Abilities of intelligent sensors include self-diagnosis and self-calibration, signal conditioning and important decision making. You can achieve high quality and high availability products for complex manufacturing systems.

1.4.3 Virtual instrument

Virtual instruments consist of industry-standard computers or workstations with mighty applications, economical and efficient hardware (such as acquisition board and driver package), which together realize the functions of traditional instrument and equipment as shown in Figure 1.16 (National Instruments, USA). Virtual instrument has realized the fundamental transformation from traditional instrument centered on hardware to modern instrument system centered on software. Through the virtual instrument system, scientists and engineers can construct different measurement and automation systems (user definitions) that fully meet the needs of all kinds of users and are not limited by the fixed functions of traditional instruments (vendor definitions). There are many advantages of virtual instruments, such as

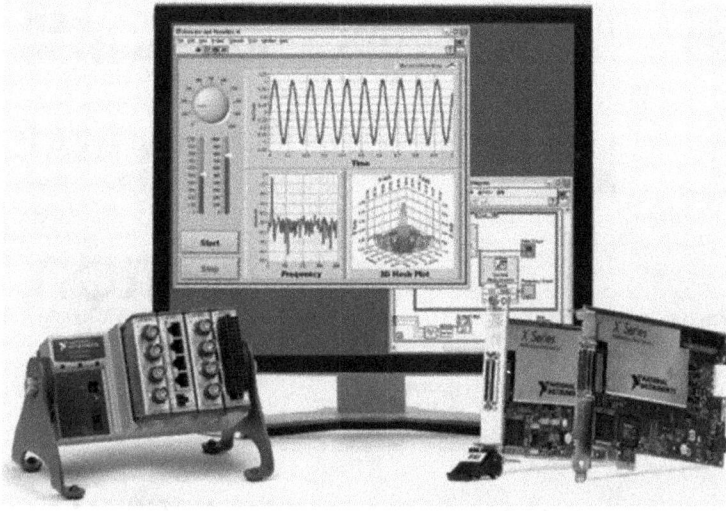

Figure 1.16: Virtual instrument (National Instruments, USA).

reducing the cost of instruments, realizing portability among various computer platforms, providing easy-to-use graphical user interfaces, using graphical representation of program structures, compiling code to separate .Exe or .Dll files and facilitating transmission control protocol/Internet protocol connection.

1.5 Manufacturers

This book introduces the world's leading manufacturers of instruments or sensors. In order to complete the measurement work, the relevant instruments and equipment must be used. Choosing the right measuring instrument can help improve the accuracy and accuracy of your measurement system. Figure 1.17 shows the trademarks of some well-known manufacturers. These manufacturers include ABB, Endress+Hauser, Emerson, GE, Honeywell, Magnetrol, Okazaki, VEGA Controls, Siemens, Yokogawa Rockwell Automation, Schneider Electric, Omron Automation, Danaher and National Instruments (NI).

(1) ABB
ABB is a leading technology and automation company providing innovative digital connectivity and enabling industrial equipment and systems. The company makes robots very popular. This company is popular for manufacturing robotics. ABB invented many measuring devices, including the first commercial online gas chromatograph, the first guided wave radar level transducer and pressure transducer.

Figure 1.17: Trademarks of some well-known manufacturers.

ABB's products include sensors for process liquids and gases, transducers and related equipment, analyzers, emission analysis, temperature measurement, pressure measurement, level measurement, flow measurement and liquid analysis. ABB provides the measurement technology needed for safe, efficient and profitable downstream refinery and petrochemical plant operations.

(2) Endress+Hauser
Endress+Hauser is a well-known Swiss company for instrumentation and process automation. Endress+Hauser offers its customers a wide range of instrumentation products for measuring volume flow, liquid level, density, pressure mass, flow and temperature. The company also produces a series of analytical instruments that measure water quality variables such as dissolved oxygen, conductivity and turbidity. In addition, Endress+Hauser also produces many instruments for hazardous environments and extreme harsh conditions.

(3) Emerson
Emerson is an American multinational group specializing in electrical equipment. Emerson provides a wide range of process measurement technologies to its customers. A full range of instrumentation solutions are offered for flow, temperature, pressure, level, analyzer and sensors. Emerson offers measurements in Coriolis, eddy current, magnetic flow and differential pressure for flow measurement technology. Rosemount pressure sensor is also manufactured by Emerson, which is

considered to be the leading devices in market share and measurement performance of the world.

(4) GE Measurement & Control

GE Measurement & Control is a department of General Electric, which specializes in designing and manufacturing sensing components, instruments, devices and systems to enable customers to control, protect, monitor and verify the security of their key processes and applications. GE measurement and control provides many innovative technologies, including sensor-based measurement, asset status monitoring, control, inspection technology, advanced software and radiation measurement solutions, as well as global-oriented professional services and skills. GE Measurement & Control uses thermal verification, ultrasonic and gas flow measurement to develop wet and clip ultrasonic flow meters for liquid, temperature sensors, gas and steam applications, negative temperature coefficient and positive temperature coefficient thermistors.

(5) Honeywell

Honeywell is a US-based company providing aerospace engineering services that supply everything from instrumentation to advanced process control. For more than 40 years, it has been a pioneer in process automation control, from traditional process control system to current leading innovation. Honeywell products include inventory level measurement, flow meters, wired and wireless pressure and temperature transmitters, radar (wired and wireless) and servo level gauges for custody transfer, as well as for pH, conductivity and oxidation reduction potential liquid analysis measuring equipment. Honeywell also offers distributed control systems that monitor and collect data from a variety of devices. The Enraf FlexLine radar gauge is one of Honeywell's recent innovations. Enraf FlexLine radar gauge ensures the level of precision required for custody transfer by combining Enraf's planar antenna technology with new software algorithms.

(6) Magnetrol

Headquartered in Aurora, Illinois, USA, Magnetrol produces flow and level instruments for monitoring and controlling the flow rates of fluid chemicals, water, slurries, gases and bulk solids. One of the typical products, the ECLIPSE 706, is a liquid level sensor based on guided wave radar technology that supplies 24 V DC. As the first innovator of the magnetic level switch and the pioneer of the technological breakthroughs in liquid level and flow, Magnetrol manufactures smarter, more reliable, easier to install and operate instruments. Magnetrol flow switches are often used to detect flow in horizontal pipes in oil and gas field processing.

(7) Okazaki

Established in 1954 in Kobe, Japan, Okazaki designed and manufactured a range of reliable, high precision temperature sensors, heaters and components. In 1963,

Okazaki developed the first metal-sheathed MgO insulation resistance thermometer for temperature measurement applications in petrochemical processes. Okazaki has also successfully developed a mineral insulated thermocouple with a 0.1 mm sheath-outer diameter, which has a response time of 1 ms. Therefore, it is suitable for measuring the temperature of any small object. In addition, Okazaki produces thermowells that are used for providing isolation between the environment and the temperature sensor.

(8) Danaher
Danaher Industrial Co., Ltd. designs and manufactures industrial and consumer products. It provides services in the fields of industry, life sciences, testing, environment and control.

(9) Siemens
Siemens, based in Germany, is one of the largest engineering companies in Europe. The company's main businesses include infrastructure, energy, automotive and industry. Siemens can be seen as a leader in infrastructure and energy solutions as well as industrial automation and software. As one of the largest energy-saving and resource-saving technology manufacturers in the world, the company provides medical imaging equipment, laboratory diagnosis and clinical IT. The company's products include turbosteam turbines, wheel compressors, high-voltage switchgear (disconnectors, circuit breakers and gas insulated switchgear), remote monitoring system and distribution panels. For pressure measurement, Siemens offers a wide range of pressure transmitters in its SITRANS P range for maximum precision, robustness and ease of use.

(10) Yokogawa
Yokogawa is a Japanese electrical and software company. It provides excellent technical services and engineering solutions, project management, operation and maintenance. Yokogawa is a technologically advanced supplier of test and measurement, and industrial automation solutions and provides field-proven system security, system operation efficiency, product quality and system reliability. For most downstream petrochemical applications, state-of-the-art field instrumentation solutions are offered by Yokogawa from open loops to critical, fast-response tight control loops, from simple primary components to transmitters. The company's product portfolio includes EJX/EJA series of high-performance pressure transmitters, high-reliability temperature transmitters, wireless transmitters, gas or liquid analyzers and other instruments.

(11) Rockwell Automation
Rockwell Automation can be called one of the top industrial automation companies in the world. It is located in Milwaukee and has offices in more than 80 countries around the world. Rockwell Automation products include condition monitoring and

I/O, advanced process control, distributed control system, design and operation software, human–machine interface, drive system, driver, industrial sensors, industrial network products, industrial control products, manufacturing actuators and motion controllers.

(12) Schneider Electric

Schneider Electric is a French multinational company specializing in energy management and automation systems by providing software and hardware. Schneider Electric has developed networking technologies and solutions to manage energy and processes in a reliable, safe, efficient and sustainable manner.

(13) Omron Automation

Omron Automation Company is one of the world's famous industrial automation companies, which was founded 80 years ago. The company's main business is the manufacture and service of automated equipment and systems. The company provides all kinds of switches, sensors, relays, safety components, logic control elements and so on.

(14) VEGA

Headquartered in the Black Forest, Germany, VEGA offers solutions for demanding measurement tasks: for food industry, pharmaceutical and chemical plant, drinking water purification and supply system, waste landfill, petroleum platform, sewage treatment plant, power generation, mining, ships and aircraft. VEGA is a leading provider of process radar level measurement and radar transmitter technology. VEGA develops and manufactures various sensors for measuring liquid level and pressure, and integrates equipment and software into process control systems. VEGA manufactures a wide range of professional-grade measuring instruments for wide applications from the tank to liquid level indication and level detection. It can measure a wide range of products from gas to coarse solids.

(15) National Instruments

National Instruments Corporation (NI) is a well-known multinational corporation in the United States. Headquartered in Austin, United States, the company is an advanced manufacturer of professional automatic testing equipment and graphical virtual instrument software. Common applications of the company's products include field instrument and equipment control, data acquisition and machine vision. NI's engineering software includes LabVIEW, LabWindows/CVI, Measurement Studio, NI Ultiboard for PCB design and NI Multisim for circuit design. NI's hardware platforms include NI CompactRIO, NI roboRIO, NI CompactDAQ and PXI Platform.

2 General description of signal

2.1 Introduction

"Signal" has different meanings in different fields. In this book, signals are usually functions of one or more independent variables containing information about the behavior or nature of certain physical phenomena. It is represented in both analog and digital forms by means of electrical and electronic components. Usually, the independent variable is time and the signal will describe the phenomenon that changes over time. Signals exist in audio, video, speech, language, images, sensors, communications, multimedia, radar, sonar, geophysics, chemistry, biology, molecules, music and medicine.

In this chapter, students will learn about the (1) classification of signals; (2) signal synthesis; (3) Fourier decomposition of a sawtooth wave; (4) spectrum of a periodic signal; (5) characteristics of a periodic signal; (6) aperiodic transient signal; (7) random signals.

2.2 Classification of signal

The nonelectricity is converted into an electrical signal by the sensor and the measuring circuit. These electrical signals are mainly embodied in the form of voltage or current. This is because electrical signals such as voltage and current signals are easy to observe, record and analyze. For example, a reciprocating displacement signal contains various information such as displacement amplitude, displacement frequency and displacement phase information. In addition, based on the differential operation, speed and acceleration information can also be obtained. The form of the signal is varied, which has a different classification.

Signals can be classified into time domain according to their properties and characteristics. They are roughly classified as
(1) Continuous-time/continuous signals;
(2) Discrete-time signals.

Definition of the continuous-time signal
A continuous-time signal or a continuous signal is a mathematically continuous function defined continuously in the time domain. That is, the domain of the function is an uncountable set. The function itself does not need to be continuous. Figure 2.1 shows the continuous-time signal.

https://doi.org/10.1515/9783110624397-002

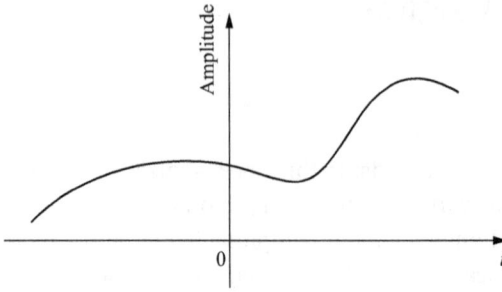

Figure 2.1: A continuous-time signal.

Definition of the discrete-time signal

Discrete-time signals are specified only on specific discrete values. The amplitude of the discrete-time signal between two discrete values is not defined. A discrete time signal has a countable domain, such as natural numbers. Figure 2.2 shows a discrete-time signal.

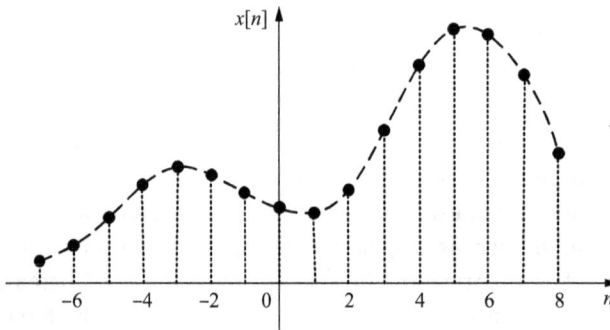

Figure 2.2: A discrete-time signal.

Continuous-time and discrete-time signals can be further divided into:
(1) Deterministic and nondeterministic signals
(2) Periodic and aperiodic signals
(3) Even and odd signals
(4) Energy and power signals

Definition of the deterministic signal

A deterministic signal is a function that is fully deterministic in time. You can predict the nature and amplitude of the signal at any time. The signal pattern is regular and can be described mathematically. For the signals in Figure 2.1 and 2.2, the amplitude can be predicted in advance at any time.

Definition of the nondeterministic signal
In contrast, a nondeterministic signal is a signal whose appearance is essentially random and whose pattern is very irregular as shown in Figure 2.3. The nondeterministic signal is also referred to as a random signal. Thermal noise in the circuit is a typical example of a nondeterministic signal.

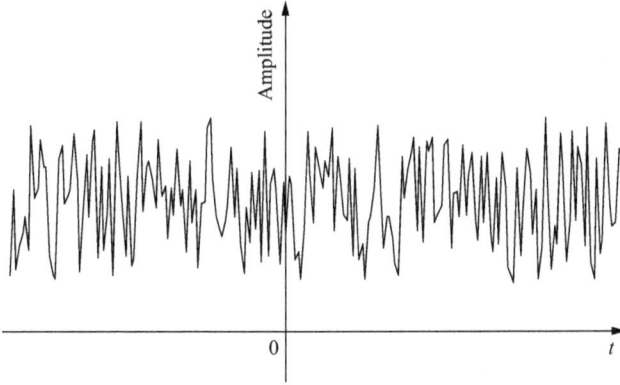

Figure 2.3: A nondeterministic signal.

Definition of the periodic signal
If the signal completes a certain output within a certain time (T_0) and repeats the output during the same subsequent period, the signal is a periodic signal as shown in Figure 2.4. Periodic signals should have a clear mathematical relationship. If the continuous-time signal satisfies the following mathematical definition, it is considered to be periodic:

$$g(t) = g(t + kT_0) \tag{2.1}$$

where k is an integer and T_0 is called the period.

Similarly, if the discrete-time signal satisfies the following mathematical definition, it is considered to be periodic:

$$x[n] = x[n + kN_0] \tag{2.2}$$

where k is an integer and N_0 is called the period.

The minimum period of the periodic signal is called the fundamental period. Besides, the fundamental frequency is the reciprocal of the fundamental period. There are some common periodic signals, including sine, square, triangle and sawtooth, as shown in Figure 2.5.

There is also a periodic signal called a composite periodic signal, also named complex periodic signal. A composite periodic signal is a signal composed of a number of sinusoidal components, wherein the frequencies of all the frequency

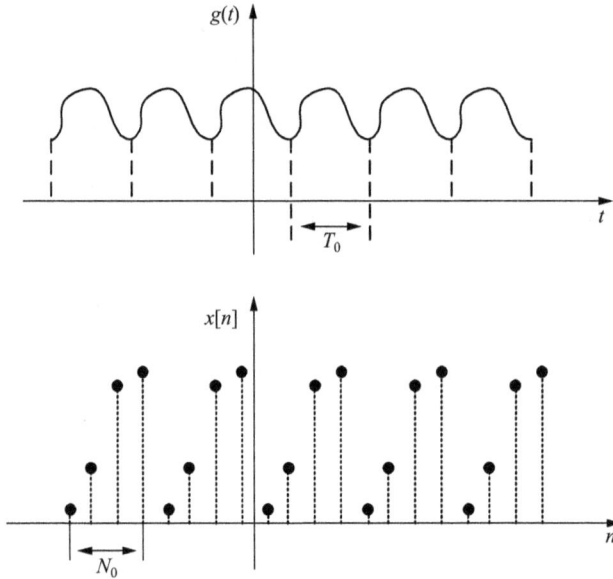

Figure 2.4: Periodic signals (mathematical definition).

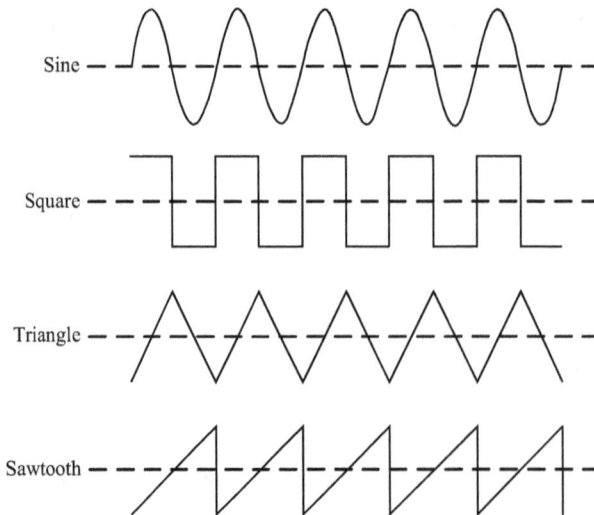

Figure 2.5: Common periodic signals.

components are integral multiples of $1/T$. (A frequency ratio is a rational number.) Figure 2.6 shows the composite periodic signal $x(t)$ for the following functions:

$$x(t) = \sin t + \frac{1}{2}\sin 3t + \frac{1}{4}\sin 5t \qquad (2.3)$$

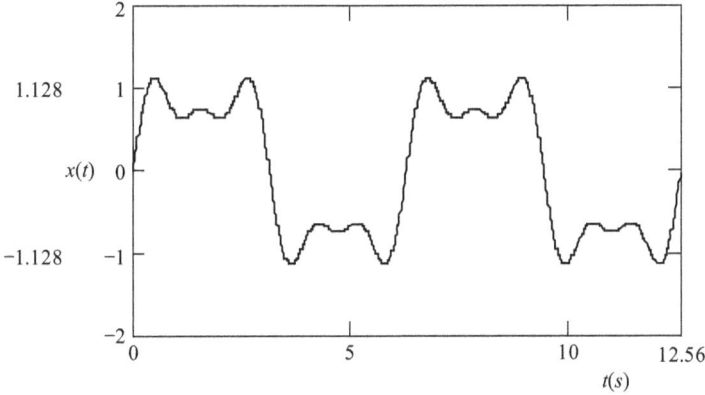

Figure 2.6: Composite periodic signal.

Definition of the aperiodic signal (nonperiodic signal)

Aperiodic signals are signals that are not repeated for a defined time interval. Aperiodic signals also have a clear mathematical relationship. However, it does not have a period. Aperiodic signals can be divided into two categories: almost periodic signals and transient signals. Almost periodic function is a function of a real number that is "almost periodic" within the certain level of precision. A good example is a planetary system, in which the planets in orbit move in an almost period. Transient signals last for a very short period of time and a broad spectrum that is often present in abrupt acoustic pressure changes, switch bounce, the shock wave generated from an impact test, seismic signals, electrical discharges and so on.

Definition of the even signal

A signal is sometimes classified by its symmetry along the time axis relative to the origin, $t = 0$. Even signal is symmetric around the vertical (could be the Y-axis.) axis. A signal $x(t)$ is said to be even, which must satisfy the following equation:

$$x(t) = x(-t) \qquad (2.4)$$

The discrete-time signal $x[n]$ is an even signal, which must satisfy the following equation:

$$x[n] = x[-n] \forall n \qquad (2.5)$$

Definition of the odd signal

The odd signal is symmetric about the origin. A signal $x(t)$ is said to be odd, which must satisfy the following equation:

$$x(t) = -x(-t) \tag{2.6}$$

The odd signal should be zero at $t = 0$, which means the odd signal passes the origin.

The discrete-time signal is said to be an odd signal if it satisfies the following condition:

$$\forall n \quad x[n] = -x[-n] \tag{2.7}$$

A real signal can always be considered as a sum of two parts (shown in Figure 2.7): even and odd:

$$x(t) = x_{even}(t) + x_{odd}(t) \tag{2.8}$$

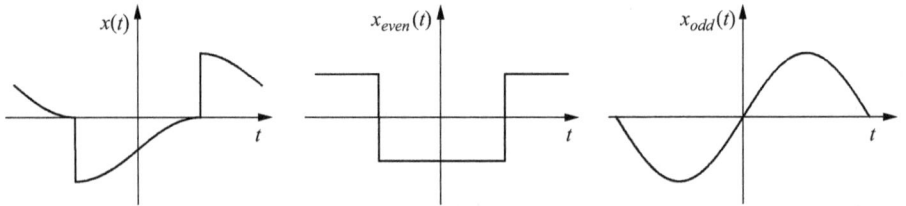

Figure 2.7: Even and odd parts.

where $x_{even}(t)$ is the even part and $x_{odd}(t)$ is the odd part defined as follows:

$$x_{even}(t) = \frac{1}{2}\{x(t) + x(-t)\} \tag{2.9}$$

$$x_{odd}(t) = \frac{1}{2}\{x(t) - x(-t)\} \tag{2.10}$$

Therefore,

$$x_{even}(-t) = \frac{1}{2}\{x(-t) + x(t)\} = x_{even}(t) \Rightarrow x_{even}(t) \text{ is even.} \tag{2.11}$$

$$x_{odd}(-t) = \frac{1}{2}\{x(-t) - x(t)\} = -x_{odd}(t) \Rightarrow x_{odd}(t) \text{ is odd.} \tag{2.12}$$

It is an important fact that it is the relative concept of Fourier series. Periodic signals can be regarded as the sum of sinusoidal and cosine signals in Fourier series. As we all know, cosine function is an even signal and sinusoidal function is an odd signal.

Definition of the energy signal

The energy of a signal such as a continuous-time signal or a discrete-time signal is defined as follows:

The energy of a continuous-time signal:

$$E_{ctx} = \int_{-\infty}^{\infty} |x(t)|^2 dt \tag{2.13}$$

The energy of a discrete-time signal:

$$E_{dtx} = \sum_{n=-\infty}^{\infty} |x[n]|^2 \tag{2.14}$$

$x(t)/x[n]$ is a continuous/discrete energy signal if it satisfies the following condition:

$$0 < E_{ctx} < \infty \quad \text{or} \quad 0 < E_{dtx} < \infty \tag{2.15}$$

Definition of the power signal

The power of a signal, that is, a continuous-time signal or a discrete-time signal is defined as follows:

Power of a continuous-time signal:

$$P_{ctx} = \lim_{T \to \infty} \frac{1}{2T} \int_{-T}^{T} |x(t)|^2 dt \tag{2.16}$$

Power of a discrete-time signal:

$$P_{dtx} = \lim_{N \to \infty} \frac{1}{2N+1} \sum_{n=-N}^{N} |x[n]|^2 \tag{2.17}$$

$x(t)/x[n]$ is a continuous/discrete power signal if it satisfies the following condition:

$$0 < P_{ctx} < \infty \quad \text{or} \quad 0 < P_{dtx} < \infty \tag{2.18}$$

The energy of the power signal will be infinite. For example, a periodic sinusoidal signal has limited nonzero power but has infinite energy.

2.3 Signal synthesis

Figure 2.8 shows a basic sine wave. The vertical axis can be force, or energy, or pressure, depending on what the wave is. The horizontal axis usually represents the time or distance through which the wave passes.

Figure 2.8 illustrates a 5 Hz sine wave with 2 N amplitude. The horizontal axis shows time expressed in seconds. The number of cycles per second is 5, so the frequency is 5 Hz. According to the frequency, it can be easily converted to the period. If a periodic wave is cycled 5 times per second, the period is 1/5 s. Similarly,

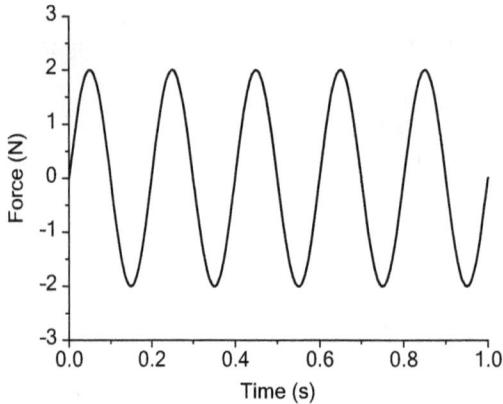

Figure 2.8: A basic sine wave of 5 Hz.

knowing the period, the frequency could be calculated. Figure 2.9 shows a sine wave of 15 Hz with 1 N amplitude. Compared with Figure 2.8, the wave period in Figure 2.9 becomes smaller as the frequency increases.

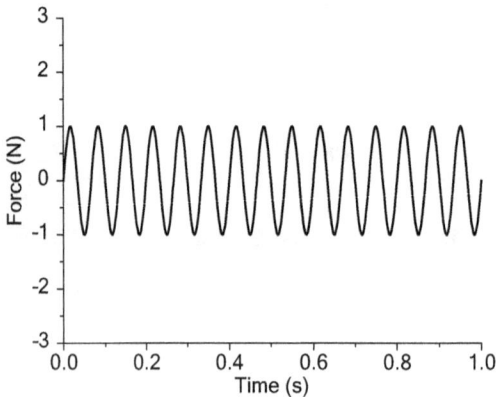

Figure 2.9: A basic sine wave of 15 Hz.

Figure 2.10 illustrates a synthesized wave by adding 5 Hz sine wave and 15 Hz sine wave together. The result of combining the two periodic signals is still a periodic signal. The synthesized frequency is 5Hz, and it is the same as the low-frequency signal.

There are two different ways to describe a signal in this book which are in the time domain and the frequency domain. Both time domain analysis and frequency domain analysis are widely used in measurement, telecommunications, electronics and other fields. Time domain analysis gives a picture of how the signal changes

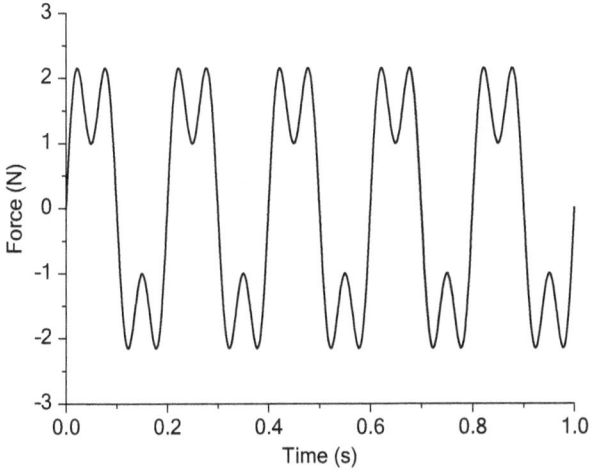

Figure 2.10: A synthesized wave.

over time. Frequency domain analysis is used for applications that require filtering, amplification and mixing. Especially when dealing with multiple sine waves, the frequency domain is more compact and more useful.

2.4 Fourier decomposition of a sawtooth wave

The sawtooth wave is a typical nonsinusoidal periodic waveform. The sawtooth waveform first inclines upward and then falls sharply. Figure 2.11 shows a sawtooth wave with an amplitude of A_0 and a period of T_0 in the time domain.

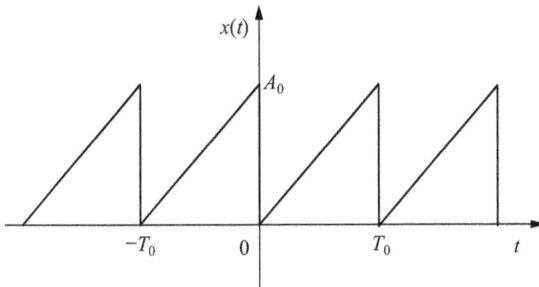

Figure 2.11: The periodic sawtooth wave.

Its mathematical function of the sawtooth wave could be written as follows:

$$\begin{cases} x(t) = x(t + nT_0) \\ \\ x(t) = \dfrac{A_0}{T_0} t, \qquad 0 < t < T_0 \end{cases} \qquad (2.19)$$

An ideal sawtooth wave with an amplitude of A_0 can be expressed as an infinite sum of sinusoidal waves by using Fourier expansion of cycle frequency f_0 varying with time t:

$$x(t) = \frac{A_0}{2} - \frac{A_0}{\pi}\left(\sin 2\pi \cdot f_0 t + \frac{1}{2}\sin 2\pi \cdot 2f_0 t + \frac{1}{3}\sin 2\pi \cdot 3f_0 t + \cdots\right) \qquad (2.20)$$

The ideal sawtooth wave contains components of integer harmonic frequencies. In order to simplify the Fourier expansion, let $T_0 = 1$ s$/f_0 = 1$ Hz, $\omega_0 = 1$ rad/s and $A_0 = 2\pi$. The above equation can be written as

$$x(t) = \pi - 2\left(\sin 2\pi t + \frac{1}{2}\sin 2\pi \cdot 2t + \frac{1}{3}\sin 2\pi \cdot 3t + \frac{1}{4}\sin 2\pi \cdot 4t + \frac{1}{5}\sin 2\pi \cdot 5t \ldots\right) \qquad (2.21)$$

Equation (2.20) shows that periodic sawtooth function can be written as an infinite sum of sinusoidal functions.

In Figure 2.12, curve B represents π, curve C represents $-2\sin 2\pi t$, curve D represents $-2\sin 2\pi \cdot 2t/2$, curve E represents $-2\sin 2\pi \cdot 3t/3$ and curve F represents $-2\sin 2\pi \cdot 4t/4$. Curve B, Curve C, Curve D, Curve E and Curve F are six sinusoidal functions (in various colors) with appropriate amplitude and frequency.

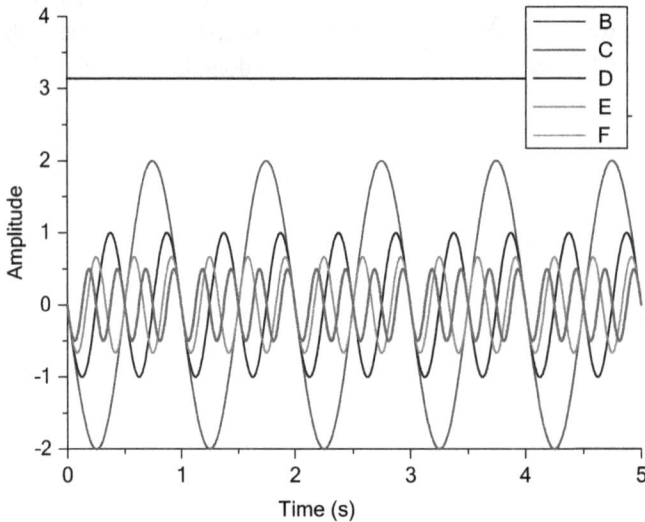

Figure 2.12: The harmonic waves.

In Figure 2.13, curve B represents $\pi - 2\sin 2\pi t$, curve C represents $\pi - 2\sin 2\pi t - 2\sin 2\pi \cdot 2t/2$, curve D represents $\pi - 2\sin 2\pi t - 2\sin 2\pi \cdot 2t/2 - 2\sin 2\pi \cdot 3t/3$, curve E represents $\pi - 2\sin 2\pi t - 2\sin 2\pi \cdot 2t/2 - 2\sin 2\pi \cdot 3t/3 - 2\sin 2\pi \cdot 4t/4$ and curve F represents $\pi - 2\sin 2\pi t - 2\sin 2\pi \cdot 2t/2 - 2\sin 2\pi \cdot 3t/3 - 2\sin 2\pi \cdot 4t/4 - 2\sin 2\pi \cdot 5t/5$. The harmonic waves with appropriate amplitude and frequency are summed up to form a sawtooth wave.

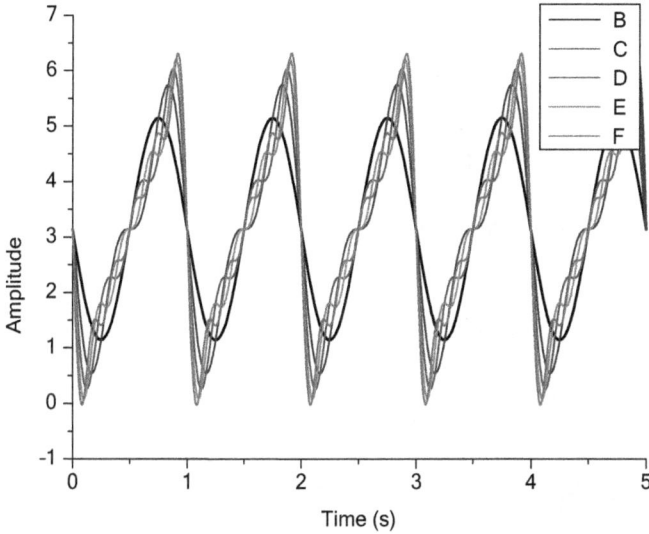

Figure 2.13: Sum of the harmonic waves.

2.5 Spectrum of periodic signal

If a periodic signal in time domain satisfies the Dirichlet condition, it can be divided into sinusoidal or complex exponents with harmonic correlation. The representation of this decomposition is called Fourier series. This section will discuss two forms of Fourier series, namely trigonometric Fourier series representation and exponential Fourier series representation.

2.5.1 Trigonometric Fourier series representation

Fourier series is the expansion of the infinite sum of sine and cosine by periodic function $f(t)$. Since sine and cosine have no mean terms, periodic signals with nonzero mean can have a constant component. The trigonometric Fourier Series representation of a periodic signal $x(t)$ with fundamental frequency f_0 is given by

$$x(t) = a_0 + \sum_{n=1}^{\infty} (a_n \cos 2\pi \cdot nf_0 t + b_n \sin 2\pi \cdot nf_0 t) \tag{2.22}$$

where a_n and b_n are Fourier coefficients for the nth harmonic given by

$$a_n = \frac{2}{T_0} \int_{t_0}^{t_0 + T_0} x(t) \cos 2\pi \cdot nf_0 t dt, \quad n = 1, 2, 3, \ldots \tag{2.23}$$

$$b_n = \frac{2}{T_0} \int_{t_0}^{t_0 + T_0} x(t) \sin 2\pi \cdot nf_0 t dt, \quad n = 1, 2, 3, \ldots \tag{2.24}$$

a_0 is the DC (average value) component of the signal, which is given by

$$a_0 = \frac{1}{T_0} \int_{t_0}^{t_0 + T_0} x(t) dt \tag{2.25}$$

If $x(t)$ is an even function, in other words, $x(t) = x(-t)$, then

$$\begin{cases} b_n = 0 \\ a_0 = \dfrac{1}{T_0} \displaystyle\int_{t_0}^{t_0 + T_0} x(t) dt \\ a_n = \dfrac{2}{T_0} \displaystyle\int_{t_0}^{t_0 + T_0} x(t) \cos 2\pi \cdot nf_0 t dt, \quad n = 1, 2, 3, \ldots \end{cases} \tag{2.26}$$

If $x(t)$ is an odd function, that is, $x(t) = -x(-t)$, then

$$\begin{cases} a_0 = 0 \\ a_n = 0 \\ b_n = \dfrac{2}{T_0} \displaystyle\int_{t_0}^{t_0 + T_0} x(t) \sin 2\pi \cdot nf_0 t dt, \quad n = 1, 2, 3, \ldots \end{cases} \tag{2.27}$$

Since $a_n \cos 2\pi \cdot nf_0 t + b_n \sin 2\pi \cdot nf_0 t = c_n \cos(2\pi \cdot nf_0 t + \theta_n)$, compact trigonometric Fourier series can be derived:

$$x(t) = c_0 + \sum_{n=1}^{\infty} c_n \cos(2\pi \cdot nf_0 t + \theta_n) \tag{2.28}$$

where $c_n \cos(2\pi \cdot nf_0 t + \theta_n)$ is nth harmonic; $c_0 = a_0$; $c_n = \sqrt{a_n^2 + b_n^2}$; f_0 is the fundamental frequency; $\theta_n = \arctan(-b_n/a_n)$.

From the compact trigonometric Fourier series representation, it can be seen that periodic signals can be described by a series of harmonically correlated (in other words, integer multiples of fundamental frequency) sine/cosine waves.

Question 2.1:
Square wave is a kind of nonsinusoidal periodic wave, whose amplitude varies between the fixed minimum and maximum, and the duration of the minimum and maximum is the same (duty ratio is 50%). The conversion between the minimum and maximum is instantaneous, so the ideal square

wave does not exist. Figure 2.14 shows a square wave with an amplitude of A_0 and a frequency of T_0 in the time domain.

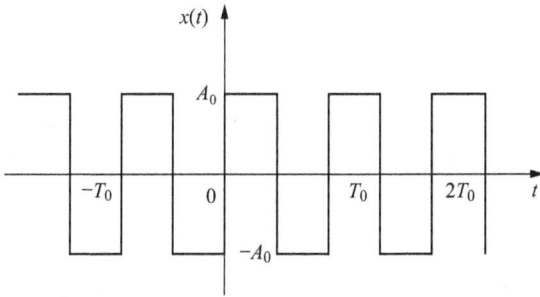

Figure 2.14: A square wave.

Find the trigonometric Fourier series for a square wave $x(t)$ and sketch its amplitude and phase spectrum.

Solution:
Its mathematical function of the square wave can be written as follows:

$$\begin{cases} x(t) = x(t + nT_0) \\ x(t) = \begin{cases} A_0 & 0 < t < \dfrac{T_0}{2} \\ -A_0 & \dfrac{T_0}{2} < t < T_0 \end{cases} \end{cases} \tag{2.29}$$

Step 1: Find DC offset

$$a_0 = 0$$

Step 2: Find the Fourier coefficient a_n
$x(t)$ is an odd function, so

$$a_n = \frac{2}{T_0} \int_{t_0}^{t_0 + T_0} x(t) \cos 2\pi \cdot n f_0 t \, dt$$
$$= 0 \tag{2.30}$$

Step 3: Find the Fourier coefficient b_n

$$b_n = \frac{2}{T_0} \int_{t_0}^{t_0 + T_0} x(t) \sin 2\pi \cdot n f_0 t \, dt$$

$$= \frac{4}{T_0} \int_0^{T_0/2} A_0 \sin 2\pi \cdot n f_0 t \, dt$$

$$= \begin{cases} \dfrac{4A_0}{\pi} \cdot \dfrac{1}{n}, & n = 1, 3, 5, \dots \\ 0, & n = 2, 4, 6, \dots \end{cases} \tag{2.31}$$

Step 4: Find the trigonometric Fourier series of the square wave function

$$x(t) = a_0 + \sum_{n=1}^{\infty} (a_n \cos 2\pi \cdot n f_0 t + b_n \sin 2\pi \cdot n f_0 t)$$

$$= \frac{4A_0}{\pi} \sum_{n=1}^{\infty} \frac{\sin 2\pi \cdot n f_0 t}{n}, \quad n = 1, 3, 5, \ldots \qquad (2.32)$$

An ideal square wave $x(t)$ with an amplitude of A_0 can be expressed as the sum of infinite sinusoidal waves. The waves contain only odd integer harmonic frequencies. In order to specify the Fourier series, let $T_0 = 1$ s, $\omega_0 = 2\pi$ rad/s and $A_0 = \pi/4$. The above equation can also be written as

$$x(t) = \sin 2\pi \cdot t + \frac{1}{3}\sin 2\pi \cdot 3t + \frac{1}{5}\sin 2\pi \cdot 5t + \frac{1}{7}\sin 2\pi \cdot 7t + \frac{1}{9}\sin 2\pi \cdot 9t + \frac{1}{11}\sin 2\pi \cdot 11t \ldots$$

$$(2.33)$$

Any enough regular periodic function can be written as the sum of infinite sinusoidal functions, according to the Fourier's theorem. Six sinusoidal functions (in various colors) with appropriate amplitude and frequency will sum up to form a square wave.

In Figure 2.15, curve B represents $\sin 2\pi \cdot t$ (first harmonic), curve C represents $\sin 2\pi \cdot 3t/3$ (second harmonic), curve D represents $\sin 2\pi \cdot 5t/5$ (third harmonic), curve E represents $\sin 2\pi \cdot 7t/7$ (fourth harmonic), curve F represents $\sin 2\pi \cdot 9t/9$ (fifth harmonic) and curve G represents $\sin 2\pi \cdot 11t/11$ (sixth harmonic).

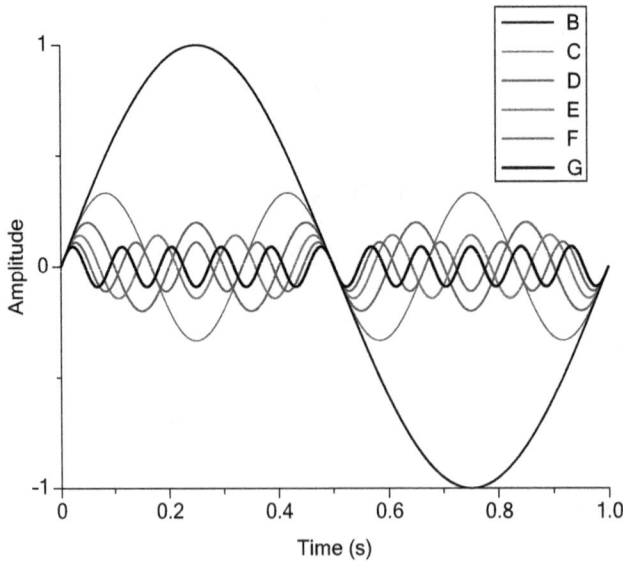

Figure 2.15: Odd-integer harmonic waves.

In Figure 2.16, curve B represents $\sin 2\pi \cdot t + \sin 2\pi \cdot 3t/3$, curve C represents $\sin 2\pi \cdot t + \sin 2\pi \cdot 3t/3 + \sin 2\pi \cdot 5t/5$, curve D represents $\sin 2\pi \cdot t + \sin 2\pi \cdot 3t/3 + \sin 2\pi \cdot 5t/5 + \sin 2\pi \cdot 7t/7$ and curve E represents $\sin 2\pi \cdot t + \sin 2\pi \cdot 3t/3 + \sin 2\pi \cdot 5t/5 + \sin 2\pi \cdot 7t/7 + \sin 2\pi \cdot 9t/9$. Based on the above analysis, it can be seen that the square signal can be expressed as a sum of many sinusoidal signals with different frequencies, amplitudes and phases.

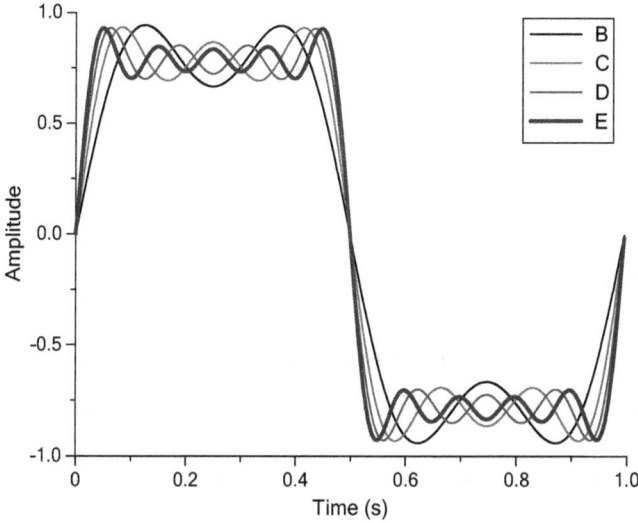

Figure 2.16: Sum of the odd-integer harmonic waves.

Step 5: Sketch its amplitude and phase spectrum
The spectrum of a signal is the amplitude and phase spectrum of different frequency components. The spectrum of the square signal $x(t)$ is shown in Figure 2.17. Each spectrum line represents a frequency harmonic. The amplitude (height) of the spectral line indicates the contribution of the frequency to the signal.

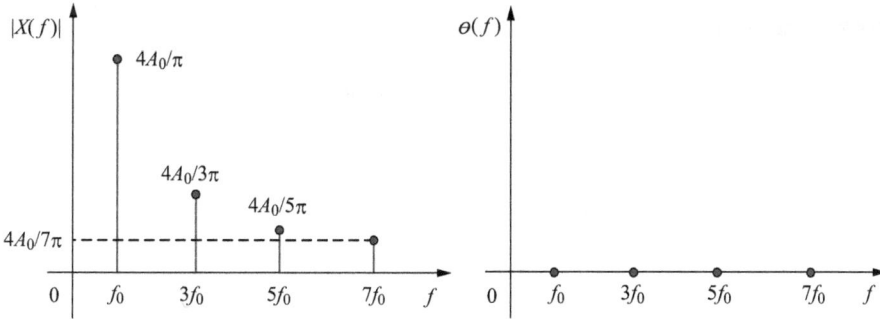

Figure 2.17: Spectrum of the square signal.

Question 2.2:
A triangle wave is the nonsinusoidal waveform, that is, the periodic piecewise linear continuous real function. Figure 2.18 shows a triangle wave with an amplitude of A_0 and a period of T_0 in the time domain.

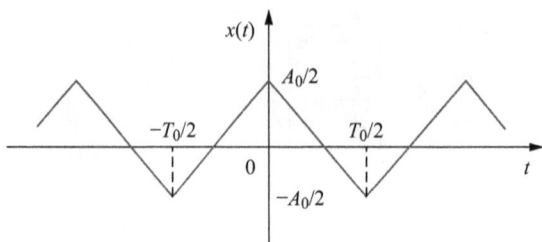

Figure 2.18: The periodic triangle wave.

Find the trigonometric Fourier series for a triangle wave $x(t)$ and sketch its amplitude and phase spectrum.

Solution:
The mathematical function of the triangle wave can be written as follows:

$$x(t) = \begin{cases} \frac{A_0}{2} + \frac{2A_0}{T_0}t, & -\frac{T_0}{2} \leq t < 0 \\ \frac{A_0}{2} - \frac{2A_0}{T_0}t, & 0 \leq t < \frac{T_0}{2} \end{cases} \tag{2.34}$$

Step 1: Find DC offset a_0

$$a_0 = \frac{1}{T_0} \int_{t_0}^{t_0 + T_0} x(t)dt$$
$$= 0 \tag{2.35}$$

Step 2: Find the Fourier coefficient a_n

$$a_n = \frac{2}{T_0} \int_{t_0}^{t_0 + T_0} x(t) \cos 2\pi \cdot nf_0 t \, dt$$

$$= \frac{4}{T_0} \int_0^{\frac{T_0}{2}} \left(\frac{A_0}{2} - \frac{2A_0}{T_0}t \right) \cos 2\pi \cdot nf_0 t \, dt$$

$$= \frac{4A_0}{n^2\pi^2} \sin^2 \frac{n\pi}{2}$$

$$= \begin{cases} \frac{4A_0}{n^2\pi^2}, & n = 1, 3, 5, \ldots \\ 0, & n = 2, 4, 6, \ldots \end{cases} \tag{2.36}$$

Step 3: Find the Fourier coefficient b_n
$x(t)$ is an even function, so

$$b_n = \frac{2}{T_0} \int_{t_0}^{t_0 + T_0} x(t) \sin 2\pi \cdot nf_0 t \, dt$$
$$= 0 \tag{2.37}$$

Step 4: Find the Fourier series of the triangle wave function

$$x(t) = \frac{4A_0}{\pi^2}\left(\cos 2\pi \cdot f_0 t + \frac{1}{3^2}\cos 2\pi \cdot 3f_0 3t + \frac{1}{5^2}\cos 2\pi \cdot 5f_0 t + \cdots\right)$$

$$= \frac{4A_0}{\pi^2}\sum_{n=1}^{\infty}\frac{1}{n^2}\cos 2\pi \cdot nf_0 t, \qquad n = 1,3,5,\ldots$$

(2.38)

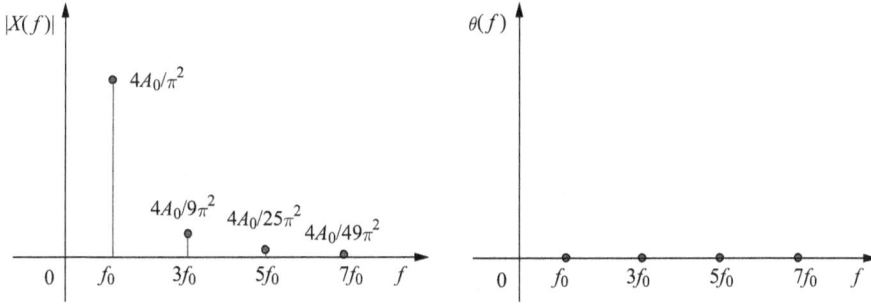

Figure 2.19: Spectrum of the triangle wave.

Step 5: Sketch its amplitude and phase spectrum as shown in Figure 2.19.
The ideal triangle waveform contains only odd harmonic components. In order to specify the Fourier expansion, let $f_0 = 1$ Hz and $A_0 = \pi^2/4$. The above equation can be written as follows:

$$x(t) = \cos 2\pi \cdot t + \frac{1}{3^2}\cos 2\pi \cdot 3t + \frac{1}{5^2}\cos 2\pi \cdot 5t + \cdots$$

$$= \sum_{n=1}^{\infty}\frac{1}{n^2}\cos 2\pi \cdot nt, \qquad n = 1,3,5,\ldots$$

(2.39)

In Figure 2.20, curve B represents $\cos 2\pi \cdot t$ (first harmonic), curve C represents $\cos 2\pi \cdot 3t/3^2$ (second harmonic), curve D represents $\cos 2\pi \cdot 5t/5^2$ (third harmonic),

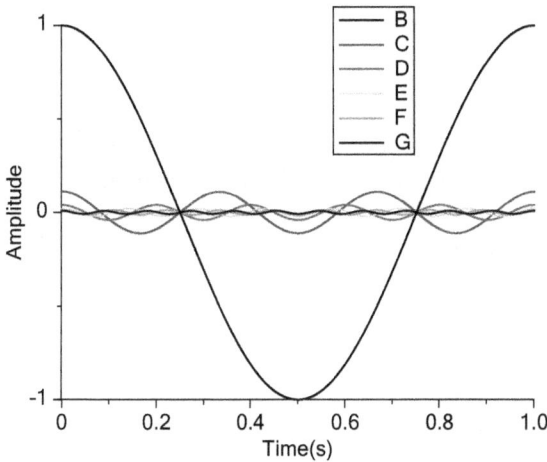

Figure 2.20: The odd-integer harmonic waves.

curve E represents $\cos 2\pi \cdot 7t/7^2$ (fourth harmonic), curve F represents $\cos 2\pi \cdot 9t/9^2$ (fifth harmonic) and curve G represents $\cos 2\pi \cdot 11t/11^2$ (sixth harmonic).

In Figure 2.21, curve B represents $\cos 2\pi \cdot t + \cos 2\pi \cdot 3t/3^2$, curve C represents $\cos 2\pi \cdot t + \cos 2\pi \cdot 3t/3^2 + \cos 2\pi \cdot 5t/5^2$, curve D represents $\cos 2\pi \cdot t + + \cos 2\pi \cdot 3t/3^2 + \cos 2\pi \cdot 5t/5^2 + \cos 2\pi \cdot 7t/7^2$ and curve E represents $\cos 2\pi \cdot t + \cos 2\pi \cdot 3t/3^2 + \cos 2\pi \cdot 5t/5^2 + \cos 2\pi \cdot 7t/7^2 + \cos 2\pi \cdot 9t/9^2$.

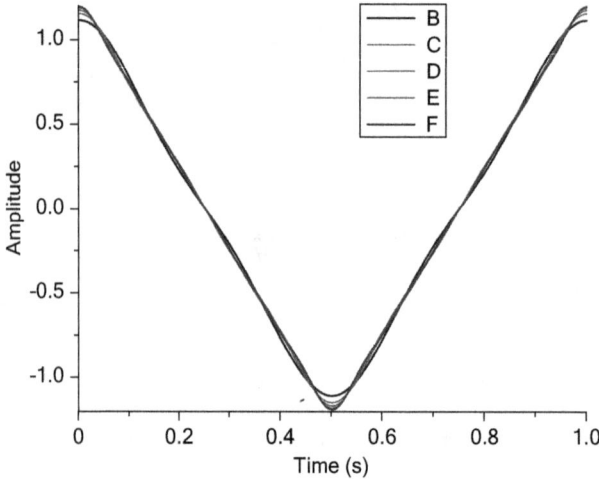

Figure 2.21: Sum of the odd-integer harmonic waves.

Like the square wave of the previous example, the triangle wave contains only odd harmonics. However, the roll-off speed of higher harmonics is much faster than that in square waves.

2.5.2 Exponential Fourier series representation

The exponential Fourier series is another type of Fourier series representation, which is equivalent to a triangular Fourier series. These two different representations give the same result. You can choose the appropriate representation according to your convenience and the type of signal. The complex exponential Fourier series representation of a periodic signal $x(t)$ with fundamental frequency f_0 is given by

$$x(t) = \sum_{n=-\infty}^{\infty} c_n e^{j2\pi \cdot nf_0 t} \tag{2.40}$$

where c_n is called the complex Fourier coefficient and is given by

$$C_n = \frac{1}{T_0} \int_{t_0}^{t_0 + T_0} x(t) e^{-j2\pi \cdot n f_0 t} dt \tag{2.41}$$

The above equation can be derived from the trigonometric Fourier series representation. The relationship between exponential Fourier series and the previous trigonometric Fourier series representation is derived from a very important identity, commonly referred to as the Euler equation:

$$e^{\pm j2\pi f_0 t} = \cos 2\pi f_0 t \pm j \sin 2\pi f_0 t \tag{2.42}$$

Hence

$$\begin{cases} \cos 2\pi \cdot n f_0 t = \frac{1}{2}\left(e^{-j2\pi \cdot n f_0 t} + e^{j2\pi \cdot n f_0 t}\right) \\ \sin 2\pi \cdot n f_0 t = j\frac{1}{2}\left(e^{-j2\pi \cdot n f_0 t} - e^{j2\pi \cdot n f_0 t}\right) \end{cases} \tag{2.43}$$

The trigonometric Fourier series representation becomes simpler if you use $e^{-j2\pi \cdot n f_0 t}$ or $e^{j2\pi \cdot n f_0 t}$ instead of $\cos 2\pi \cdot n f_0 t$ and $\sin 2\pi \cdot n f_0 t$. So

$$x(t) = a_0 + \sum_{n=1}^{\infty} (a_n \cos 2\pi \cdot n f_0 t + b_n \sin 2\pi \cdot n f_0 t)$$

$$= a_0 + \sum_{n=1}^{\infty} \left(\frac{a_n}{2}\left(e^{-j2\pi \cdot n f_0 t} + e^{j2\pi \cdot n f_0 t}\right) + j\frac{b_n}{2}\left(e^{-j2\pi \cdot n f_0 t} - e^{j2\pi \cdot n f_0 t}\right) \right)$$

$$= a_0 + \sum_{n=1}^{\infty} \left(\frac{1}{2}(a_n + jb_n)e^{-j2\pi \cdot n f_0 t} + \frac{1}{2}(a_n - jb_n)e^{j2\pi \cdot n f_0 t} \right)$$

$$= \sum_{n=1}^{\infty} \frac{1}{2}(a_n - jb_n)e^{j2\pi \cdot n f_0 t} + a_0 + \sum_{n=1}^{\infty} \frac{1}{2}(a_n + jb_n)e^{-j2\pi \cdot n f_0 t}$$

$$= \sum_{n=-\infty}^{\infty} C_n e^{j2\pi \cdot n f_0 t} \quad (n = 0, \pm 1, \pm 2, \ldots) \tag{2.44}$$

where, in this case, C_n is given by

$$C_n = \begin{cases} \frac{1}{2}(a_n - jb_n), & n = 1, 2, 3, \ldots \\ a_0, & n = 0 \\ \frac{1}{2}(a_{-n} + jb_{-n}), & n = -1, -2, -3, \ldots \end{cases} \tag{2.45}$$

C_n is normally complex except C_0 and satisfies:

$$C_n = C^*_{-n} \tag{2.46}$$

In eq. (2.46), the * represents the complex conjugate.
Exponential Fourier series can be written as

$$x(t) = \sum_{n=-\infty}^{\infty} C_n e^{j2\pi \cdot n f_0 t} \quad (n = 0, \pm 1, \pm 2, \ldots) \tag{2.47}$$

Similarly, the complex Fourier coefficient c_n can also be derived from a_n and b_n. Thus, for positive n

$$c_n = \frac{1}{2}(a_n - jb_n)$$

$$= \frac{1}{2} * \frac{2}{T_0} \int_{t_0}^{t_0+T_0} x(t) \cos 2\pi \cdot nf_0 t \, dt - \frac{1}{2} * \frac{2}{T_0} *j* \int_{t_0}^{t_0+T_0} x(t) \sin 2\pi \cdot nf_0 t \, dt$$

$$= \frac{1}{T_0} \int_{t_0}^{t_0+T_0} x(t)(\cos 2\pi \cdot nf_0 t - j\sin 2\pi \cdot nf_0 t)\, dt$$

$$= \frac{1}{T_0} \int_{t_0}^{t_0+T_0} x(t)e^{-j2\pi \cdot nf_0 t}\, dt \tag{2.48}$$

For negative n

$$c_n = \frac{1}{2}(a_{-n} + jb_{-n})$$

$$= \frac{1}{2} * \frac{2}{T_0} \int_{t_0}^{t_0+T_0} x(t) \cos 2\pi \cdot -nf_0 t \, dt + \frac{1}{2} * \frac{2}{T_0} *j* \int_{t_0}^{t_0+T_0} x(t) \sin 2\pi \cdot -nf_0 t \, dt$$

$$= \frac{1}{T_0} \int_{t_0}^{t_0+T_0} x(t)(\cos 2\pi \cdot nf_0 t - j\sin 2\pi \cdot nf_0 t)\, dt$$

$$= \frac{1}{T_0} \int_{t_0}^{t_0+T_0} x(t)e^{-j2\pi \cdot nf_0 t}\, dt \tag{2.49}$$

For $n = 0$

$$c_n = a_0$$

$$= \frac{1}{T_0} \int_{t_0}^{t_0+T_0} x(t)\, dt$$

$$= \frac{1}{T_0} \int_{t_0}^{t_0+T_0} x(t)e^{-0}\, dt \tag{2.50}$$

So

$$c_{\pm n} = c_{nR} + jc_{nI} = \frac{a_{|n|} \mp jb_{|n|}}{2} \tag{2.51}$$

$$a_n = 2c_{nR}, \quad c_{nR} - \text{real part} \tag{2.52}$$

$$b_n = -2c_{nI}, \quad c_{nI} - \text{imaginary part} \tag{2.53}$$

Then, you can plot the double-sided (or two-sided) amplitude spectrum $|C_n| - f$ or $C_{nR} - f/C_{nI} - f$ (real/imaginary spectrum) and phase spectrum $\phi_n - f$.

Question 2.3:

Sketch the real spectrum, imaginary spectrum, single-sided and double-sided amplitude spectrum of $x(t) = 3 \sin 2\pi t + 5 \cos 8\pi t$.

Solution:

Based on Euler's equation:

$$e^{\pm j 2\pi \cdot f_0 t} = \cos 2\pi \cdot f_0 t \pm j \sin 2\pi \cdot f_0 t \tag{2.54}$$

you will get

$$x(t) = 3 \cdot j \frac{1}{2}\left(e^{-j 2\pi t} - e^{j 2\pi t}\right) + 5 \cdot \frac{1}{2}\left(e^{-j 8\pi t} + e^{j 8\pi t}\right)$$

$$= \frac{5}{2}\left(e^{-j 2\pi \cdot 4t} + e^{j 2\pi \cdot 4t}\right) + j\frac{3}{2}\left(e^{-j 2\pi t} - e^{j 2\pi t}\right) \tag{2.55}$$

A plot of real spectrum is shown in Figure 2.22.

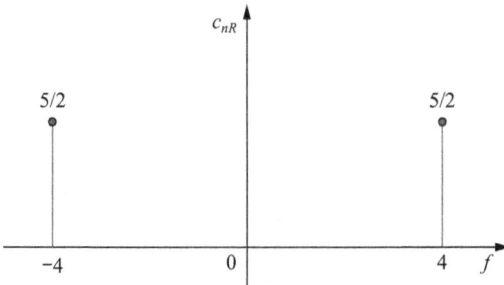

Figure 2.22: The real spectrum.

A plot of imaginary spectrum is shown in Figure 2.23.

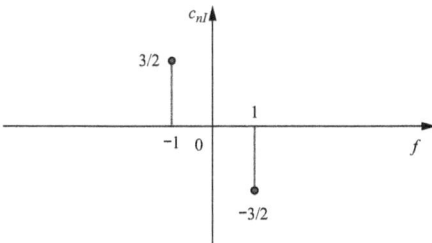

Figure 2.23: The imaginary spectrum.

A plot of two-sided amplitude spectrum is shown in Figure 2.24.

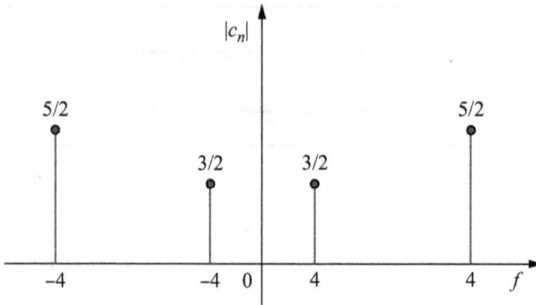

Figure 2.24: Two-sided amplitude spectrum.

A plot of single-sided amplitude spectrum is shown in Figure 2.25.

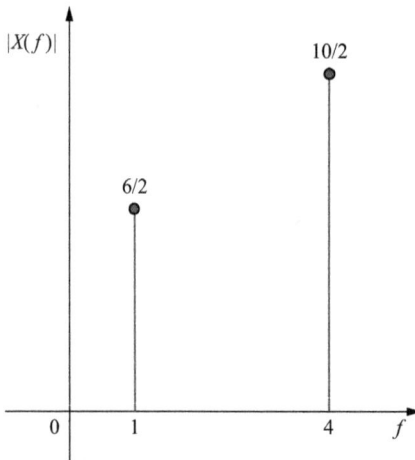

Figure 2.25: Single-sided amplitude spectrum.

2.6 Characteristics of periodic signal

Peak value, peak-to-peak value, average value, average absolute value, R.M.S (root mean square) value and average power are the most important characteristics of a periodic wave. This book uses a sine wave as an example to illustrate the characteristics of the periodic signal shown in Figure 2.26.

Definition of the peak value

The maximum value obtained by the alternating amount in a period is called its peak value x_p. It is also called the maximum, amplitude or peak value. The sinusoidal cross variable gets a peak at 90° shown in Figure 2.26:

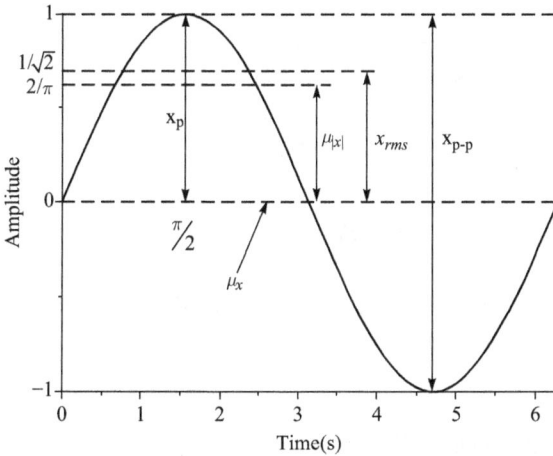

Figure 2.26: A sine signal.

$$x_p = |x(t)|_{max} \tag{2.56}$$

Definition of the peak-to-peak value

The peak-to-peak value is described as the vertical distance between the highest amplitude value (peak) and the lowest amplitude value (trough) of the wave. It will be measured in volts as a voltage waveform and can be marked as x_{p-p}:

$$x_{p-p} = |x(t)_{max}| + |x(t)_{min}| \tag{2.57}$$

Definition of the average value

The average value μ_x is the mean value of all the instantaneous values of an alternating voltage and current in a complete period. The average value, also called the mean value, is the most commonly used parameter:

$$\mu_x = \frac{1}{T_0} \int_0^{T_0} x(t)dt \tag{2.58}$$

For example, the average value μ_x of a sine wave is 0.637 times that of the peak value.

Definition of the average absolute value

If a symmetric wave is considered, such as a sinusoidal wave, the positive half period equals exactly the negative half period. The average value over the entire period will be zero. Therefore, the average absolute value $\mu_{|x|}$ is determined without considering the symbol:

$$\mu_{|x|} = \frac{1}{T_0} \int_0^{T_0} |x(t)|dt \tag{2.59}$$

Definition of the R.M.S value

The R.M.S value x_{rms} of a set of values is the arithmetic mean's square root of the squares of the instantaneous values. The R.M.S value of all times of the periodic signal is equal to the R.M.S value of one cycle of the signal. The RMS current is the "direct current value that dissipates power in a resistor" in physics:

$$x_{rms} = \sqrt{\frac{1}{T_0} \int_0^{T_0} x(t)^2 dt} \tag{2.60}$$

For example, the R.M.S value of a sine wave is 0.707 times that of the peak value, and similarly, 0.354 times that of the peak-to-peak value.

Definition of the average power

The average power P_{av} of the signal $x(t)$ is defined as

$$P_{av} = \frac{1}{T_0} \int_0^{T_0} x(t)^2 dt \tag{2.61}$$

2.7 Aperiodic transient signal

Aperiodic transient signal can be defined as a signal that has a short duration compared with the observed time. These signals include damped oscillating signals, switch bounce, exponential decay signals, rectangular pulse signals, single pulse signals, shock waves generated by impact tests and seismic signals.

2.7.1 Definition of the Fourier transform

Fourier transform is a tool that decomposes a signal into an alternative representation, characterized by sine and cosine. The Fourier transform shows that every signal can be rewritten as a sum of sinusoidal functions. The definition of the trigonometric Fourier series representation is closely related to the Fourier transform for the function $x(t)$. For such a function, the Fourier series can be calculated over any interval containing points where $x(t)$ is not equal to zero. You can also define Fourier transform as such a function. When the interval length used to calculate the Fourier series is increased, the coefficients of the Fourier series begin to look more like the Fourier transform, while the sum of the Fourier series looks more like the inverse Fourier transform.

In order to explain this point more accurately, this textbook assumes that T_0 is large enough to contain an interval $[t_0, t_0 + T_0]$ in which $x(t)$ is not equal to zero. $x(t)$ is given by

$$x(t) = \sum_{n=-\infty}^{\infty} c_n e^{j2\pi \cdot nf_0 t}$$

$$= \sum_{n=-\infty}^{\infty} \left(\frac{1}{T_0} \int_{t_0}^{t_0+T_0} x(t) e^{-j2\pi \cdot nf_0 t} dt \right) e^{j2\pi \cdot nf_0 t} \qquad (2.62)$$

If T_0 tends to infinity, we will get:

$$\Delta f = f_0 = 1/T_0 \to 0 \qquad (2.63)$$

Therefore

$$\Delta f \to df \qquad (2.64)$$

$$\int_{t_0}^{t_0+T_0} \to \int_{t_0-\frac{T_0}{2}}^{t_0+\frac{T_0}{2}} \to \int_{-\infty}^{\infty} \qquad (2.65)$$

$$nf_0 \to f \qquad (2.66)$$

$x(t)$ can be written as

$$x(t) = \int_{-\infty}^{\infty} df \left(\int_{-\infty}^{\infty} x(t) e^{-j2\pi ft} dt \right) e^{j2\pi \cdot ft}$$

$$= \int_{-\infty}^{\infty} \left(\int_{-\infty}^{\infty} x(t) e^{-j2\pi \cdot ft} dt \right) e^{j2\pi \cdot ft} df \qquad (2.67)$$

If

$$X(f) = \int_{-\infty}^{\infty} x(t) e^{-j2\pi \cdot ft} dt \qquad (2.68)$$

So

$$x(t) = \int_{-\infty}^{\infty} X(f) e^{j2\pi \cdot ft} df \qquad (2.69)$$

$X(f)$ is the Fourier transform of $x(t)$. The Fourier transform converts the time function $x(t)$ into a frequency function $X(f)$. $x(t)$ is the inverse Fourier transform of $X(f)$. The inverse Fourier transform converts a frequency function $X(f)$ into a time function $x(t)$. Fourier transform and inverse Fourier transform are a pair of transformations. This is because the Fourier transform and the inverse Fourier transform are only different in the symbols of the exponential parameters. The relationship between them is usually written more simply as follows:

$$x(t) \underset{\text{IFT}}{\overset{\text{FT}}{\Longleftrightarrow}} X(f) \qquad (2.70)$$

where $x(t)$ and $X(f)$ are called Fourier transform pairs.

Question 2.4:
Figure 2.27 shows a periodic rectangular pulse train $x(t)$. Sketch the double-sided amplitude spectrum of the periodic signal.

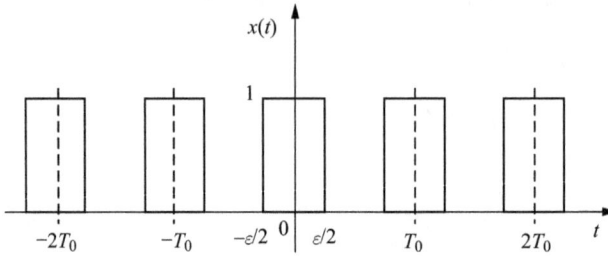

Figure 2.27: The periodic rectangular pulse train.

Solution:
Its mathematical function can be written as follows:

$$x(t) = \begin{cases} 1 & 0 \le |t| \le \varepsilon/2 \\ 0 & \varepsilon/2 < |t| < \dfrac{T_0}{2} \end{cases} \tag{2.71}$$

Based on the exponential Fourier series representation, you will get:

$$C_n = \frac{1}{T_0} \int_{t_0}^{t_0 + T_0} x(t) e^{-j 2\pi \cdot n f_0 t} dt$$

$$= \frac{1}{T_0} \int_{-\varepsilon/2}^{\varepsilon/2} e^{-j 2\pi \cdot n f_0 t} dt$$

$$= \frac{1}{T_0 \, 2\pi \cdot n f_0 j} \left[-e^{-j 2\pi \cdot n f_0 t} \right]\Big|_{-\varepsilon/2}^{\varepsilon/2}$$

$$= \frac{1}{T_0 \, 2\pi \cdot n f_0 j} \left[e^{\frac{j 2\pi \cdot n f_0 \varepsilon}{2}} - e^{-\frac{j 2\pi \cdot n f_0 \varepsilon}{2}} \right]$$

$$= \frac{1}{T_0 \pi \cdot n f_0} \sin \frac{2\pi \cdot n f_0 \varepsilon}{2}$$

$$= \frac{\varepsilon}{T_0} \frac{\sin \pi \cdot n f_0 \varepsilon}{\pi \cdot n f_0 \tau}$$

$$= \frac{\varepsilon}{T_0} \sin c(\pi \cdot n f_0 \varepsilon) \tag{2.72}$$

$$|C_n| = \left| \frac{\varepsilon}{T_0} \sin c(\pi \cdot n f_0 \varepsilon) \right|, n = 0, \pm 1, \pm 2, \dots \tag{2.73}$$

Plot amplitude spectrum
Case 1: $T_0 = 1$ s and pulse ratio = 20%, that is, $T_0 = 5\varepsilon = 1$ s.

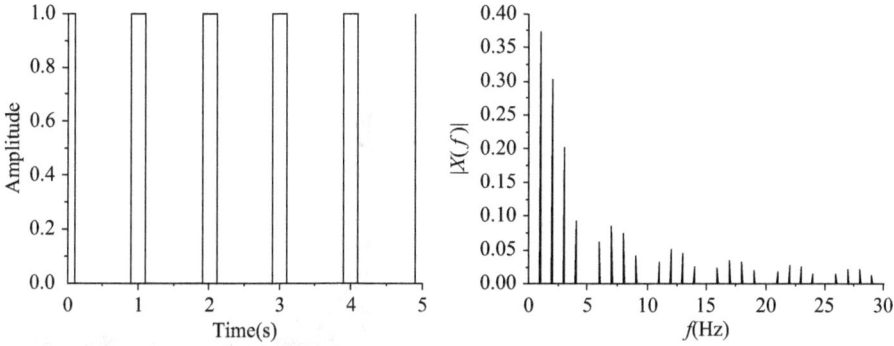

Figure 2.28: The amplitude spectrum under $T_0 = 5\varepsilon = 1$ s..

Case 2: $T_0 = 1$ s and pulse ratio = 10%, that is, $T_0 = 10\varepsilon = 1$ s.

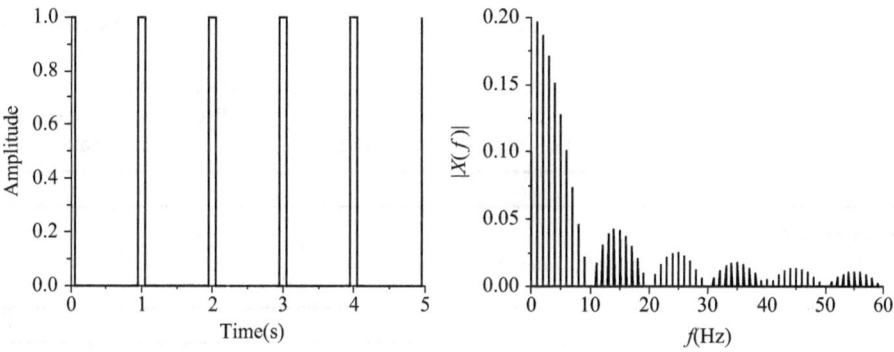

Figure 2.29: The amplitude spectrum under $T_0 = 10\varepsilon = 1$ s..

Case 3: $T_0 = 1$ s and pulse ratio = 5%, that is, $T_0 = 20\varepsilon = 1$ s.

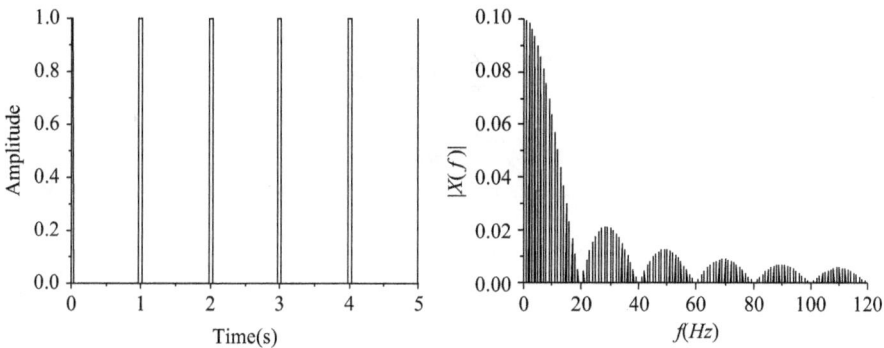

Figure 2.30: The amplitude spectrum under $T_0 = 20\varepsilon = 1$ s..

Case 4: $T_0 = 1$s and pulse ratio = 2.5%, that is, $T_0 = 40\varepsilon = 1$s.

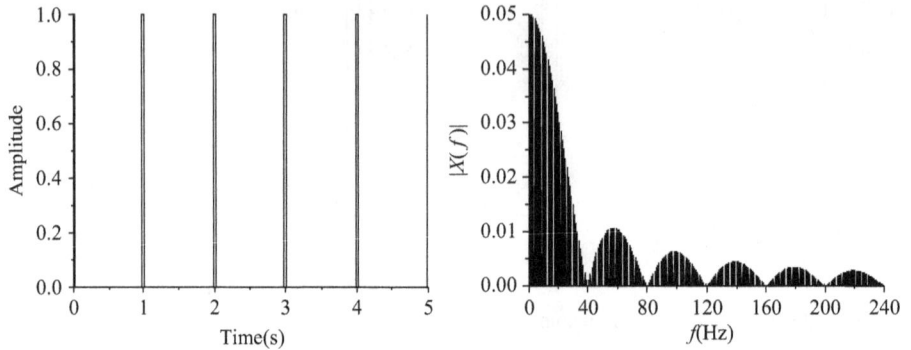

Figure 2.31: The amplitude spectrum under $T_0 = 40\varepsilon = 1$s.

From Figures 2.28, 2.29, 2.30 and 2.31, it is shown that the lower the pulse ratio, the larger the relative period. The density of the short lines (spikes) increases as the relative period increases, while the amplitude shows a decreasing trend. Therefore, it is conceivable that when the relative period is infinite, the envelope of the spectral short lines changes from a discrete point to a continuous curve. The spectrum achieves a conversion from a periodic signal to a transient signal.

Question 2.5:
Find the spectrum of the rectangular window function $x(t)$, $x(t) = \begin{cases} A_0, & -T_0/2 < t < T_0/2 \\ 0, & \text{else} \end{cases}$.

Solution:
Based on the Fourier transform, you will get:

$$X(f) = \int_{-\infty}^{\infty} x(t)e^{-j2\pi ft}\,dt$$

$$= \int_{-\frac{T_0}{2}}^{\frac{T_0}{2}} A_0 e^{-j2\pi ft}\,dt$$

$$= \frac{-A_0}{j2\pi f}\left(e^{-j\pi fT_0} - e^{j\pi fT_0}\right)$$

$$= A_0 T_0 \frac{\sin \pi fT_0}{\pi fT_0}$$

$$= A_0 T_0 \sin c(\pi fT_0) \tag{2.74}$$

The amplitude spectrum $X(f) - f$ is shown in Figure 2.32.

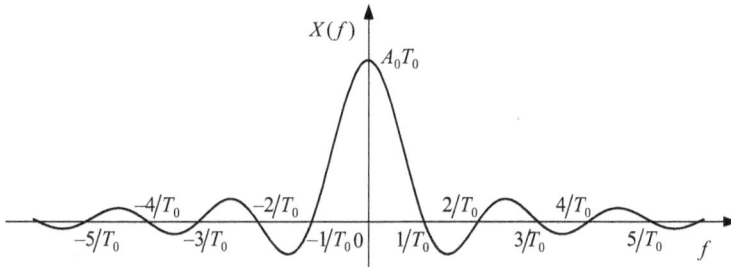

Figure 2.32: The amplitude spectrum of the rectangular pulse train.

The phase spectrum is determined by the sign of $\sin c(\pi f T_0)$ (Figure 2.33).

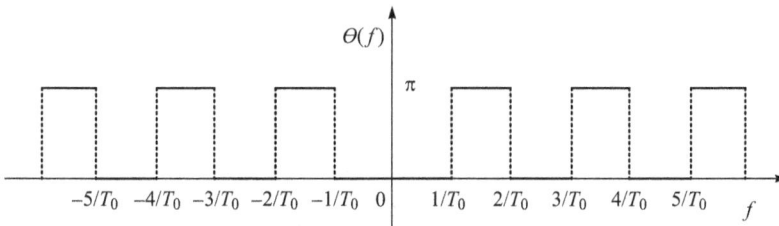

Figure 2.33: The phase spectrum of a rectangular pulse train.

2.7.2 Properties of Fourier transform

The Fourier transform is used to establish the correspondence between the time do-main and the frequency domain. The Fourier expansion property of periodic func-tions is a special case listed here. Some simple properties of the Fourier transform will be presented with even simpler proofs. The Fourier transform has the basic properties as follows:

(1) Linearity
If $x(t)$ has the Fourier transforms $X(f)$ and $y(t)$ has the Fourier transforms $Y(f)$, the sum $z(t) = x(t) + y(t)$ has the Fourier transform $X(f)+Y(f)$.

For any complex numbers a and b,

If

$$z(t) = ax(t) + by(t) \tag{2.75}$$

Then

$$Z(f) = aX(f) + bY(f) \tag{2.76}$$

A practical example is shown in Figure 2.34.

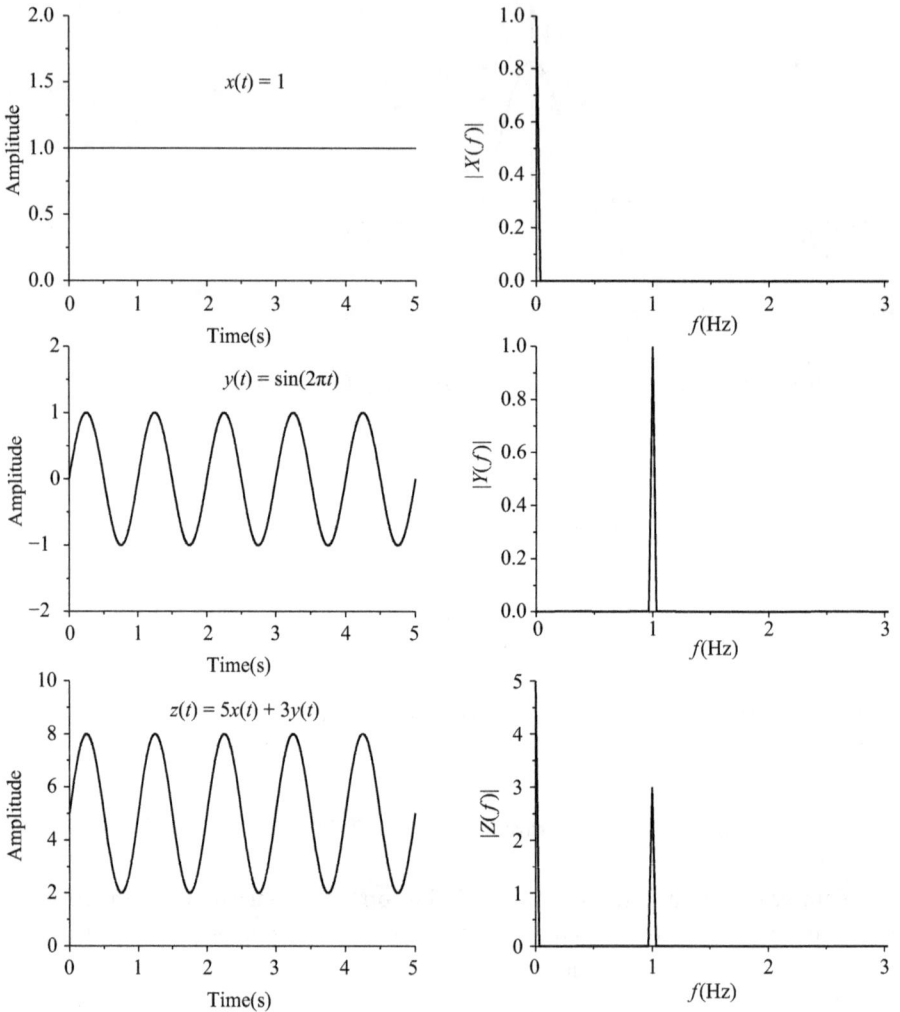

Figure 2.34: The linearity property ($z(t) = 5x(t) + 3y(t)$).

(2) Symmetry/duality

If the Fourier transform of a given time function is known, we will get the Fourier transform of the signal, which has the same functional form as the Fourier transform. In other words, the Fourier transform of the frequency function is consistent with the original signal that is inverted in time.

If

$$F[x(t)] = X(f) \qquad\qquad (2.77)$$

Then

$$F[X(t)] = x(-f) \tag{2.78}$$

Proof:

$$x(t) = \int_{-\infty}^{\infty} X(f)e^{j2\pi ft}df \tag{2.79}$$

Let $-t = t$

$$x(-t) = \int_{-\infty}^{\infty} X(f)e^{-j2\pi ft}df \tag{2.80}$$

Interchanging t and f, you will get

$$x(-f) = \int_{-\infty}^{\infty} X(t)e^{-j2\pi ft}dt \tag{2.81}$$

if the signal is even in particular

$$x(t) = x(-t) \tag{2.82}$$

Then you will have
   If

$$F[x(t)] = X(f) \tag{2.83}$$

Then

$$F[X(t)] = x(f) \tag{2.84}$$

You can get the following conclusion: the spectrum of even square wave is a standard sinc function. According to the symmetry, it is known that the spectrum of standard sinc function is an even square wave waveform, as shown in Figure 2.35.

(3) Time scaling

If the Fourier transform of $x(t)$ is $X(f)$, then, according to the time scaling property, we will get the Fourier transform of $x(kt)$. k is a real constant bigger than zero, and it can be determined by substituting $t = kt$ in the Fourier integral equation.
If

$$x(t) \Leftrightarrow X(f) \tag{2.85}$$

Then

$$x(kt) \Leftrightarrow \frac{1}{|k|}X\left(\frac{f}{k}\right) \tag{2.86}$$

Proof:

$$F[x(kt)] = \int_{-\infty}^{\infty} x(kt)e^{-j2\pi ft}dt$$

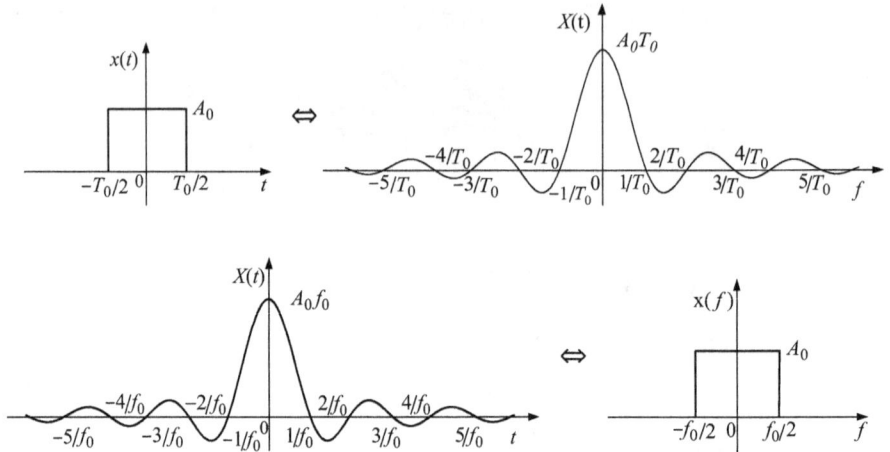

Figure 2.35: The duality property.

$$= \int_{-k*\infty}^{k*\infty} x(kt)e^{-j2\pi \frac{f}{k}(kt)}\frac{1}{k}d(kt) \qquad (2.87)$$

Now, if k is positive, the result is very simple:

$$F[x(kt)] = \int_{-\infty}^{\infty} x(kt)e^{-j2\pi \frac{f}{k}(kt)}\frac{1}{k}d(kt)$$

$$= \frac{1}{k}X\left(\frac{f}{k}\right)$$

$$= \frac{1}{|k|}X\left(\frac{f}{k}\right) \qquad (2.88)$$

If k is negative, the integration limits flip, which introduces an extra minus sign

$$F[x(kt)] = -\int_{\infty}^{\infty} x(kt)e^{-j2\pi \frac{f}{k}(kt)}\frac{1}{k}d(kt)$$

$$= -\frac{1}{k}X\left(\frac{f}{k}\right)$$

$$= \frac{1}{|k|}X\left(\frac{f}{k}\right) \qquad (2.89)$$

A practical example is shown in Figure 2.36.

(4) Time shift and frequency shift
If $x(t)$ can be shifted by a constant t_0, then by substituting $t \pm t_0$, you will get the Fourier transform

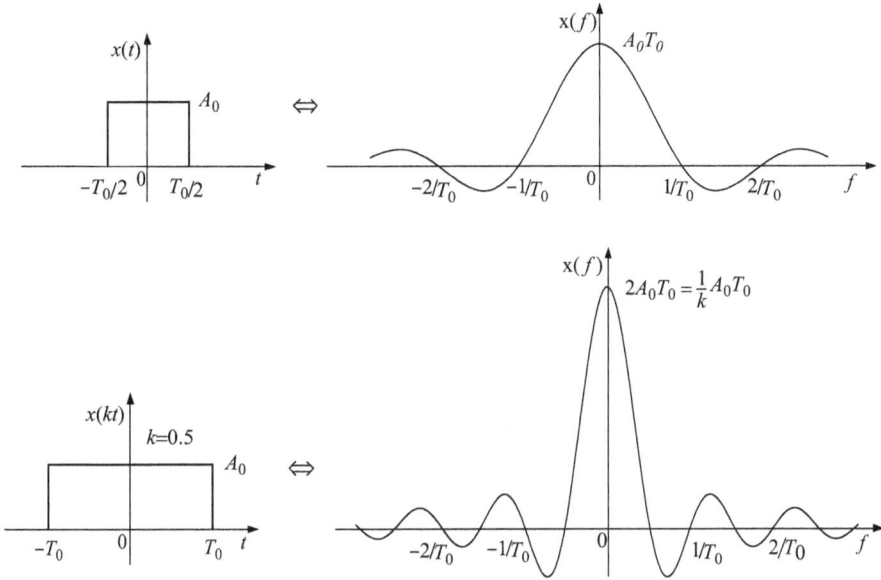

Figure 2.36: The time scaling property.

$$F[x(t \pm t_0)] = X(f)e^{\pm j2\pi f t_0} \tag{2.90}$$

This is the time shift property. A delayed signal $x(t - t_0)$ requires all the corresponding sinusoidal components $e^{j2\pi f t_0}$ for $-\infty < f < \infty$ delay t_0, thus changing their individual absolute phases.

Proof:

$$F[x(t - t_0)] = \int_{-\infty}^{\infty} x(t - t_0)e^{-j2\pi f(t - t_0 + t_0)}d(t - t_0)$$

$$= \int_{-\infty}^{\infty} x(t - t_0)e^{-j2\pi f(t - t_0)}e^{-j2\pi f t_0}d(t - t_0)$$

$$= e^{-j2\pi f t_0}\int_{-\infty}^{\infty} x(t - t_0)e^{-j2\pi f(t - t_0)}d(t - t_0)$$

$$= e^{-j2\pi f t_0}X(f) \tag{2.91}$$

If $X(f)$ can be shifted by a constant f_0, you will get its inverse transform

$$F^{-1}[X(f \pm f_0)] = x(t)e^{\mp j2\pi f_0 t} \tag{2.92}$$

This is the frequency shift property.

Proof:

$$F^{-1}[X(f-f_0)] = \int_{-\infty}^{\infty} X(f-f_0)e^{j2\pi t(f-f_0+f_0)}d(f-f_0)$$

$$= \int_{-\infty}^{\infty} X(f-f_0)e^{j2\pi t(f-f_0)}e^{j2\pi tf_0}d(f-f_0)$$

$$= e^{j2\pi tf_0}\int_{-\infty}^{\infty} X(f-f_0)e^{j2\pi f(f-f_0)}d(f-f_0)$$

$$= e^{j2\pi f_0 t}x(t) \tag{2.93}$$

(5) Convolution

The convolution theorem indicates that convolution in the time domain corresponds to multiplication in the frequency domain, and vice versa:

If

$$F[x_1(t)] = X_1(f) \tag{2.94}$$

$$F[x_2(t)] = X_2(f) \tag{2.95}$$

Then

$$F[x_1(t)*x_2(t)] = X_1(f)X_2(f) \tag{2.96}$$

$$F^{-1}[X_1(f)*X_2(f)] = x_1(t)x_2(t) \tag{2.97}$$

Proof:

$$F[x_1(t)*x_2(t)] = \int_{-\infty}^{\infty}\left[\int_{-\infty}^{\infty} x_1(\tau)x_2(t-\tau)d\tau\right]e^{-j2\pi ft}dt$$

$$= \int_{-\infty}^{\infty} x_1(\tau)\left[\int_{-\infty}^{\infty} x_2(t-\tau)e^{-j2\pi ft}dt\right]d\tau$$

$$= \int_{-\infty}^{\infty} x_1(\tau)X_2(f)e^{-j2\pi f\tau}d\tau$$

$$= X_1(f)X_2(f) \tag{2.98}$$

(6) Differentiation and integration

If

$$F[x(t)] = X(f) \tag{2.99}$$

Then

$$F\left[\frac{d^n x(t)}{dt^n}\right] = (j2\pi f)^n X(f) \tag{2.100}$$

$$F^{-1}\left[\frac{d^n X(f)}{df^n}\right] = (-j2\pi t)^n x(t) \tag{2.101}$$

$$F\left[\int_{-\infty}^{t} x(t)dt\right] = \frac{1}{j2\pi f}X(f) \tag{2.102}$$

Proof:

Differentiating the inverse Fourier transform $X(f)$ with respect to t we get:

$$\begin{aligned}
\frac{dx(t)}{dt} &= \frac{d}{dt}\left[\int_{-\infty}^{\infty} X(f)e^{j2\pi ft}df\right] \\
&= \int_{-\infty}^{\infty} X(f)\frac{d}{dt}e^{j2\pi ft}df \\
&= \int_{-\infty}^{\infty} j2\pi f X(f)e^{j2\pi ft}df \\
&= F^{-1}[j2\pi f X(f)]
\end{aligned} \tag{2.103}$$

Repeating this process we get:

$$F\left[\frac{d^n x(t)}{dt^n}\right] = (j2\pi f)^n X(f) \tag{2.104}$$

2.7.3 Spectrums of typical aperiodic transient signals

In this section, rectangular window function, δ function, sine/cosine function and sampling function are introduced.

2.7.3.1 Rectangular window function

Rectangular window (sometimes called a boxcar or Dirichlet window) is the simplest window. It is a function that remains constant throughout the interval, and zero elsewhere. This kind of function is called a rectangular window. When a waveform is multiplied by a rectangular window function, you will see that the product is zero outside the interval: all that remains is the portion of their overlap. The rectangular window function can be used to intercept the signal. When analyzing transient signals in modal analysis, such as impulse responses, pulses, chirped bursts, sinusoidal bursts and noise bursts, where the energy versus time distribution is extremely nonuniform, the rectangular window may be most suitable. The convolution in the time domain using a rectangular function is equivalent to multiplication in the frequency domain with a sinc function.

A rectangular window is described in the time domain or the frequency domain, as shown in Figure 2.37. It can be seen from the figure that it has a sinc-shaped frequency response, with a main lobe in the center of the frequency

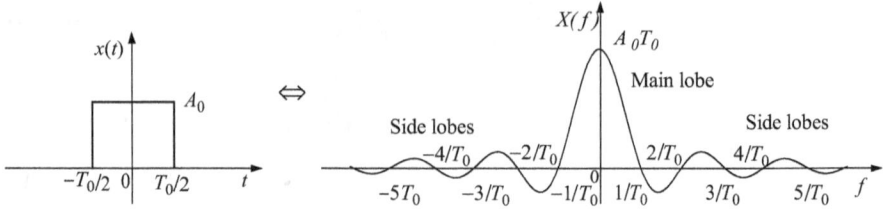

Figure 2.37: A rectangular window.

response and smaller side lobes on both sides. The main lobe is defined as a lobe containing higher power. The width of the main lobe is $2/T_0$, which is inversely proportional to the window width in the time domain. As T_0 increases, the main lobe narrows (in other words, the windows have a better frequency resolution). T_0 has no effect on the height of the side lobes.

2.7.3.2 Dirac delta function

(1) Definition of the Dirac delta function

In mathematics, the Dirac delta function, or δ function, or unit impulse function, is a generalized function on the real number line that is zero everywhere except at zero, with an integral of one over the entire real line. In the context of signal processing it is often referred to as the unit impulse symbol. The Dirac delta function is a nonphysical singularity function with the following definition:

$$\delta(t) = \begin{cases} \infty, & t = 0 \\ 0, & t \neq 0 \end{cases} \tag{2.105}$$

But with the requirement that

$$\int_{-\infty}^{\infty} \delta(t)dt = \lim_{\varepsilon \to 0} \int_{-\infty}^{\infty} \delta_\varepsilon(t)dt$$
$$= 1 \tag{2.106}$$

That is, the function has a unit area.

(2) δ sampling property/shifting property

The important property of the delta function is the following relation:

$$\int_{-\infty}^{\infty} \delta(t)x(t)dt = \int_{-\infty}^{\infty} \delta(t)x(0)dt$$
$$= x(0) \int_{-\infty}^{\infty} \delta(t)dt$$
$$= x(0) \tag{2.107}$$

This is easy to see for any function $x(t)$. First, except for $t = 0$, $\delta(t)$ disappears anywhere. Therefore, except for $t = 0$, the value of the function $x(t)$ is irrelevant. It can be seen that $\delta(t)x(t) = \delta(t)x(0)$. Then $f(0)$ can be pulled out of the integral, because it does not depend on t. By the change of the variable $t = t - t_0$, it follows that:

$$\int_{-\infty}^{\infty} \delta(t - t_0)x(t)dt = \int_{-\infty}^{\infty} \delta(t - t_0)x(t_0)dt$$

$$= x(t_0)\int_{-\infty}^{\infty} \delta(t)dt$$

$$= x(t_0) \tag{2.108}$$

This property is sometimes called the shifting property of the delta function. When you multiply any signal $x(t)$ by the time-shifting delta function, it basically filters out all amplitudes except for $t = t_0$ ($\delta(t - t_0)$ is zero except for $t = t_0$). $\delta(t - t_0)$ is a time-shifted delta function. $x(t_0)$ can be considered as a constant and can be placed outside the integral.

(3) δ function and convolution

The convolution of $x(t)$ function and $\delta(t)$ function is

$$x(t)*\delta(t) = \int_{-\infty}^{\infty} x(\tau)\delta(t - \tau)d\tau$$

$$= \int_{-\infty}^{\infty} x(\tau)\delta(\tau - t)d\tau$$

$$= x(t) \tag{2.109}$$

This is a key technique in signal processing. Similarly, when Dirac delta function is a time-shifted function $\delta(t \pm t_0)$,

$$x(t)*\delta(t \pm t_0) = \int_{-\infty}^{\infty} x(\tau)\delta(t \pm t_0 - \tau)d\tau$$

$$= \int_{-\infty}^{\infty} x(\tau)\delta(\tau - (t \pm t_0))d\tau$$

$$= x(t \pm t_0) \tag{2.110}$$

The convolution result shows that the $x(t)$ is redrawn at $t = \pm t_0$. The convolution of the any function and delta function is the simplest convolution. The convolution results of $x(t)$ and $\delta(t)$, and $x(t)$ and $\delta(t \pm t_0)$ are shown in Figure 2.38.

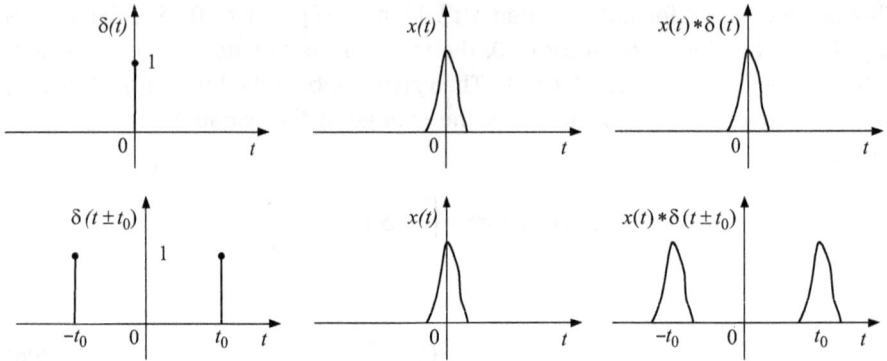

Figure 2.38: The convolution results.

(4) Dirac delta's spectrum/spectral density function

Fourier transform

$$\Delta(f) = \int_{-\infty}^{\infty} \delta(t)e^{-j2\pi ft}\,dt$$

$$= e^0$$

$$= 1 \tag{2.111}$$

Inverse Fourier transform

$$\delta(t) = \int_{-\infty}^{\infty} 1e^{j2\pi ft}\,df \tag{2.112}$$

Delta function has an infinitely broad spectrum, and 1 indicates that the spectrum of the delta is a uniform spectral density.

(5) Properties
The delta function satisfies the following scaling property for a nonzero scalar k:

$$\int_{-\infty}^{\infty} \delta(kt)\,dt = \int_{-\infty}^{\infty} \frac{\delta(kt)}{|k|}\,dkt$$

$$= \frac{1}{|k|} \tag{2.113}$$

Time shift property

$$F[\delta(t \pm t_0)] = e^{\pm j2\pi ft_0} \tag{2.114}$$

Frequency shift property

$$F^{-1}[\delta(f \pm f_0)] = e^{\mp j2\pi ft_0} \tag{2.115}$$

2.7.3.3 Sine and cosine functions

The sine/cosine function in the time domain can be considered as inverse Flourier transform of the differences of the two δ functions with different frequency shift in the frequency domain:

$$F[\sin 2\pi f_0 t] = F\left[j\frac{1}{2}(e^{-j2\pi f_0 t} - e^{j2\pi f_0 t})\right]$$

$$= j\frac{1}{2}[\delta(f + f_0) - \delta(f - f_0)] \tag{2.116}$$

$$F[\cos 2\pi f_0 t] = F\left[\frac{1}{2}(e^{-j2\pi f_0 t} + e^{j2\pi f_0 t})\right]$$

$$= \frac{1}{2}(\delta(f + f_0) + \delta(f - f_0)) \tag{2.117}$$

2.7.3.4 Dirac comb

The Dirac comb is an infinite delta pulse sequence that is evenly distributed over time. It is also known as the sampling function or impulse trains in electrical engineering:

$$\text{III}(t, T_0) \overset{\text{def}}{=} \sum_{n=-\infty}^{\infty} \delta(t - nT_0), \quad n = 0, \pm 1, \pm 2, \ldots \tag{2.118}$$

where T_0 is the interval between two neighboring impulses, and $f_0 = 1/T_0$ is the sampling rate.

As $\text{III}(t, T_0)$ is periodic (period T_0), it can be Fourier expanded:

$$\text{III}(t, T_0) = \sum_{n=-\infty}^{\infty} c_n e^{j2\pi n f_0 t} \tag{2.119}$$

where the Fourier series coefficients are

$$c_n = \frac{1}{T_0} \int_{t_0}^{t_0 + T_0} \text{III}(t, T_0) e^{-j2\pi \cdot n f_0 t} dt$$

$$= \frac{1}{T_0} \int_{-\frac{T_0}{2}}^{\frac{T_0}{2}} \text{III}(t, T_0) e^{-j2\pi k f_0 t} dt \quad , \quad n = 0, \pm 1, \pm 2, \ldots$$

$$= \frac{1}{T_0} \int_{-\frac{T_0}{2}}^{\frac{T_0}{2}} \delta(t) e^{-j2\pi k f_0 t} dt$$

$$= \frac{1}{T_0} \tag{2.120}$$

Now the expression for $\text{III}(t, T_0)$ can be written as

$$\text{III}(t, T_0) = \frac{1}{T_0} \sum_{n=-\infty}^{\infty} e^{j2\pi n f_0 t} \tag{2.121}$$

Fourier transform from the time domain to the frequency domain can be obtained by

$$F[III(t, T_0)] = F\left[\sum_{n=-\infty}^{\infty} \frac{1}{T_0} e^{j2\pi n f_0 t}\right]$$

$$= \frac{1}{T_0} F\left[\sum_{n=-\infty}^{\infty} e^{j2\pi n f_0 t}\right]$$

$$= \frac{1}{T_0} \sum_{n=-\infty}^{\infty} \delta(f - n f_0)$$

$$= \frac{1}{T_0} \sum_{n=-\infty}^{\infty} \delta\left(f - \frac{n}{T_0}\right) \qquad (2.122)$$

The spectrum of the Dirac comb and its still impulse sequence are shown in Figure 2.39. It should be noted that the sampling rate $f_0 = 1/T_0$ is the interval between two adjacent pulses. That is to say, the high sampling rate (the small interval between two adjacent sampling pulses) of $III(t, T_0)$ corresponds to the large gap f_0 in its spectrum.

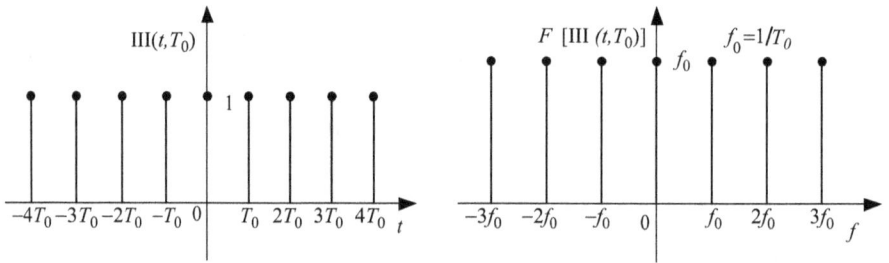

Figure 2.39: The spectrum of the Dirac comb.

2.8 Random signals

In engineering, mathematical descriptions of signals without precise signals are often encountered because they are random time functions. This kind of random development is usually caused by a single random variable, and it is usually the result of many random variables. Random signals are those that cannot be accurately predicted. Using the signal's past history and the magnitude of the amplitude it uses, it is not possible to precisely predict what particular value it will take on at certain instants in the future. The value of the signal can only be predicted based on certain probabilities. Many of the signals processed by a computer can be considered random, such as voice, video, audio, digital communications, biological and economic signals. Many systems have random signals and noise, and signal processing techniques enable engineers to extract useful signals.

The random time function can be the desired signal, or it can be an unnecessary signal, accidentally adding the desired signal and interfering with the desired signal. Unwanted signal that is inadvertently added to the desired (information) signal disturbs the desired signal. We call the signal we need as random signal and the signal we don't need as noise. Noise is a kind of interference. In signal processing, the required signal can be minimized to minimize interference and suppress interference as much as possible. In all cases, the effect of the interfering signal on system performance needs to be described.

Understanding the nature of random signals and noise is critical to detecting signals and reducing and minimizing the effects of noise. The concept of random variables is sufficient to handle unpredictable signals, which allows us to propose probability descriptions of random quantities. In the real world, signals not only vary in magnitude but also in terms of time parameters. We should develop new and feasible mathematical tools for probability representation of random signals. In this way, the resulting theory extends the mathematical model of probability to include temporal parameters, often referred to as stochastic or stochastic process theory.

We consider a sample space S with sample points (related to some random experiments) s_1, s_2, ... For each one $s_i \in S$, a really valuable time function will be assigned. In this case, as shown in Figure 2.40, a sample space with n points and n waveforms is shown, marked as $x_i(t)$, $i = 1$, 2, 3, 4, 5, ..., n.

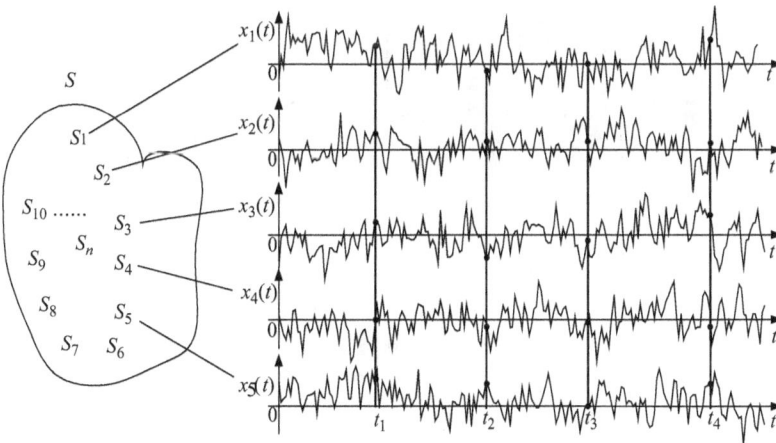

Figure 2.40: A simple random process.

Let's look for some point $t = t_1$ in Figure 2.40 to observe the waveforms. Each point s_i of S in the graph is associated with a probability P_i and a number $x_i(t_1)$. In this way, the set of numbers $\{x_i(t_1), i = 1, 2, 3, 4, 5, \ldots, n\}$ forms a random variable. We can observe the waveform at the second time $t = t_2$ and generate different sets of numbers $\{x_i(t_2),\ i = 1,\ 2,\ 3,\ 4,\ 5,\ \ldots,\ n\}$, thus producing different random variables. In practical terms, this set of n waveforms defines a random variable for the selection of each observation time. You can also easily extend the above situation to an infinite number of sampling points.

Generally, a probabilistic system consisting of a set of time functions, a sample space and a probability measure is called a stochastic process and expressed as $X(t)$. Stochastic processes are functions of two variables, such as $-\infty < t < \infty$ and $s \in S$. Therefore, a better representation will be $X(s, t)$. As a matter of convenience, a simplified representation $X(t)$ is used to represent a stochastic process. Each waveform is referred to as a sample function, and a probability is assigned to any meaningful event associated with these sample functions. For a stochastic process $X(t)$, the following quantities can be determined:

$X(t)$ is the stochastic process;

$x_i(t)$ is a sampling function, which is associated with the sampling point s_i;

$X(t_i)$ is the random variables at $t = t_i$, which is obtained from observation processes $t = t_i$;

$x_i(t_i)$ is the real number, which is the value of $x_i(t)$ at $t = t_i$.

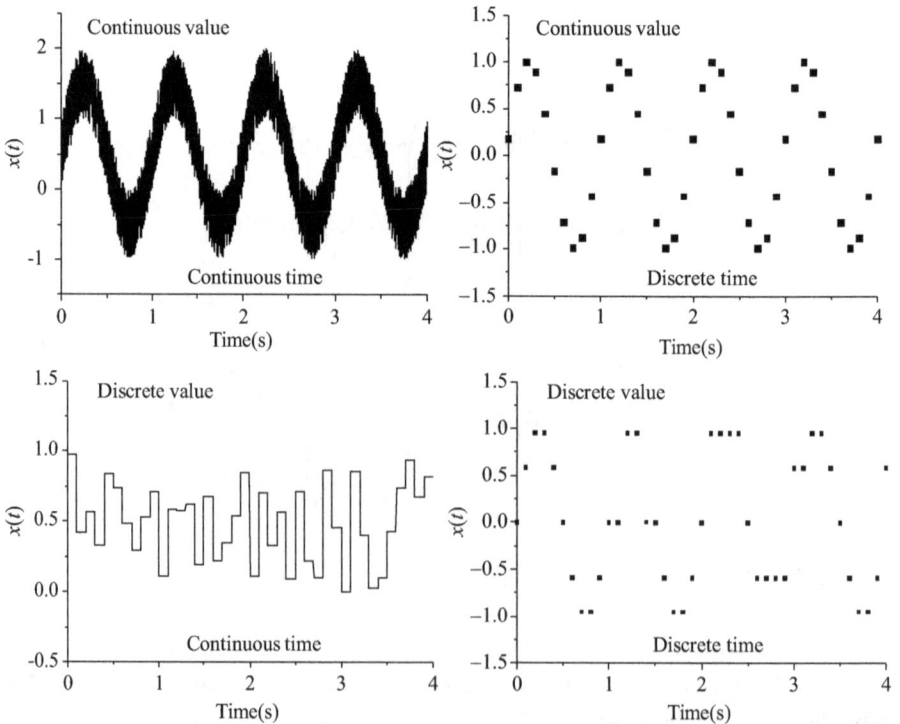

Figure 2.41: Types of stochastic processes.

The values of time t and $X(t)$ are continuous. This process is defined as a continuous stochastic process. The values of $X(t)$ are continuous, while time t is discrete. These processes are called continuous random sequences. Supposing that the values of $X(t)$ are discrete and the time t is continuous, the process can be called a discrete stochastic process. Last, if the time t and the values of $X(t)$ are discrete, it can be said that the process is a discrete random sequence. Types of stochastic processes are shown in Figure 2.41.

3 Characteristics of measuring systems

3.1 Introduction

The purpose of an experiment is to understand physical systems or phenomena better. When students are prepared to measure certain physical variables or attributes of the system, they must decide which measuring device is more suitable for their experiment. Students must be sure that their measurement system accurately reflects the measured value or change in measuring. The characteristics of the measuring device can be divided into two categories: static characteristics and dynamic characteristics.

In this chapter, students will learn: (1) static characteristics; (2) dynamic characteristics; (3) step responses of measuring system; (4) nondistorting measurement; (5) identification of dynamic characteristics; (6) loading effect; (7) anti-interference technology.

3.1.1 Classification of measuring instruments

Various classification methods for measuring instruments are as follows:

(1) Active and passive instruments

Instruments can be divided into active instruments and passive instruments. This classification is based on whether the output of instrument is produced totally by the measurand, or whether the measurand requires additional external power.

Active instruments: These measurement equipment require an external power supply to operate like strain gauges, RTD (Resistance Temperature Detector), sonar, radar and more. They only work when they are supplied with external power. Active instrument power supplies are usually electrical, but in some cases, it can also be other forms of energy, such as hydraulic and pneumatic. The world's largest radio telescope (China) is probably the world's biggest active sensor (Figure 3.1). It is an active measurement equipment.

Passive instruments: These instruments do not require power to work like a piezoelectric sensor, a thermocouple, a moving coil generator and a photovoltaic cell. They convert the physical change into a voltage signal.

(2) Null-type and deflection-type instruments

Null-type instruments: These instruments attempt to keep the deflection at zero by appropriately applying a known effect opposite to that produced by the measured amount. The deadweight meter shown in Figure 3.2 is a typical null-type instrument.

https://doi.org/10.1515/9783110624397-003

Figure 3.1: The world's largest radio telescope (Pingtang County, China).

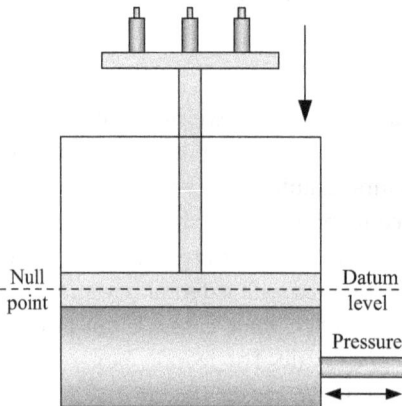

Figure 3.2: A typical null-type instrument.

In Figure 3.2, weights need to be placed at the top of the piston to balance the downward force. Weights are added here until the piston reaches the null point. The pressure measurement is based on the weight value required to reach the zero position. The fluid pressure can be calculated based on the added weight and the cross-sectional area of the piston. The advantage of a null-type instrument is that it minimizes the measurement load error and the interaction between the measurement system and the measurand. However, it is only used for static measurements.

Deflection-type instruments: The output of a deflection-type instrument is deflected in proportion to the magnitude of the measurand, which is the most common measuring instrument. The deflection-type instrument is obviously more convenient to use. It can be seen that reading the pointer position is much simpler than adding and subtracting weights (up to the null point). The biggest advantage of the deflection-type instrument is its high dynamic response.

(3) Analogue and digital instruments

Analog instrument: The output signal provided by analog instruments is continuous in amplitude and time. The output signal of the instrument is proportional to the input signal.

Digital instrument: A digital instrument provides an output signal that varies over discrete times. The output signal has an infinite number of values within the range designed for measurement.

The analog oscilloscope and digital oscilloscope are shown in Figure 3.3. Many digital instruments combine analog sensors and digital readouts.

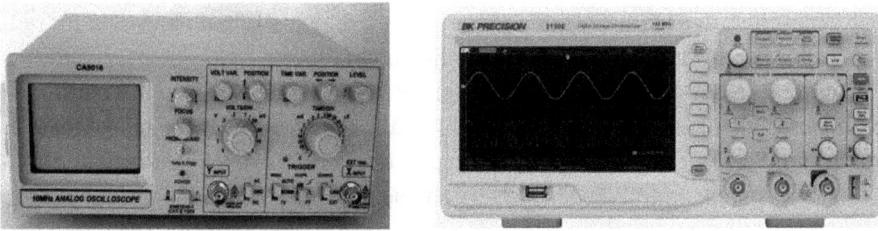

Figure 3.3: Analog oscilloscope (SMT MAX, Canada) and digital oscilloscope (B&K Precision, USA).

(4) Indicating instruments and instruments with a signal output

Indicating instrument: The indicating instrument indicates the amount of the measurement. Its indication is usually moved by the pointer to the calibration dial. Ordinary voltmeters, ammeters, wattmeters and so on are typical indicating instruments.

Instrument with a signal output: An instrument with a signal output gives out the output in the form of measuring signal, whose value is proportional to the measured value. The output signal can be voltage or current.

The indicating instrument and the instrument with a signal output are given in Figure 3.4.

(5) Smart and nonsmart instruments

Smart instruments: Smart instruments use a microprocessor to measure or control a single process variable. It is very flexible due to the parameters set by the vendor or

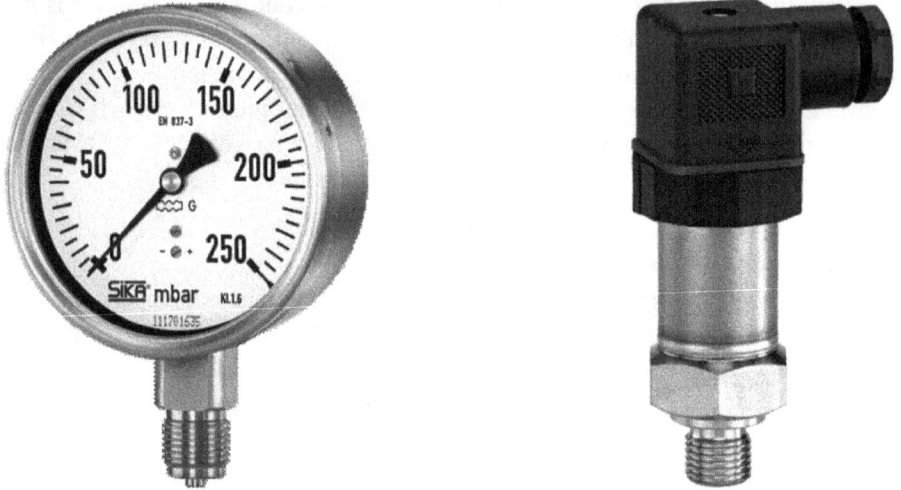

Figure 3.4: Capsule element pressure gauge and pressure sensor with a signal output (SiKA, German).

user. Smart instruments have self-calibration, self-diagnostic compensation for random errors and the ability to measure nonlinear variables.

Figure 3.5 shows a Carbo 510 smart sensor. The Carbo 510 is a smart sensor for self-diagnosis. This smart sensor delivers a value for every 15 s is hailed as the fastest carbon dioxide analyzer on the market. It offers the highest accuracy and reliability. It also detects and displays errors and service requirements.

Figure 3.5: Carbo 510 smart sensor (Anton Paar GmbH, Austria).

Nonsmart instruments: These can have a mechanical indicator without a microprocessor.

(6) Manually operated and automatic-type instruments

Manually operated type instrument: This requires manual operation. A typical manual operation instrument is a glass mercury thermometer for measuring temperature shown in Figure 3.6.

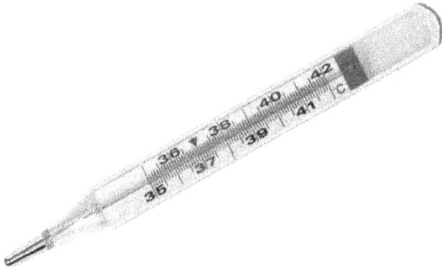

Figure 3.6: The mercury-in-glass thermometer.

Automatic-type instrument: This makes use of automation to measure and evaluate the test results quickly, which can automatically test and diagnose faults using chips and integrated circuits.

(7) Contacting- and noncontacting-type instruments

Contacting-type instrument: This instrument itself is stored in the measured medium. A clinical glass mercury thermometer is a typical example of the instrument shown in Figure 3.6.

Noncontacting-type instrument: This measures the required measurand without directly contacting the measured medium. An infrared thermometer is an example of such an instrument (Figure 3.7). The infrared thermometer measures only the surface temperature. It is ideal for moving parts and rough surfaces.

3.1.2 Description of the characteristics

To choose the device that best fits a particular measurement application, the system characteristics need to be understood. The characteristic is the relationship between the input and the output, which shows the performance of the instrument to be used (Figure 3.8). It is roughly divided into two categories: static characteristics and dynamic characteristics.

For an engineering measurement, if the input X and the output Y are known, the characteristics of the system can be inferred. This process is called system

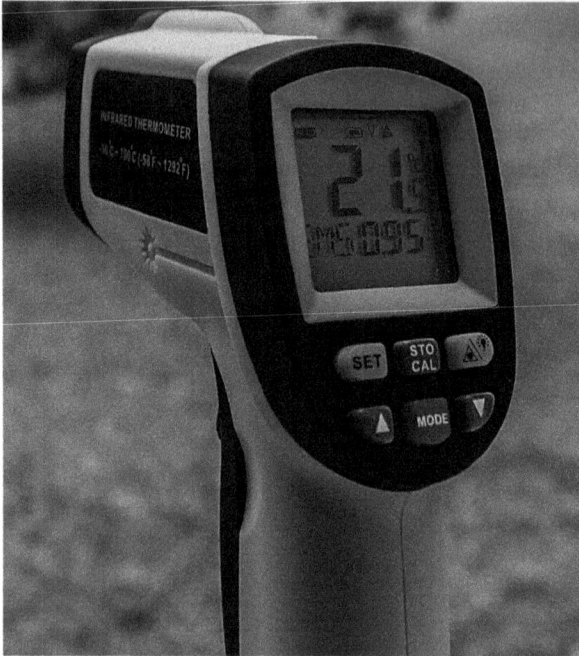

Figure 3.7: The infrared thermometer (Benetech, USA).

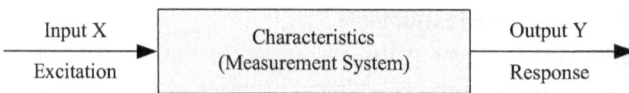

Figure 3.8: The relationship between input and output.

identification. The characteristics of the linear time-invariant (LTI) system can be described in the time domain, plural domain and frequency domain. In the time domain, you can set up the mathematical model of the system and solve the differential equations. In the plural domain, you can use Laplace transforms to get the transfer function $H(s) = Y(s)/X(s)$. The frequency response function $H(f) = Y(f)/X(f)$ can be obtained through experiments in the frequency domain.

3.1.3 Linear system and its main properties

The input $x(t)$/output $y(t)$ relationship in Figure 3.8 can be described by the linear constant–coefficient differential equation as follows:

$$a_0 y(t) + a_1 \frac{dy(t)}{dt} + \cdots + a_{n-1} \frac{d^{n-1}y(t)}{dt^{n-1}} + a_n \frac{d^n y(t)}{dt^n}$$

$$= b_0 x(t) + b_1 \frac{dx(t)}{dt} + \cdots + b_{m-1} \frac{d^{m-1}x(t)}{dt^{m-1}} + b_m \frac{d^m x(t)}{dt^m}$$

(3.1)

where t is a variable; $a_0, a_1, \ldots, a_{n-1}, a_n, b_0, b_1, \ldots, b_{m-1}, b_m$ are constant coefficients. This system is called an LTI system. You can consider many physical systems as LTI systems. A good example of an LTI system is a circuit, which consists of resistors, capacitors and inductors. LTI systems play a key role in designing many of the most dynamic systems. In discussing the properties of the LTI system, a very important linear result will be discussed. For the linear system, if the input of the system is zero for all t, the output of the system is also zero. A continuous LTI system is a system that can satisfy the principles of linearity and time invariance and process continuous time signals. The LTI system must meet the following properties:

(1) Superposition principle

Two inputs $x_1(t)$, $x_2(t)$ of an LTI system are different, and then the outputs will be $y_1(t)$ and $y_2(t)$. If the sum of individual inputs $x_1(t) \pm x_2(t)$ is applied to this system, the output will be $y_1(t) \pm y_2(t)$:

$$x_1(t) \pm x_2(t) \rightarrow y_1(t) \pm y_2(t)$$

(3.2)

(2) Scaling property

When you input $ax(t)$ to an LTI system, you will get the output $ay(t)$:

$$ax(t) \rightarrow ay(t)$$

(3.3)

(3) Differential property

When you input $(dx(t))/dt$ to an LTI system, you will get the output $(dy(t))/dt$:

$$\frac{dx(t)}{dt} \rightarrow \frac{dy(t)}{dt}$$

(3.4)

(4) Integral property

When you input $\int_0^{t_0} x(t)dt$ to an LTI system, you will get the output $\int_0^{t_0} y(t)dt$:

$$\int_0^{t_0} x(t)dt \rightarrow \int_0^{t_0} y(t)dt$$

(3.5)

(5) Frequency property

When you input a sine wave to the LTI system, you will get the output that is also a sine wave with a different amplitude and phase angle, but with the same frequency:

$$x(t) = X_0 \sin \omega t$$
$$y(t) = Y_0 \sin(\omega t + \varphi) \tag{3.6}$$

Proof:

According to the scaling property, if you input $\omega^2 x(t)$, you will get the output $\omega^2 y(t)$:

$$\omega^2 x(t) \rightarrow \omega^2 y(t) \tag{3.7}$$

According to the differential property, if the input is $(d^2 x(t))/dt^2$, the output will be $(d^2 y(t))/dt^2$:

$$\frac{d^2 x(t)}{dt^2} \rightarrow \frac{d^2 y(t)}{dt^2} \tag{3.8}$$

Since $x(t)$ is a sine signal, it can be written as

$$x(t) = X_0 e^{j\omega t} \tag{3.9}$$

The second derivative of $x(t)$

$$\frac{d^2 x(t)}{dt^2} = \frac{d^2 (X_0 e^{j\omega t})}{dt}$$
$$= (j\omega)^2 X_0 e^{j\omega t}$$
$$= -\omega^2 x(t) \tag{3.10}$$

Thus

$$\frac{d^2 x(t)}{dt^2} + \omega^2 x(t) = 0 \tag{3.11}$$

The corresponding output will be

$$\frac{d^2 y(t)}{dt^2} + \omega^2 y(t) = 0 \tag{3.12}$$

The only possible solution is

$$y(t) = Y_0 e^{j(\omega t + \phi)} \tag{3.13}$$

Superposition principle and frequency property are very important in measuring technology.

3.2 Static characteristics

In the static measurement, the measured value remains constant, or changes/varies quite slowly. For instruments that measure invariant process conditions, you need to consider the static characteristics of the instrument. The main static characteristics are accuracy, precision, range or span, linearity, sensitivity, resolution, bias, dead zone, threshold, hysteresis, drift, reproducibility, stability, tolerance and error.

3.2.1 Accuracy

Accuracy is the proximity of the instrument reading to the true value being measured. Accuracy usually depends on the limitations of the instrument itself and the unavoidable defects in the measurement process. Accuracy is expressed in the following three ways: point accuracy, the percentage of accuracy in the scale range and the percentage of accuracy in the true value. Point accuracy is specified only at a specific scale point. When an instrument is operated at a uniform scale, its accuracy can be expressed in a proportional range. The best way to imagine accuracy is to specify it based on the actual value being measured. In this book, accuracy is defined as a percentage of error, that is, the percentage of full-scale reading such as 0.1%, 0.2% or 0.3% of the actual value:

$$\text{Percentage of F.S. deflection} = \frac{(\text{Measured value} - \text{True value})}{\text{Maximum scale value}} \times 100\% \quad (3.14)$$

Question 3.1:
Find the maximum error of a displacement sensor with a range between 0 and 100 mm with an accuracy of $\pm 0.5\%$F.S.

Solution:

$$\text{Maximum error} = \pm 0.5\% \times 100 \text{ mm}$$
$$= \pm 0.5 \text{ mm} \quad (3.15)$$

Notes:
It is essential to select an instrument that has a suitable operating range.

Question 3.2:
Calculate the accuracy of a displacement sensor (has an error of ± 0.2 mm) with a range between 0 and 50 mm. Calculate the error percentage when the reading point is 20 mm.

Solution:

$$\text{Accuracy} = \pm \frac{0.2}{50} \times 100\%$$
$$= \pm 0.4\% \text{ F.S.} \tag{3.16}$$

$$\text{Error percentage} = \pm \frac{0.2}{20} \times 100\%$$
$$= \pm 1\% \text{ F.S.} \tag{3.17}$$

Notes:
This displacement sensor is not suitable for low range reading.

Question 3.3:
Two displacement sensors (A and B) have a full-scale accuracy of ±0.1%. Sensor A has a range of 0–50 mm and sensor B 0–100 mm. Which sensor is more suitable to be used if the reading is 40 mm?

Solution:
Displacement sensor A

$$\text{Maximum error} = \pm 0.1\% \times 50 \text{ mm}$$
$$= \pm 0.05 \text{ mm} \tag{3.18}$$

Error percentage at 40 mm

$$\text{Error percentage} = \pm \frac{0.05}{40} \times 100\%$$
$$= \pm 0.125\% \text{ F.S.} \tag{3.19}$$

Displacement sensor B

$$\text{Maximum error} = \pm 0.1\% \times 100 \text{ mm}$$
$$= \pm 0.1 \text{ mm} \tag{3.20}$$

Error percentage at 40 mm

$$\text{Error percentage} = \pm \frac{0.1}{40} \times 100\%$$
$$= \pm 0.25\% \text{ F.S.} \tag{3.21}$$

Notes:
Displacement sensor A is more suitable to use at a reading of 40 mm because the error percentage (±0.125%) is smaller compared to the percentage error of displacement sensor B (±0.25%).

3.2.2 Precision

Precision describes how randomly the instrument changes its output when measuring a constant amount. It is used to judge the reproducibility of the instrument. In fact, people often confuse the difference between precision and accuracy. There is little correlation between high precision and measurement accuracy. The instrument precision usually depends on the elements causing random or accidental errors, and the instrument accuracy depends on systematic errors. Precision depends on the repeatability of the device that is precise but not necessarily accurate. Figure 3.9 shows this relationship. High accuracy means that the average value is close to the true value, and high precision means that the standard deviation σ is small.

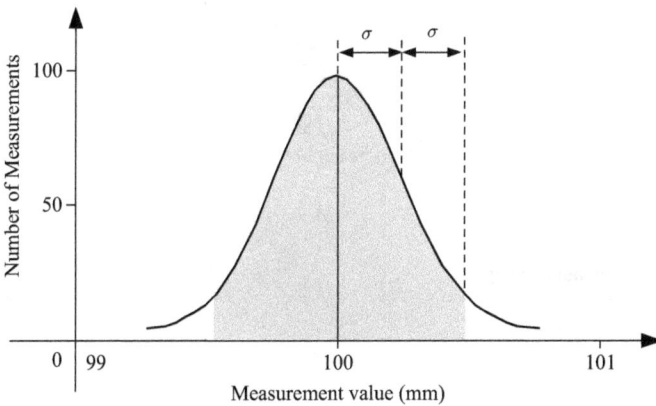

Figure 3.9: The standard deviation.

A clearer description can be found in Figure 3.10 when measuring a fixed target position. As shown in Figure 3.10, target A has a perfect accuracy with high precision, target B has high precision with low accuracy, target C has a low accuracy with low precision and target D has relatively good accuracy (average value has high accuracy) with low precision. This can be inferred that it is better to be precise or repeatable, not accurate.

3.2.3 Range or span

The range or span is defined as the reading between the minimum and maximum values, which is designed to measure for the instrument. In this book, accuracy is calculated based on the range or span of the instrument. Ranges or spans usually have positive values. For example, the reading range of a sensor used to measure pressure has a reading range of −100 to 2,000 kPa, which is 2,100 kPa.

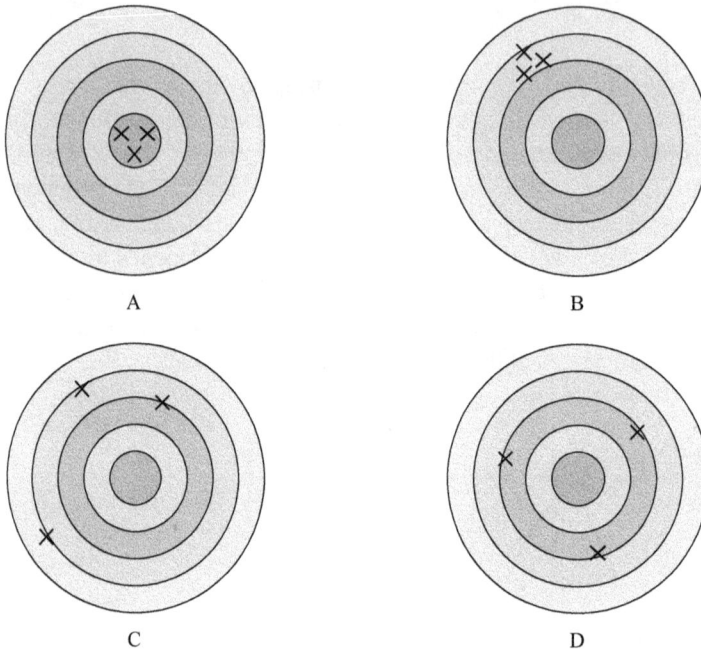

Figure 3.10: The target position measurement.

3.2.4 Linearity

Linearity is defined as the maximum deviation between the calibration points (curve) and the idealized straight line. In some books, linearity is often reported as non-linearity. The output of an instrument must be linearly proportional to the measured quantity. The linearity is usually displayed as a percentage of full-scale percentage (% F.S.). Linearity δ is defined as

$$\delta = \frac{\Delta y_{max}}{y_{max}} \cdot 100\% \tag{3.22}$$

where Δy_{max} is the maximum deviation between the calibration curve and the idealized straight line; y_{max} is the maximum range of the instrument. The maximum deviation Δy_{max} can be found in Figure 3.11. The idealized straight line can be obtained by the end-point straight-line method and the least squares method.

3.2.5 Sensitivity

The sensitivity represents the minimum variation in the measured variable to which the instrument responds. It is the ratio of instrument output changes to input changes.

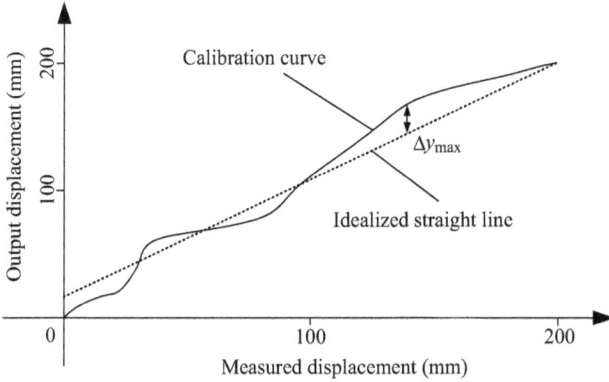

Figure 3.11: The maximum deviation.

The sensitivity S is defined as the ratio of magnitude of the measured quantity over the magnitude of the response:

$$S = \lim_{\Delta x \to 0} \left(\frac{\Delta y}{\Delta x} \right) = \frac{dy}{dx} \qquad (3.23)$$

where Δy is the change in output and Δx is the change in input. If the calibration curve has linear characteristics, as shown in Figure 3.12, the sensitivity of the instrument can be considered as the slope of the calibration curve. If the calibration curve does not have linear characteristics, as also shown in Figure 3.12, the sensitivity varies with the input.

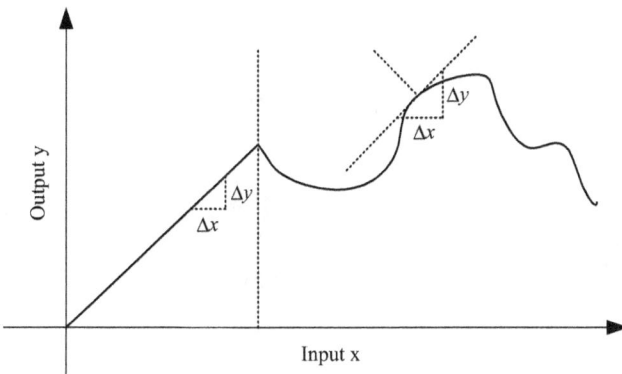

Figure 3.12: The sensitivity.

Question 3.4:
Calculate the sensitivity of the load cell. Its output is as follows:

Input (mm)	Output (kg)
0	0
0.02	10
0.04	20
0.06	30
0.08	40
0.1	50

Solution:
From the above table, the input and the output are changing linearly, so the sensitivity can be calculated as

$$S = \frac{dy}{dx} = \frac{50-0}{0.1-0} = \frac{50 \text{ kg}}{0.1 \text{ mm}} = 500 \text{ kg/mm} \qquad (3.24)$$

3.2.6 Resolution

Resolution is defined as the smallest increment of the input value that is detected by the instrument. If you slowly increase the input from an arbitrary input value, you will again find that the output will not change at all until it exceeds an increment. This increment is defined as resolution. Figure 3.13 shows different resolutions of the images.

High resolution   Low resolution

Figure 3.13: The different resolutions of the images.

3.2.7 Bias

Bias is a constant error that exists throughout the measurement range of the instrument. This error can be rectified through calibration. For example, a pressure sensor always gives a bias reading. The device always provides a 30 kPa reading even without any pressure applied. Therefore, if a standard pressure with 300 kPa is given, the given reading will be 330 kPa. This indicates that there is a constant deviation of 30 kPa that requires correction.

3.2.8 Dead zone

The dead zone is the largest change in input, and there is no output. It is defined as the range of input readings when the output remains unchanged, as shown in Figure 3.14. This may be due to friction, clearance or hysteresis in the instrument. For example, the backlash can be defined as the maximum angle or distance through which any part of the mechanical system can move in one direction without causing the next part to move. The dead zone is inevitable due to the existence of backlash.

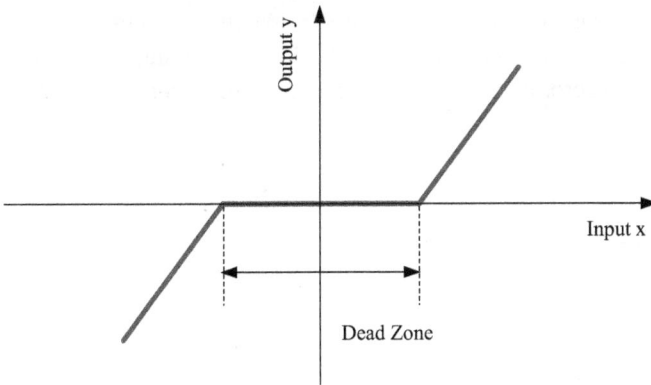

Figure 3.14: The dead zone.

3.2.9 Threshold

A threshold is defined as the smallest measurable input, and the output change is not recognized below this value. When the input reading is increased from zero, it will reach a specific value before the change occurs in the output. The minimum limit for the input reading defines the instrument's threshold shown in Figure 3.15.

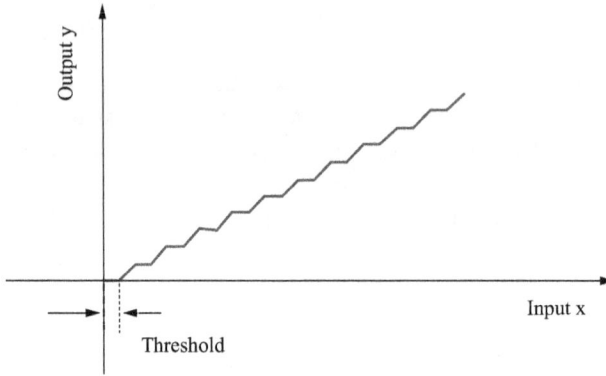

Figure 3.15: The threshold.

3.2.10 Hysteresis

Hysteresis is the largest difference between output readings for the same measurement point, one point obtained while increasing from zero and the other obtained while falling from full-scale. These points are taken on the same continuous cycle. Since all the energy loaded into the stressed parts during unloading is not recoverable, there will be a hysteresis, as shown in Figure 3.16. When the input of the instrument changes from zero to full scale, the output will be changing if the input decreases from full scale to zero. Energy storage/dissipation in systems usually results in hysteresis.

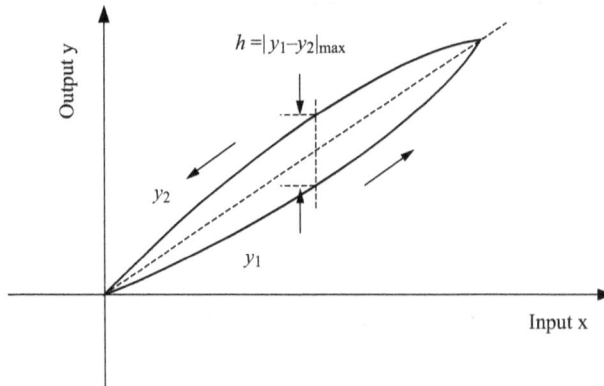

Figure 3.16: The hysteresis.

The deviation is expressed as a percent of full scale:

$$H = \frac{h}{y_{max}} \times 100\%$$

$$= \frac{|y_1 - y_2|_{max}}{y_{max}} \times 100\%$$

$$= \frac{\Delta y_{max}}{y_{max}} \times 100\% \qquad (3.25)$$

3.2.11 Drift

Drift is defined as the gradual change during a period of time when the input variable is constant. Environmental factors such as stray electric fields, stray magnetic fields, temperature changes and mechanical vibrations may cause drift. Drift is divided into three parts: zero drift, sensitivity drift and combined drift, as shown in Figure 3.17. Zero drift is the output change of the instrument itself, and the input change cannot cause zero drift. Internal temperature changes and component instability can cause zero drift. Sensitivity drift defines the amount where the sensitivity of the instrument changes as environmental conditions change. The combined drift consists of zero drift and sensitivity drift.

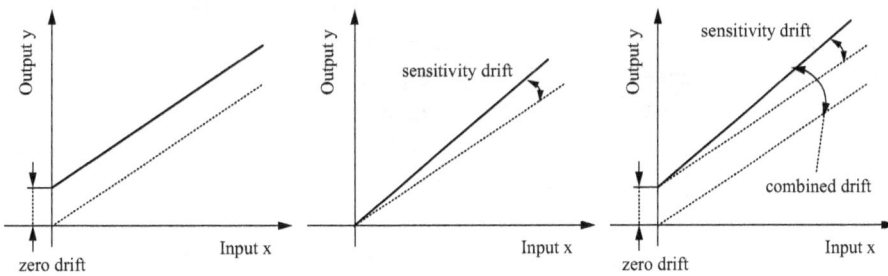

Figure 3.17: The drift.

3.2.12 Repeatability

Repeatability is the closeness of the results obtained with the same measured quantity using the same observer, same measurement procedure, same operating conditions, same measuring system and same location in a short time. It is random in nature. Repeatability is also known as nonrepeatability.

3.2.13 Stability

Stability refers to the ability of the instrument to maintain its performance during the specified storage and service life.

3.2.14 Tolerance

Tolerance is the maximum error allowed in measurement and is expressed as a value. It specifies the maximum allowable deviation between the manufacturing equipment and the mentioned values. As you can see from the definition, there are subtle differences between tolerances and accuracy.

3.2.15 Error

Error is the algebraic difference between the actual value and the measured value. Errors include schematic errors, random errors, systematic errors and gross error.

3.3 Dynamic characteristics

The output response depends on input types, sensor types, initial conditions and system characteristics. The behavior of such a system, where the input changes from instantaneous to instantaneous and the output also changes from instantaneous to instantaneous, is defined as the dynamic response of the system. In the dynamic measurement, the measured value changes rapidly over time. The dynamic characteristics of the measurement system include response speed, fidelity, measuring lag and dynamic error. The response speed is called as the rapidity of the instrument, which responds to the changes in the measured quantity. In this way, the speed and activity of the system can be obtained. Fidelity refers to the degree where the measurement system faithfully reproduces the input changes without any dynamic errors. Measuring lag refers to the extent where measurement changes are indicated by the measurement system without dynamic errors. The dynamic error is the difference between the true value of the quantity as a function of time and the indication value of the measuring system (if static error is not assumed).

3.3.1 Types of dynamic inputs

The types of dynamic inputs include periodic inputs (sinusoidal functions of constant amplitude) and transient inputs (step inputs, ramp inputs and parabolic inputs). The

periodic inputs periodically change over time or repeat itself after a constant interval
T. Transient inputs vary nonperiodically over time.

3.3.2 Physical meaning of convolution

Convolution plays a role as a bridge between the time domain and the frequency
domain. The input is $x(t)$, the output is $y(t)$ and the impulse response function
(IRF) is $h(t)$, as shown in Figure 3.18. You can get the relationship of $x(t)$ and $y(t)$:

$$y(t) = x(t)*h(t) \tag{3.26}$$

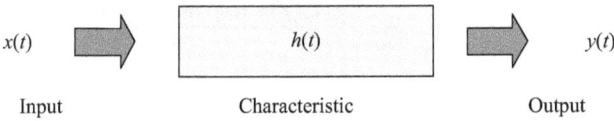

Figure 3.18: The relationship between input and output.

The signal $x(t)$ can be divided into many strips with a narrow width W as shown in
Figure 3.19. The height of the strip is $x(nW)$ at time $t = nW$. You can see that when t
tends to 0, the narrow strip can be considered as δ function. The area of the strip is
$x(nW) \times W$.

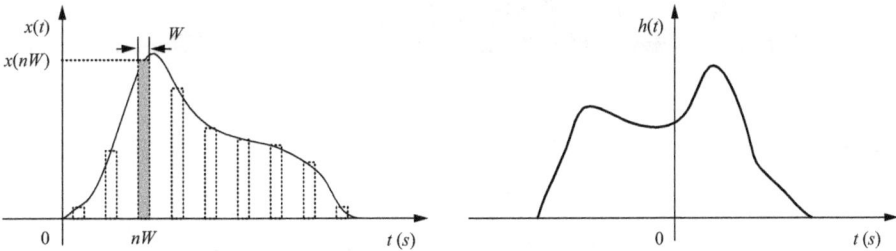

Figure 3.19: The input signal and the impulse response function.

The response $y(nW)$ caused by the narrow pulse at time $t = nW$ as shown in
Figures 3.20 and 3.21:

$$y(nW) = x(nW) \times W \times h(t - nW) \tag{3.27}$$

Figure 3.20 indicates $y(2W)$, $y(8W)$, $y(14W)$ and $y(20W)$ caused by the narrow pulse
at time $t = 2W$, $t = 8W$, $t = 14W$ and $t = 20W$. Figure 3.21 indicates the $y(nW)$ caused
by the narrow pulse at time $t = nW$, $n = 1, 2, 3, \ldots, 26$.

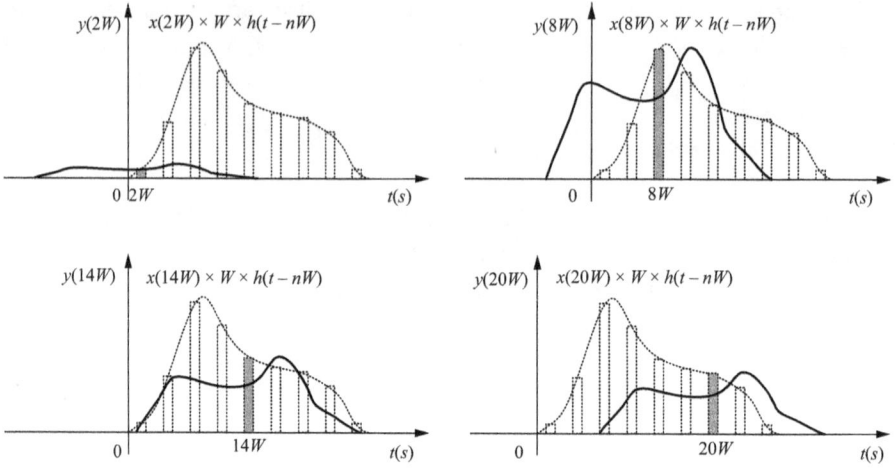

Figure 3.20: The input signal and the impulse response function.

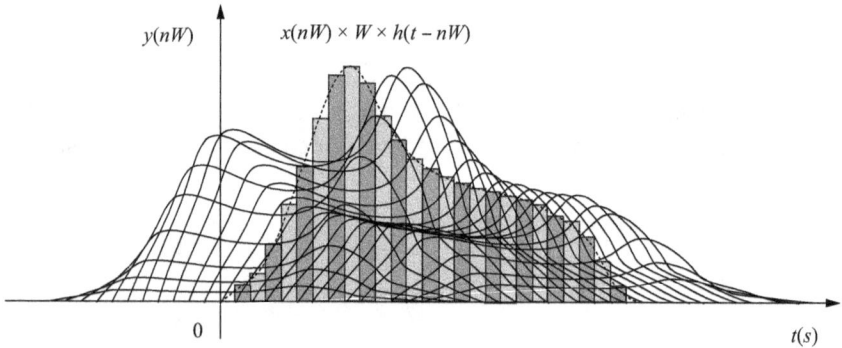

Figure 3.21: The response caused by the narrow pulses.

Finally, you will get the output signal $y(t)$ as shown in Figure 3.22:

$$y(t) = \sum_{n=0}^{\infty} x(nW) \times W \times h(t - nW) \tag{3.28}$$

3.3.3 Mathematical description

The characteristic of the instrument can be described by the transfer function and frequency response function.

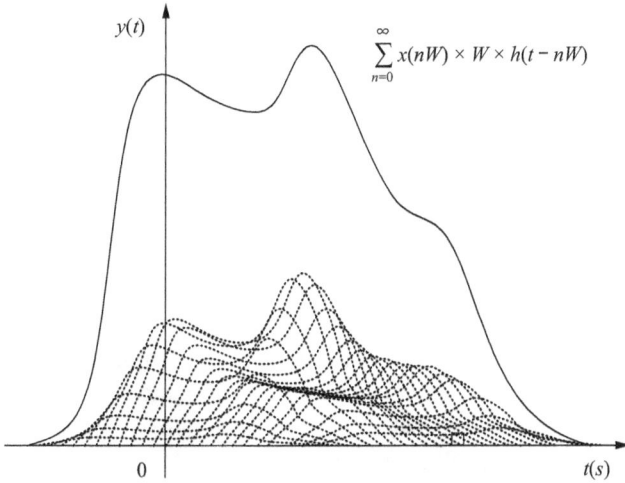

Figure 3.22: The output signal.

3.3.3.1 Transfer function

By Laplace transform, you will get the transfer function:

$$H(s) = \frac{Y(s)}{X(s)} + G_h(s) \tag{3.29}$$

$$H(s) = \frac{b_m S^m + b_{m-1} S^{m-1} + \cdots + b_1 S + b_0}{a_n S^n + a_{n-1} S^{n-1} + \cdots + a_1 S + a_0} \tag{3.30}$$

where $G_h(s)$ depends on initial conditions and input. $H(s)$ is independent of the initial conditions and the input.

3.3.3.2 Frequency response function

The system has different input responses to different frequencies. Some systems can amplify the magnitude of components of some frequencies and attenuate the magnitude of components of other frequencies. The output of the system is associated with the input of the system at different frequencies and is referred to as the frequency response of the system. In the Fourier domain, the relationship between system input and output is called frequency response. In fact, the frequency response is a complex function, so each frequency component has an amplitude and phase angle. Amplitude response is defined as the magnitude of frequency response. The phase response is the relationship between the phase of a sine input and the output signal passing through a device that accepts input and produces an output signal, which is the phase of the frequency response.

If transfer function $H(s)$ is known, let $s = j\omega$, you will find the frequency response function:

$$H(j\omega) = \frac{b_m(j\omega)^m + b_{m-1}(j\omega)^{m-1} + \cdots + b_1(j\omega) + b_0}{a_n(j\omega)^n + a_{n-1}(j\omega)^{n-1} + \cdots + a_1(j\omega) + a_0} \tag{3.31}$$

$$H(j\omega) = |H(j\omega)|e^{i\angle H(j\omega)} \tag{3.32}$$

$$A(\omega) = |H(j\omega)|$$

$$= \sqrt{\mathrm{Re}[H(j\omega)]^2 + \mathrm{Im}[H(j\omega)]^2} \tag{3.33}$$

$$\phi(\omega) = \angle H(j\omega)$$

$$= \mathrm{arctg}(\mathrm{Im}[H(j\omega)]/\mathrm{Re}[H(j\omega)]) \tag{3.34}$$

$A(\omega)$ and $\phi(\omega)$ directly reflect the distortions for the input signals with the different frequencies. The frequency response is characterized by the magnitude of the system's response, usually measured in decibels (dB) or decimal, phase in radians or degrees and frequency in radians/seconds or Hertz (Hz).

By inverse Fourier transform, you will get

$$h(t) = F^{-1}[H(\omega)] \tag{3.35}$$

3.3.3.3 Swept-sine method and its application

It can be considered that the measuring instrument can be regarded as a "black box" shown in Figure 3.23. You need to know the effect of this black box on the spectrum of $x(t)$, which represents the entire input signal.

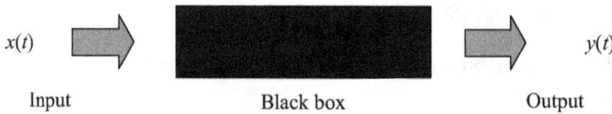

Figure 3.23: The black box.

Figure 3.24 shows an input–output pair of an instrument, at the frequency ω_0. $x(t)$ shows the input signal and $y(t)$ shows the output signal:

$$\text{Input signal } x(t) = A\sin(\omega_0 t + \phi_x) \tag{3.36}$$

$$\text{Output signal } y(t) = B\sin(\omega_0 t + \phi_y) \tag{3.37}$$

The instrument gain is the ratio of the peak output amplitude over the peak input amplitude at that frequency. You can also say that the amplitude response is B/A at the frequency ω_0 (rad/s). The phase response of the instrument is the phase of the output signal minus the phase of the input signal at that frequency. Figure 3.24 shows that the instrument has a phase response equal to $\phi_y - \phi_x$ at the frequency ω_0. In this way, a sinusoidal wave can be input at each frequency (from 0 to ω_{max}),

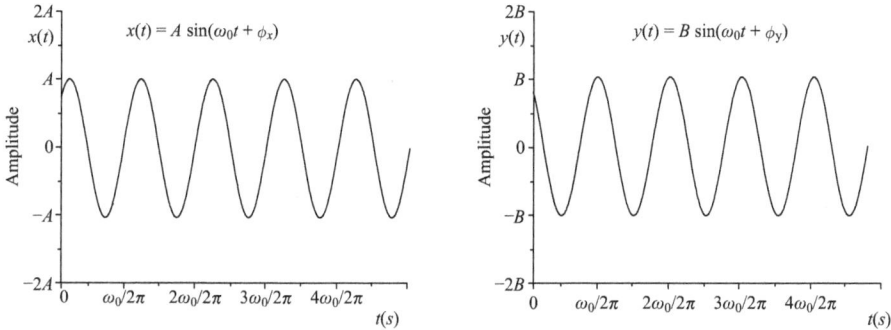

Figure 3.24: The input and output signals.

the input and output waveforms can be checked as shown in Figure 3.24, and the peak-to-amplitude ratio (gain) and phase shift of each frequency can be recorded on the graph. The resultant pair of plots is called the frequency response.

Swept-sine methods use two signal-generation techniques: continuous swept-sine method and stepped swept-sine method. A continuous swept-sine method sweeps through a set of frequencies to produce a chirp pattern. In previous chirp signals, the frequency linearly sweeps with time. In audio applications, the frequency is scanned exponentially over time. A stepped swept-sine method steps through a set of frequencies. Figure 3.25 shows an excitation signal for a continuous swept-sine method and an excitation signal for a stepped swept-sine method. The swept-sine method requires an excitation signal. The excitation signal is always a single signal. The swept-sine method measures both harmonic distortion and linear response, and compares the response signal to the excitation signal to calculate the frequency response function. The amplitude of the frequency response function is equal to the gain indicating the ratio of the output level to the input level for each test frequency. The phase is equal to the phase lag introduced by each test frequency. This method is robust against noise, time variance and weak nonlinearities.

First typical application:
A typical application is to measure natural vibration frequency by the swept-sine method and the internal signal from an exciter controller is used. A measuring natural vibration frequency system (ONO SOKKI Co., Ltd, Japan) is shown in Figure 3.26. It is a good tool for measuring the natural frequencies of parts. Using sine sweep method, the frequency response function of the vibration on the excitation table and acceleration of measurement object can be calculated by the sweep averaging function.

For example, you can set the controller of the vibration exciter to a sine sweep of 1–100 Hz, and then measure the frequency response function of the vibration of the exciter and the acceleration of the component. The sine sweep time must be

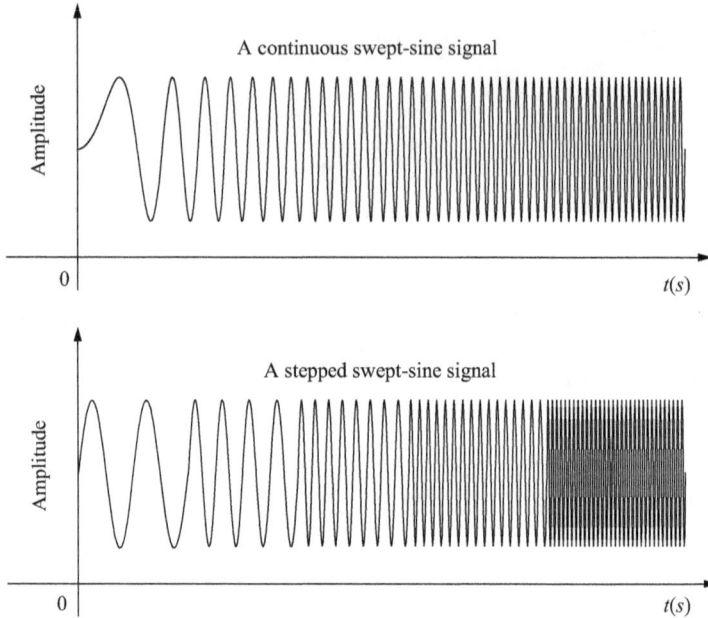

Figure 3.25: Swept-sine signals.

guaranteed to be long enough. The swept-sine method can also help you to fine the damping ratio. In addition, force/velocity (mechanical impedance), displacement/force (compliance), velocity/force (mobility), force/acceleration (dynamic mass) and force/ displacement (dynamic rigidity) can be displayed and evaluated by integral functions.

Second typical application

The advantage of the swept-sine method is its simpler form. However, it is inefficient. Figure 3.27 shows the schematic diagram of a laboratory instrument. It is an electrohydraulic force excitation instrument whose frequency response can be found by the swept-sine method.

Parts 1 and 5 are the position sensors; part 2 is the servo cylinder; part 3 is the load cell; part 4 is the soil specimen; parts 6 and 7 are the pressure transducers; part 8 is the servo valve; part 9 is the DAQ board; part 10 is the PC.

The frequency response function of the excitation instrument can be obtained directly by experimental approaches. The inputs here are sine signals with different frequencies. Meanwhile, the outputs are measured to find the dynamic characteristics. In the excitation instrument, a series of sine waves with different frequencies (0.5–30.5 Hz) and amplitudes of 50 kg are obtained. The interval frequency is set to 0.5 Hz. The frequency response of the electrohydraulic force excitation control instrument is shown in Figure 3.28.

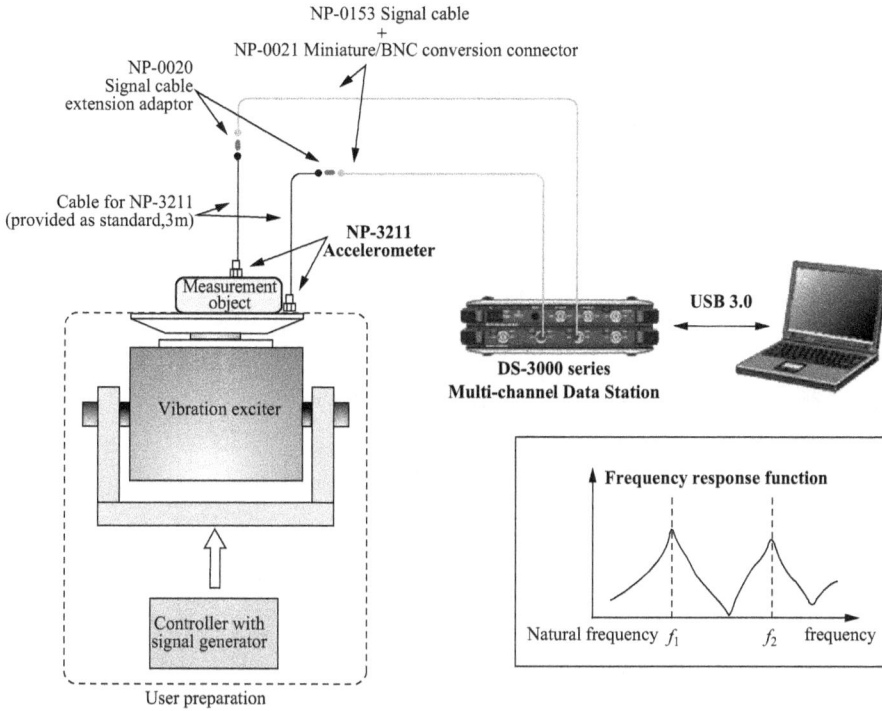

Figure 3.26: Measuring natural vibration frequency system (ONO SOKKI Co., Ltd, Japan).

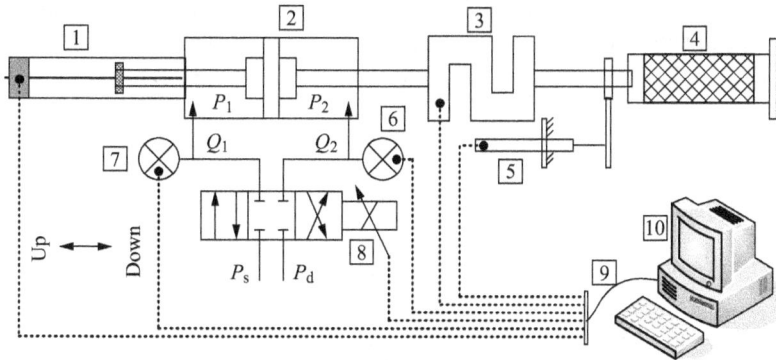

Figure 3.27: Schematic diagram of the laboratory instrument.

3.3.3.4 Impulse response method and its application

The impulse response method or the IRF method is used to get the output of a dynamic system when presented with a brief input signal. An impulse response is the reaction of any dynamic system in response to some external change.

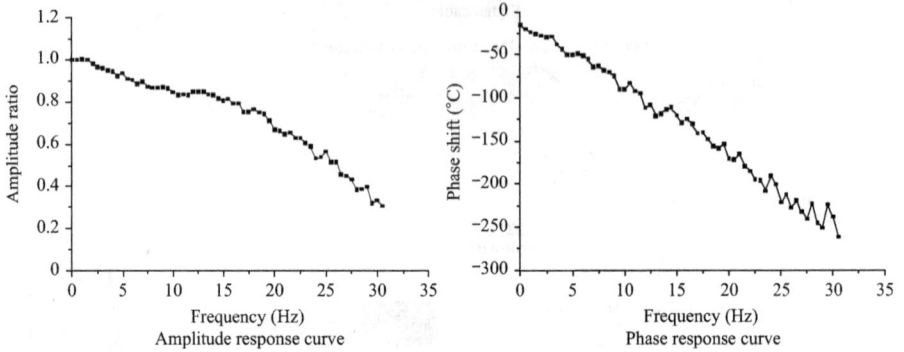

Figure 3.28: The frequency response of the excitation instrument.

If the input signal is Dirac delta $\delta(t)$, its Fourier transform is

$$F[x(t)] = x(\omega) = F[\delta(t)] = 1 \tag{3.38}$$

Then,

$$H(\omega) = \frac{Y(\omega)}{X(\omega)}$$

$$= Y(\omega)$$
$$= F[y(t)] \tag{3.39}$$

where $h(t) = y(t)$ is the IRF. An example of the input signal $x(t)$ and the output signal $y(t)$ is shown in Figure 3.29. The impulse response for a system can be obtained by experiment, transfer function, state space matrix and other techniques.

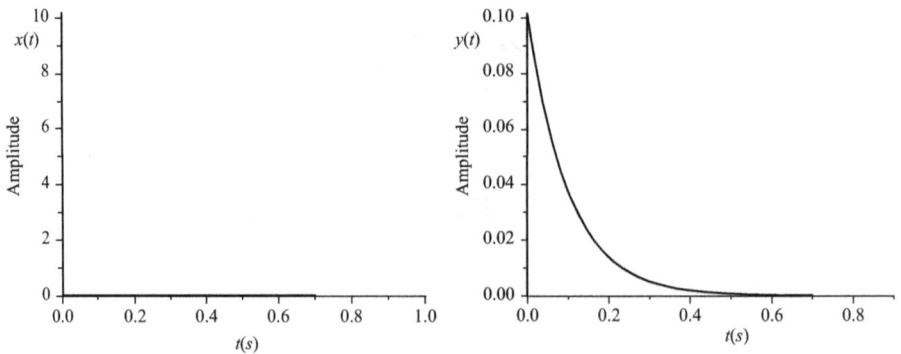

Figure 3.29: The input and output signals.

First typical application

The impulse response method is a transient method. The method can generate a pulse signal by means of a device such as a starting pistol and record the response result without distortion. This is achieved by playing back the entire frequency range of the bursts in space and recording them.

The Sonic echo-impulse response system (Model: SE/IR-1 Model, Olson Instruments, USA) shown in Figure 3.30 is to determine the length and integrity of foundations, and the upper part or part of the foundation is accessible. This system includes an accelerometer and a hammer with interchangeable plastic to rubber head for data acquisition in one channel and can only be processed in the time domain. The system and its related methods can be used in new and existing foundations, and can be achieved by impacting the foundations and utilizing nearby receivers to record echoes from defects or base bottoms. This method not only is best suited for pile and borehole foundation, but also has been successfully applied to cushion foundation, abutment wall and other similar structures. According to the hammer force spectrum (obtained by fast Fourier transform (FFT)) and velocity spectrum, the surface mobility map is calculated by dividing the velocity spectrum by the force spectrum (the signal from the hammer force sensor).

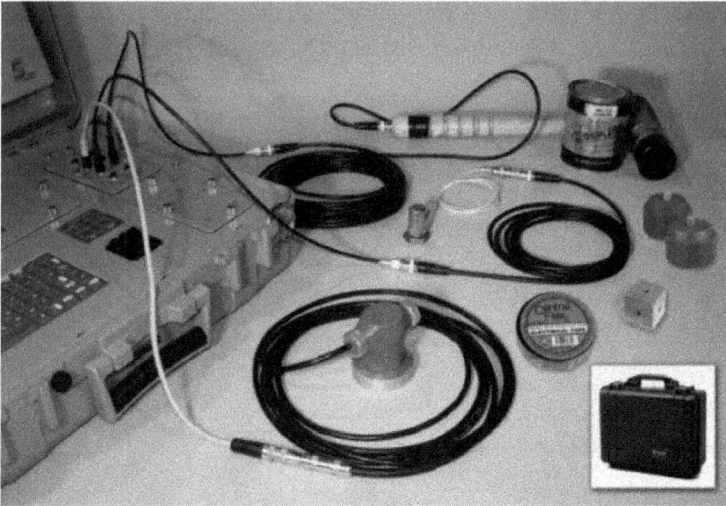

Figure 3.30: Sonic echo/impulse response instrument (Model: SE/IR-1, Olson Instruments, USA).

The impulse response method is a reliable non-destructive testing method for evaluating concrete structure such as tunnel lining and bridge deck, enabling the faster study of large areas and a greater understanding of the internal conditions of the structure. The impulse response method is used to quickly screen a large area of the structure to identify possible defective areas for further detailed analysis. On bridges, the impulse response method can determine delamination and voids in the deck or wear layer. A hammer with a built-in force sensor on the hammer head hits the surface. This effect is about 100 times stronger than the impact-echo method. The low strain shock sends a stress wave through the device under test. Use a speed sensor to record the motion or speed of the surface. Records of impact and surface speed are stored in the notebook.

Second typical application
The impulse response of the room is the sound signal that you can hear at a specific location in the room when someone makes a very short impulse in another part of the room. Strictly speaking, the room response is the sum of the acoustic reflections of the Dirac pulse. The easiest way is to use a starting pistol, balloon or clapper (especially at the shooting position) to generate a pulse signal. Any loud sound will produce broadband noise. The advantage of using transient sounds (starting a pistol, clapper, balloon) is that no postprocessing is required. Pulse recording can be used directly in convolution software. Although sinusoidal scanning is the most preferred method, it may not always be convenient to carry the speaker and play the sinusoidal scan back. The basic idea of room response estimation is to use the speaker to generate long self-orthogonal signals and calculate the cross-correlation between the playback and the measurement signals. The measurement principle diagram is shown in Figure 3.31.

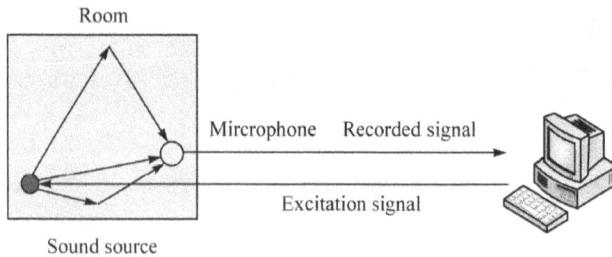

Figure 3.31: The measurement of the room impulse response.

A sound card that has a high-quality microphone preamplifier and digital step gain control and the AC output of the sound level meter can be used in Figure 3.31. In any case, it must be possible to maintain controlled input and output gains to calibrate the measurement system. In the basic arrangement, the measurements are taken synchronously, so the same PC will play the test sound and simultaneously record the signal from the microphone. This means that the system must be fully wired and use a cable that connects the sound card output to the speaker. The recording is processed using a mathematical process called "deconvolution" that reconstructs the impulse response from the nonpulsed recorded signal.

Third typical application

The application shows how to measure the natural vibration frequency and damping ratio. The measurement principle is shown in Figure 3.32 (ONO SOKKI Co., Ltd, Japan). The measurement object is suspended in free vibration (or placed on a soft object) and subjected to a hammer test by an impulse force hammer. Then, the freely damped vibration is detected by an accelerometer. The frequency response function of the acceleration (a) and the striking force (force) is calculated, and the resonance frequency at the peak of (a)/(force) is obtained. Information such as force/acceleration (dynamic mass), displacement/force (compliance), speed/power (mobility), force/displacement (dynamic stiffness) or force/speed (mechanical impedance) can be measured in the system.

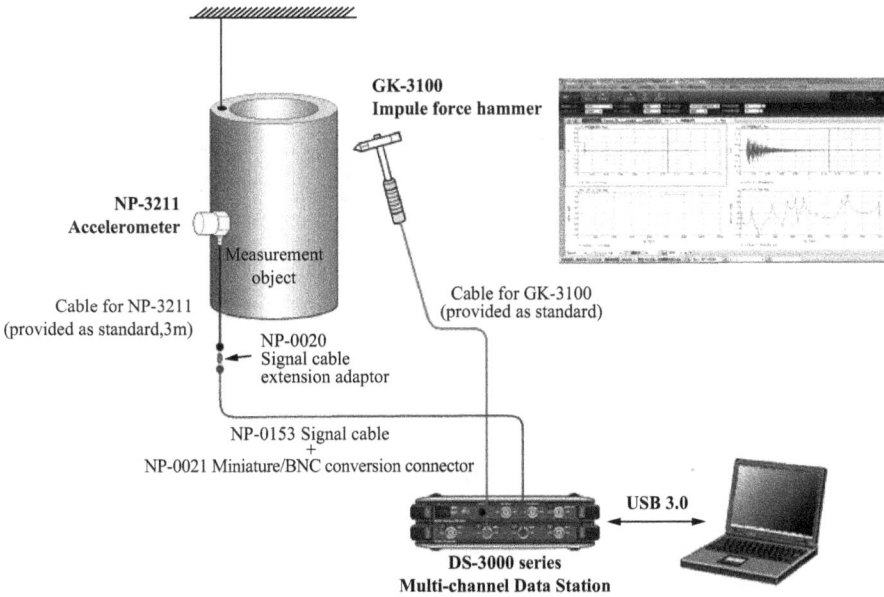

Figure 3.32: Measuring natural vibration frequency and damping ratio by hammering test (ONO SOKKI Co., Ltd, Japan).

3.3.3.5 Step response method and its application

The Dirac Delta represents an extreme case of pulses, which are very short (resulting in an infinitely high peak) while maintaining their area or integral unchanged. It is a useful idealization, although it is impossible in any real system. However, the step signal is easier to generate. Step response recognition is one of the most important subjects in process control. Step input may be the most commonly used excitation signal in process industry. To deal with different practical problems in the recognition based on step response, many new methods have been proposed.

If the input $x(t)$ is the unit step signal,

$$x(t) = 1 \tag{3.40}$$

The input and output signals are shown in Figure 3.33.

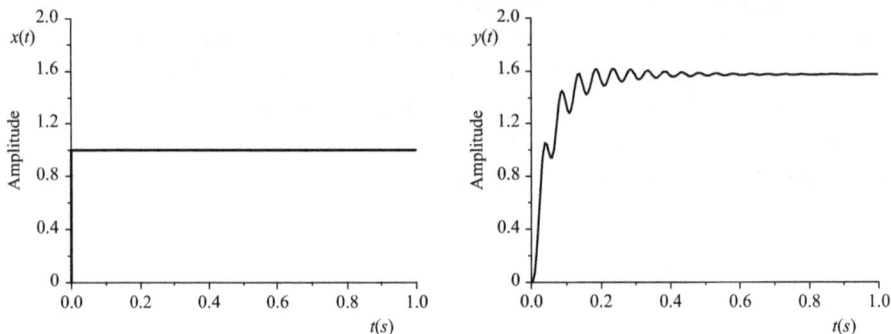

Figure 3.33: The input and output signals.

Then

$$X(s) = 1/s \tag{3.41}$$

The dynamic characteristic $Y(s)$:

$$H(s) = \frac{Y(s)}{X(s)}$$
$$= Y(s) \cdot s \tag{3.42}$$

Let $s = j\omega$, then

$$H(\omega) = Y(j\omega) \cdot j\omega \tag{3.43}$$

$H(\omega)$ consists of a real part and an imaginary part:

$$H(\omega) = R(\omega) + jI(\omega) \tag{3.44}$$

$R(\omega)$ and $I(\omega)$ are real functions over ω. You can plot $R(\omega)$-ω or $I(\omega)$-ω and get real or imaginary frequency characteristics. In Cartesian coordinates, the real part $R(\omega)$ is plotted on the X-axis and the imaginary part $I(\omega)$ is plotted on the Y-axis. This plot is called the Nyquist plot. Another useful plot is the Bode plot, which is a plot of either the magnitude or the phase of a transfer function as a function of ω. The magnitude plots are more common because they represent the gain of the system.

Until now, dynamic characteristics can be described as $h(t)$ in the time domain, $H(\omega)$ in the frequency domain and $H(s)$ in the plural domain. The relationship between $h(t)$, $H(s)$ and $H(\omega)$ can be written as

$$h(t) \leftrightarrow H(s) \quad \text{Laplace transform pairs} \tag{3.45}$$

$$h(t) \leftrightarrow H(\omega) \, \text{Fourier transform pairs} \tag{3.46}$$

A typical application of a step response method is used to find the damping ratio and natural frequency of a second-order measurement system by experiment.

Typical application

The 1940 Tacoma Strait Suspension Bridge was built in Washington, DC. The new bridge collapsed 4 months after its completion. It was the third longest suspension bridge in the world at the time. As we know, any system will vibrate at its natural frequency. The eddy currents of the fluid are formed at a specific frequency, which cause vibrations in the structure. If the vibration is consistent with the natural frequency or resonant frequency of the structure, it will cause failure. If the natural frequency of the structure is achieved due to any factor, more vibration and more energy storage will be caused in the system. When the energy exceeds the load limit of the object, the structure loses its integrity. In the case of Tacoma Narrows, when the natural frequency encounters the frequency of wind blowing, it causes an increase in amplitude and eventually collapses. Structural vibration characteristics are very important for the bridges. A photo in Figure 3.34 shows the Xinghai Bay Cross-Sea Bridge in Dalian, northeast China's Liaoning Province. It has a total length of

Figure 3.34: Xinghai Bay Cross-Sea Bridge in Dalian (Dalian, China).

6.8 km with eight lanes in two ways and special pavement for pedestrians. Once the construction was completed, a series of vibration characteristics were also measured.

The step response method is an effective and applicable method for detecting structural vibration characteristics. Determining damage or fault detection by changes in dynamic characteristics or structural response is the subject of current structural vibrational resistance. Changes in structural stiffness, whether local or distributed, can cause changes of the modal parameters (especially modal shapes, natural frequencies, etc.), and these changes can determine the location and severity of damage in the structure. The structure is tested and evaluated through improved system identification.

The process of the step response method consists of three steps: vibrating the structure, extracting the acceleration response through the accelerometer and analyzing the obtained response. Using modal analysis method, the structure is vibrated by moving the car crash barrier on the bridge. This will create a step shock. Using an accelerometer installed on the laptop, data extraction is performed. Finally, the samples were analyzed by FFT.

3.3.4 Series connection and parallel connection

Figure 3.35 shows the interconnection of two LTI systems.

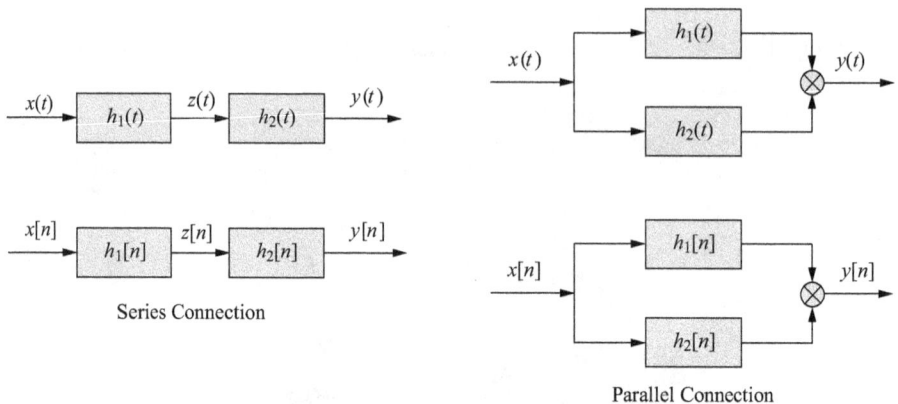

Series Connection

Parallel Connection

Figure 3.35: Interconnection of two LTI systems.

The impulse response of two LTI systems in series connection is the convolution of their individual impulse responses:

$$y(t) = z(t)*h_2(t)$$
$$= x(t)*h_1(t)*h_2(t) \tag{3.47}$$

$$y[n] = z[n]*y_2[n]$$
$$= x[n]*h_1[n]*h_2[n] \qquad (3.48)$$

$$H(\omega) = \prod_{i=1}^{n} H_i(\omega) \qquad (3.49)$$

$$A(\omega) = \prod_{i=1}^{n} A_i(\omega)$$
$$\qquad (3.50)$$
$$\phi(\omega) = \sum_{i=1}^{n} \phi_i(\omega)$$

The impulse response of two LTI systems in parallel is the sum of their individual impulse responses:

$$y(t) = y_1(t) + y_2(t)$$
$$= x(t)*h_1(t) + x(t)*h_2(t)$$
$$= x(t)*(h_1(t) + h_2(t)) \qquad (3.51)$$

$$y[n] = y_1[n] + y_2[n]$$
$$= x[n]*h_1[n] + x[n]*h_2[n]$$
$$= x[n]*(h_1[n] + h_2[n]) \qquad (3.52)$$

$$H(\omega) = \sum_{i=1}^{n} H_i(\omega) \qquad (3.53)$$

3.3.5 Mathematical models of the measuring systems

Three mathematical models of the measuring system including zeroth-order, first-order and second-order systems are discussed in this section.

3.3.5.1 Zeroth-order system

The scales, proving ring, mechanical levers, amplifiers and potentiometer can be considered as zeroth-order systems, as is shown in Figure 3.36. You can imagine a thermometer measuring the temperature, which will indicate the current temperature of the installation location. The output of the sensor follows the input "exactly" is the defining characteristic of the zeroth-order systems.

Mathematically speaking, if you put $x(t)$ into the system as a time function and the output is $y(t)$, then the relationship between the input and the output is

$$y(t) = kx(t) \qquad (3.54)$$

where $x(t)$ is the actual temperature, $y(t)$ is the indicated temperature and k is a constant, which multiplies the input to produce the output. For example, if the output of the potentiometer is an electrical signal, k will be a constant in volts/degree.

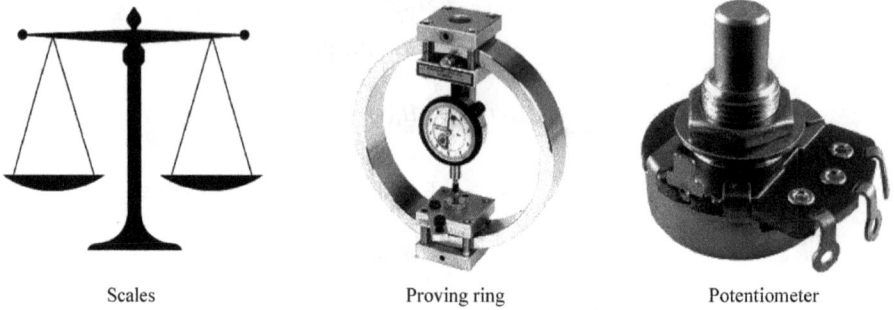

| Scales | Proving ring | Potentiometer |

Figure 3.36: The typical zeroth-order systems.

k is often referred to as the static sensitivity. The value of static sensitivity is obtained by a static calibration process. You need to note that the output of the zeroth-order system is not affected by the speed at which $x(t)$ changes. Regardless of how the input changes, the output is proportional to the input. The output is a faithful reproduction of input without any distortion or time lag.

3.3.5.2 First-order system

Glass column thermometer, RC integrator and direct drive valves (Moog Industrial Controls Division, USA) as Figure 3.37 shows can be considered as the typical first-order systems.

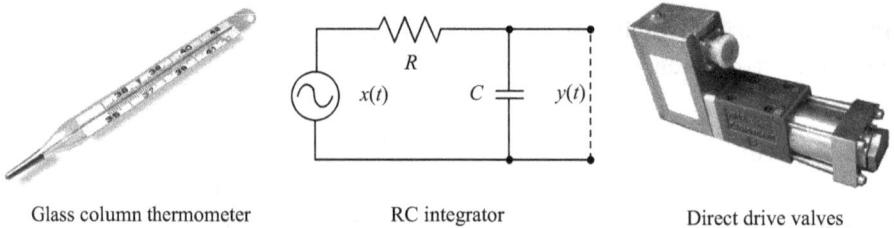

| Glass column thermometer | RC integrator | Direct drive valves |

Figure 3.37: The typical first-order systems.

The simplest RC circuit is a resistor and a capacitor in series as shown in Figure 3.37. When a circuit consists of only a resistor R and a charged capacitor C, the capacitor releases its stored energy through the resistor. The voltage on the capacitor is related to time. It can be calculated by Kirchhoff's current law, in which the current charged by the capacitor must be equal to the current through the resistor. This process produces a linear differential equation as follows:

$$RC\frac{dy(t)}{dt} + y(t) = x(t) \qquad (3.55)$$

$$\tau = RC \qquad (3.56)$$

where $\tau = RC$ is the time constant of the instrument:

$$H(s) = \frac{1}{\tau s + 1} \qquad (3.57)$$

Frequency response function is given by

$$H(f) = \frac{1}{j\tau 2\pi f + 1}$$
$$= \frac{1}{1 + (2\pi f\tau)^2} - j\frac{2\pi f\tau}{1 + (2\pi f\tau)^2} \qquad (3.58)$$

The magnitude of the gains across the two components is

$$M(f) = \frac{1}{\sqrt{1 + (2\pi f\tau)^2}} \qquad (3.59)$$

The phase angle is

$$\phi(f) = -\text{arctg}(2\pi f\tau) \qquad (3.60)$$

The magnitude response and phase response can be plotted in the Cartesian coordinate system and logarithmic coordinate system, as shown in Figure 3.38.

In fact, these responses can be drawn in the following three ways: (1) using frequency as a parameter to plot magnitude and phase on a single rectangular plot, the Nichols plot can be obtained. (2) using frequency as a parameter to plot the magnitude and phase angle on a single polar plot, the Nyquist plot can be obtained; (3) plotting the magnitude and phase measurements on two rectangular plots as functions of frequency, the Bode plot can be obtained.

Engineer Hendrik Wade Bode (1905–1982) designed a simple method, which was also very precise. Working at Bell Labs (the United States) in the 1930s, it can be used to plot gain and phase shift plots. You can express the magnitude axis of the Bode plot as power decibels by the 20 logarithmic rule: 20 times the common (base 10) logarithm of the amplitude gain. Since the amplitude gain is logarithmic, the Bode plot makes the multiplication of the amplitude very simple, simply adding the distance (in decibels) to the graph such as

$$\log(xy) = \log(x) + \log(y) \qquad (3.61)$$

A Bode phase plot describes the relationship between phase and frequency, which is also drawn on the logarithmic frequency axis. It is usually used together with the magnitude plot to evaluate the phase offset of the signal. The Bode plot of the RC

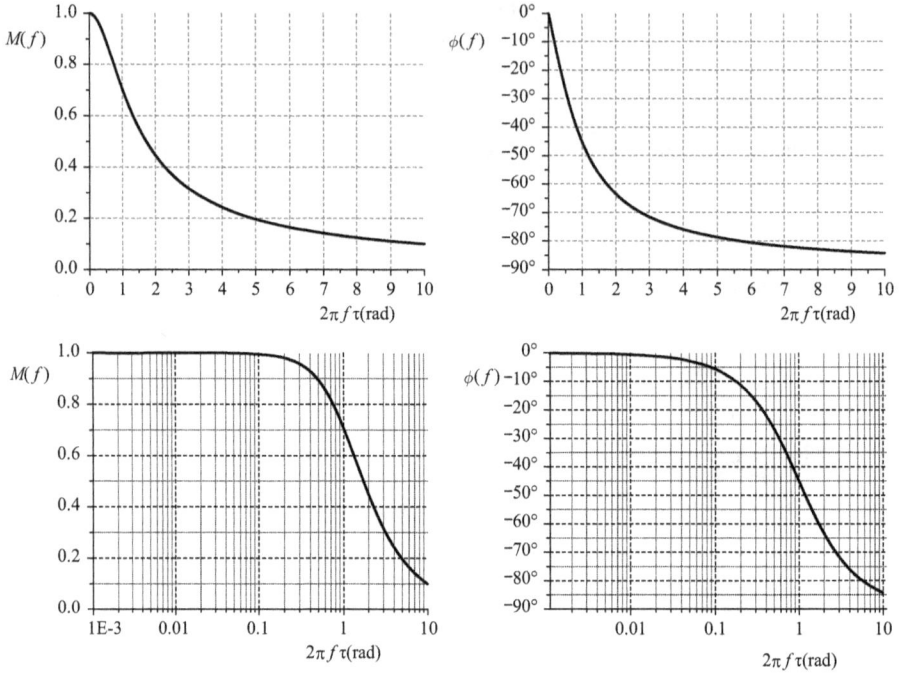

Figure 3.38: The magnitude response and phase response.

first-order system is shown in Figure 3.39. Low-frequency sine waves are transmitted by the *RC* integrator, and high-frequency sine waves are suppressed. Figure 3.39 shows that the amplitude of the output is approximately equal to the amplitude of the input, when $2\pi f\tau$ is much less than 1. When $2\pi f\tau$ is bigger than 2–3, the output is proportional to the integral of the input. The system is equivalent to an integrator. $M(f)$ is almost inversely proportional to the frequency, and the phase lag is about 90°.

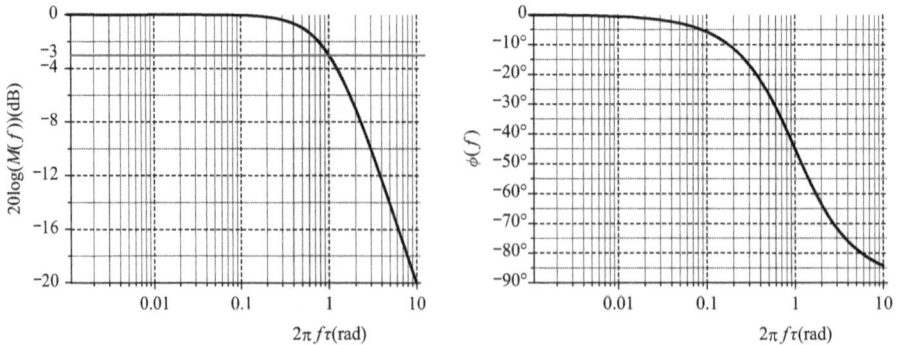

Figure 3.39: The Bode plot of the first-order system.

Nyquist's plot is named after Harry Nyquist. He is a former engineer at Bell Labs. The Nyquist plot is a parameter map of the frequency response for signal processing and automatic control. Usually, the most common use of Nyquist diagrams is to evaluate the stability of systems with feedback. In the Cartesian coordinate system, the imaginary part is drawn on the Y-axis, and the real part of the transfer function is plotted on the X-axis. The frequency is scanned as a parameter to obtain a graph of each frequency. You can describe the same graph using polar coordinates. The phase of the transfer function is the corresponding angular coordinate and the gain of the transfer function is the radial coordinate. Figure 3.40 shows the Nyquist plot of the first-order system described earlier:

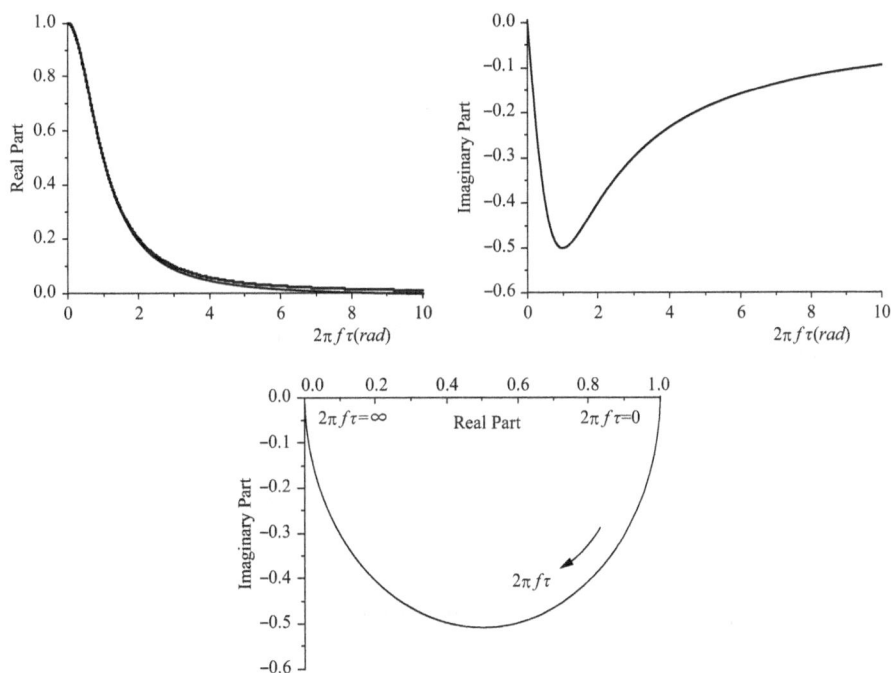

Figure 3.40: The Nyquist plot of the first-order system.

$$H(f) = \frac{1}{1 + (2\pi f \tau)^2} - j\frac{2\pi f \tau}{1 + (2\pi f \tau)^2} \tag{3.62}$$

The Nichols plot is a diagram for signal processing and control design, named after American engineer Nathaniel B. Nichols. The Nichols plot is shown with $20 \log(|M(f)|)$ versus $\phi(f)$. The Nichols plot is plotted in a Cartesian coordinate system and Figure 3.41 shows the Nichols plot of the RC first-order system.

The impulse response of the RC first-order system in the time domain is shown in Figure 3.42:

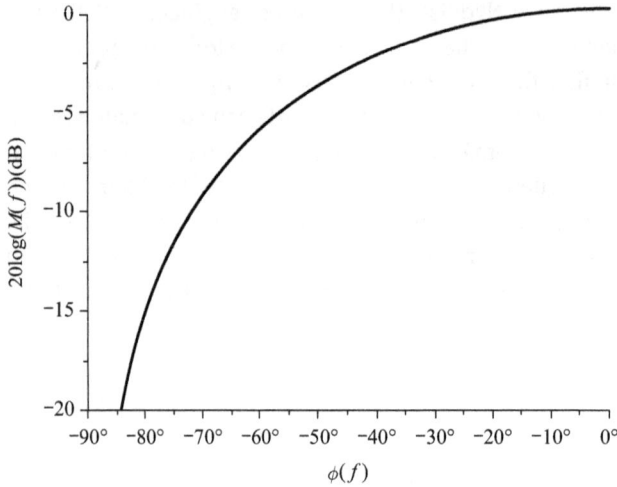

Figure 3.41: The Nichols plot of the first-order system.

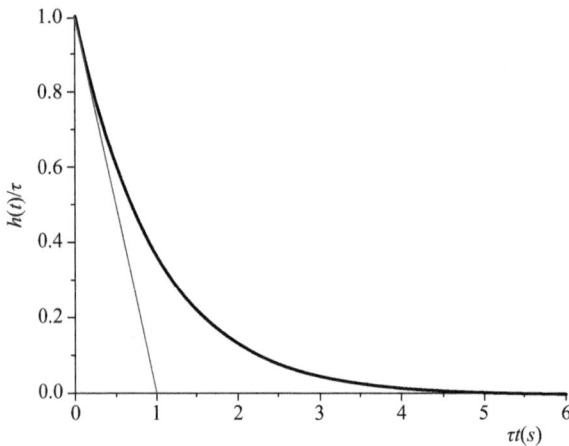

Figure 3.42: The impulse response of the first-order system.

$$h(t) = \frac{1}{\tau} e^{-t}/\tau \qquad (3.63)$$

You can find a physical interpretation of the time constant τ from the initial condition response of any output variable $y(t)$. If $\tau < 0$, the response increases exponentially, and the system is unstable. If $\tau > 0$, the response of any system variable decreases exponentially from the initial value $y(0)$ to zero, and the system is stable.

The time constant τ of first-order systems is often easy to find. For example, the translating friction-mass system's time constant is m/B (B, viscous frictional coefficient); the translating friction-spring system's time constant is b/K (K, the spring constant); the rotating friction- flywheel system's time constant is J/B_r (B_r, the rotating damping constant); the rotating friction-spring system's time constant is B_r/K_r (K_r, the rolling spring constant) and the RL system's time constant is L/R.

3.3.5.3 Second-order system
The single degree of freedom mass–spring system (the spring has mass), RLC circuit, moving coil meter and servo valve (Moog Industrial Controls Division, USA) shown in Figure 3.43 can be considered as the typical second-order systems.

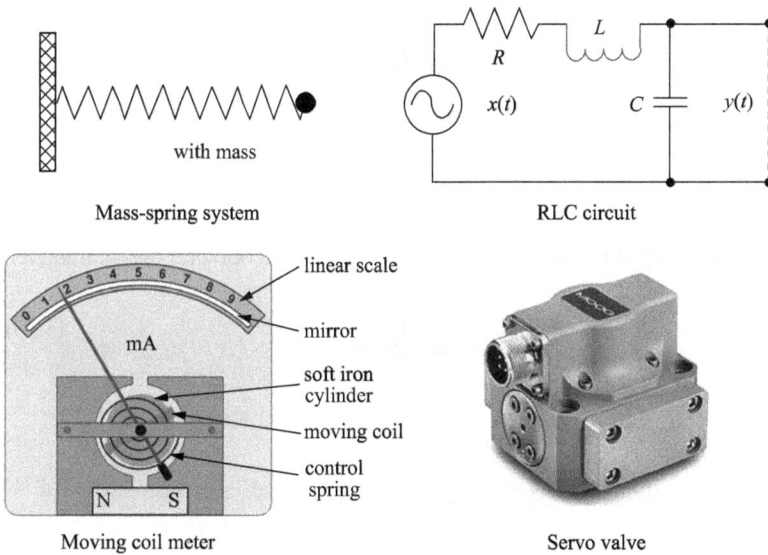

Figure 3.43: Typical second-order systems.

The light-beam oscillograph is an instrument, which is used to detect and measure very small currents. Most modern galvanometers are moving coil type. For the movable coil, a circle of thin wires wrapped around the aluminum molder is suspended by conductive tape around the soft core between the permanent magnet poles. When the current flows through the coil, a magnetic field will be generated, and it interacts with the magnetic field of the permanent magnet and generates the torque. This will cause the coil to rotate until it is completely resisted by the suspension. The resulting displacement is proportional to the current. In general, the formula for the moving circle is given by

$$J\frac{d^2y(t)}{dt^2} + C\frac{dy(t)}{dt} + Ky(t) = Gx(t) \tag{3.64}$$

where $x(t)$ is the input current, $y(t)$ is the needle position, J is the moment of inertia of the needle, C is the damping factor, G is the current gain and K is the stiffness of the spring.

The above equation can be rewritten as

$$\frac{d^2y(t)}{dt^2} + \frac{C}{J}\frac{dy(t)}{dt} + \frac{K}{J}y(t) = \frac{G}{J}x(t) \tag{3.65}$$

Let

$$\omega_n = \sqrt{K/J} \tag{3.66}$$

$$S = G/K \tag{3.67}$$

$$\zeta = C/2\sqrt{KJ} \tag{3.68}$$

You will get

$$\frac{d^2y(t)}{dt^2} + 2\zeta\omega_n\frac{dy(t)}{dt} + \omega_n^2 y(t) = S\omega_n^2 x(t) \tag{3.69}$$

where S is the sensitivity of the moving coil. The transfer function can be obtained as

$$H(s) = \frac{S\omega_n^2}{s^2 + 2\zeta\omega_n s + \omega_n^2} \tag{3.70}$$

If $S = 1$, you will get

$$H(s) = \frac{\omega_n^2}{s^2 + 2\zeta\omega_n s + \omega_n^2} \tag{3.71}$$

The characteristic equation is

$$s^2 + 2\zeta\omega_n s + \omega_n^2 = 0 \tag{3.72}$$

The roots of characteristic equation are

$$s = \frac{-2\zeta\omega_n \pm \sqrt{(2\zeta\omega_n)^2 - 4\omega_n^2}}{2}$$
$$= -\zeta\omega_n \pm \omega_n\sqrt{\zeta^2 - 1} \tag{3.73}$$

From formula (3.73), it is shown that the two roots are imaginary when $\zeta = 0$; the two roots are real and equal when $\zeta = 1$. The two roots are real but not equal when $\zeta > 1$; the two roots are complex conjugate when $0 < \zeta < 1$.

Let $s = j2\pi f$, the frequency response function can be written as

$$H(f) = \frac{(2\pi f_n)^2}{(j2\pi f)^2 + 2\zeta 2\pi f_n (j2\pi f) + (2\pi f_n)^2}$$

$$= \frac{(f_n)^2}{-f^2 + 2j\zeta f_n f + (f_n)^2}$$

$$= \frac{1}{\left[1 - \left(\frac{f}{f_n}\right)^2\right] + j2\zeta\frac{f}{f_n}} \tag{3.74}$$

The magnitude response function is

$$M(f) = \frac{1}{\sqrt{\left[1 - \left(\frac{f}{f_n}\right)^2\right]^2 + 4\zeta^2\left(\frac{f}{f_n}\right)^2}} \tag{3.75}$$

The phase response function is

$$\phi(f) = -\arctan\frac{2\zeta\left(\frac{f}{f_n}\right)}{1 - \left(\frac{f}{f_n}\right)^2} \tag{3.76}$$

Similarly, the magnitude response and phase response of the second-order system can be plotted in the Cartesian coordinate system and logarithmic coordinate system shown in Figure 3.44.

The Bode plot of the second-order system is shown in Figure 3.45. Figure 3.45 shows that: (1) resonance (maximum amplitude of response) is the greatest when the damping factor in the system is small. The effect of increasing damping is to reduce the amplitude at resonance. (2) The resonant frequency coincides with the natural frequency for an underdamped system, but as the damping is increased, the resonant frequency becomes smaller. (3) When the damping factor is greater than 0.707, there is no resonant peak. However, for a value whose damping coefficient is lower than 0.707, a resonance peak appears. (4) The phase shift characteristics depend strongly on the damping factor for all frequencies. (5) In an instrument system, the flattest possible response up to the highest possible input frequency is achieved with a damping factor of 0.707.

Figure 3.46 shows the Nyquist plot of the second-order system described earlier:

$$H(f) = \frac{1}{\left[1 - \left(\frac{f}{f_n}\right)^2\right] + j2\zeta\frac{f}{f_n}}$$

$$= = \frac{\left[1 - \left(\frac{f}{f_n}\right)^2\right]}{\left[1 - \left(\frac{f}{f_n}\right)^2\right]^2 + 4\zeta^2\left(\frac{f}{f_n}\right)^2} - j\frac{2\zeta\frac{f}{f_n}}{\left[1 - \left(\frac{f}{f_n}\right)^2\right]^2 + 4\zeta^2\left(\frac{f}{f_n}\right)^2} \tag{3.77}$$

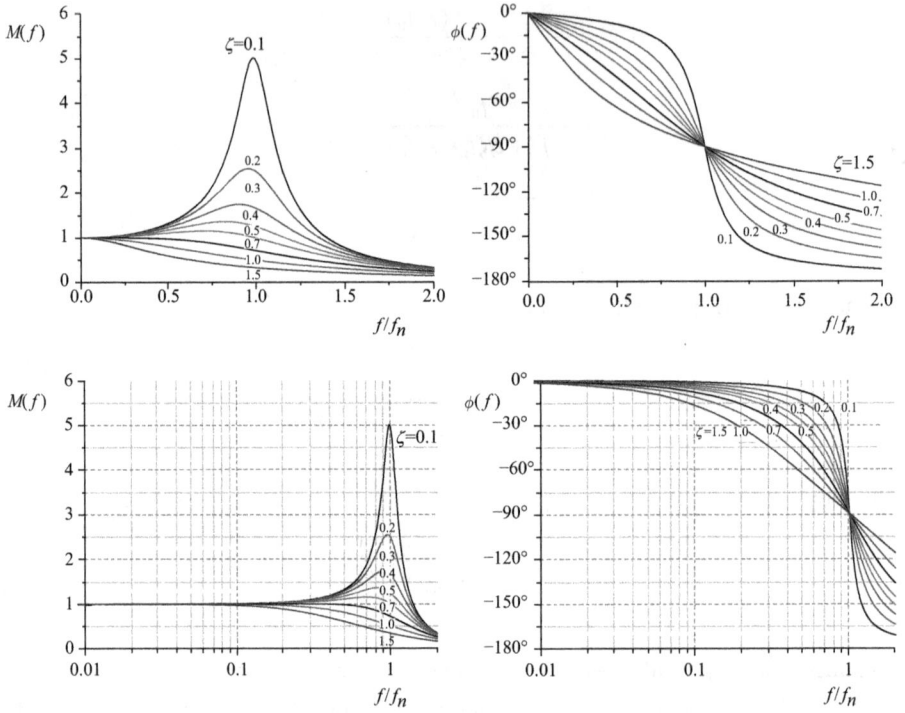

Figure 3.44: The magnitude response and phase response of the second-order system.

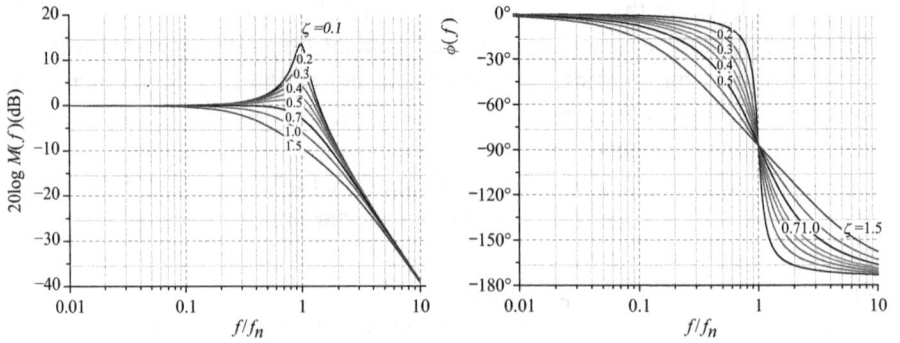

Figure 3.45: Bode plot of the second-order system.

The impulse response of the second-order system in the time domain is shown in Figure 3.47:

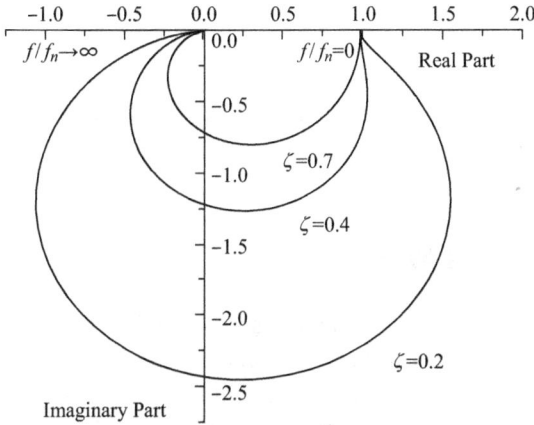

Figure 3.46: The Nyquist plot of the second-order system.

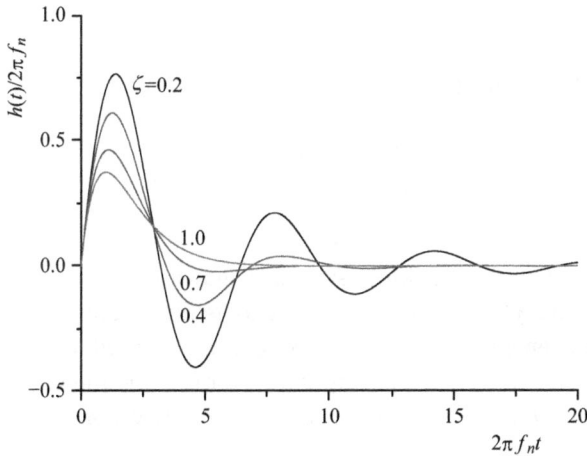

Figure 3.47: Impulse response of the second-order system.

$$h(t) = 2\pi f_n \sin 2\pi f_n t, \quad \zeta = 0 \tag{3.78}$$

$$h(t) = (2\pi f_n)^2 t e^{-2\pi f_n t}, \quad \zeta = 1 \tag{3.79}$$

$$h(t) = \frac{2\pi f_n}{2\sqrt{\zeta^2 - 1}} e^{-2\pi f_n \left(\zeta - \sqrt{\zeta^2 - 1}\right)t}, \quad \zeta > 1 \tag{3.80}$$

$$h(t) = \frac{2\pi f_n}{\sqrt{1 - \zeta^2}} e^{-2\pi f_n \zeta t} \sin 2\pi \sqrt{1 - \zeta^2} f_n t, \quad 0 < \zeta < 1 \tag{3.81}$$

In summary, we must know that the zeroth-, first- and second-order systems are mathematical ideals for system response. The actual instrument can be modeled differently under different conditions. For example, if the input changes very slowly, you can treat a given instrument as zeroth-order system. The same instrument, when used in applications, may be considered as a first-, second- or even higher-order system if the input changes rapidly. Our motivation is to use the simplest model to describe the behavior of the instrument and to ensure the accuracy required for a given application.

3.4 Step responses of measuring system

The step response method is one of the most commonly used methods for system identification, especially in process industries as described in the Section 3.3.3.5. In 1928, the idea of step response was first introduced by Küpfmüller, who also proposed the first method for estimating the parameters of the first-order plus time-delay model from the step response. A step input is used to represent the change of input from one constant to another in a very short time.

3.4.1 The response of the first-order system

If the input is a unit step signal, the output of the first-order system will be

$$y(t) = 1 - e^{-t/\tau} \tag{3.82}$$

Unit step input and the response of the first-order system is shown in Figure 3.48. From Figure 3.48, it is shown that $y(t)$ is 63.2% of the way to its final value after a time of one time constant τ. In general, $y(t)$ is considered to have achieved its final value (99.3%) after five time constants 5τ. The steady-state error is 0 and the initial slope is $1/\tau$. The first-order system with a smaller time constant τ will respond more

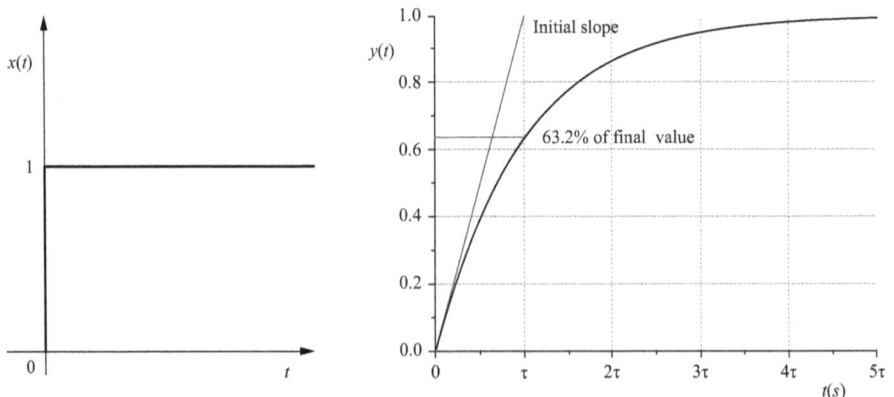

Figure 3.48: Unit step input and the response of the first-order system.

quickly to a step input or any other input. It can be considered that the smaller the time constant τ is, the better the performance will be.

3.4.2 The response of the second-order system

Similarly, if the input is a unit step signal, the output of the second-order system will be

(1) Underdamped system ($0 < \zeta < 1$)

It can be clearly seen from the above expression that when the error of the signal is 255, it is oscillatory and its amplitude decreases exponentially.

From the above expression, it can be clearly seen that when $\zeta < 1$, the error of the signal is exponentially attenuated oscillating:

$$\left.\begin{aligned} y(t) &= 1 - \frac{e^{-2\pi f_n \zeta t}}{\sqrt{1-\zeta^2}} \sin\left(2\pi f_n \sqrt{1-\zeta^2}t + \varphi\right), \quad 0 < \zeta < 1 \\ \varphi &= \arctan \frac{\sqrt{1-\zeta^2}}{\zeta} \end{aligned}\right\} \tag{3.83}$$

(2) Undamped case ($\zeta = 0$)

When the damping ratio is zero ($\zeta = 0$), the expression of the output signal can also be rewritten as follows:

$$\begin{aligned} y(t) &= 1 - \frac{e^{-0 \cdot 2\pi f_n t}}{\sqrt{1-0^2}} \sin\left(2\pi f_n \sqrt{1-\zeta^2}t + \varphi\right) \\ &= 1 - \sin\left(2\pi f_n \sqrt{1-\zeta^2}t + \frac{\pi}{2}\right) \\ &= 1 - \cos\left(2\pi f_n \sqrt{1-\zeta^2}t\right) \\ &= 1 - \cos(2\pi f_n t) \end{aligned} \tag{3.84}$$

(3) Critically damped case ($\zeta = 1$)

$$y(t) = 1 - e^{-2\pi f_n t}(1 + 2\pi f_n t) \tag{3.85}$$

Subjective unit step function has no oscillation in the expression of output signal. Therefore, the time response of the second-order control system is called critical damped.

(4) Overdamped case ($\zeta > 1$)

In this case, the solution is (Figure 3.49)

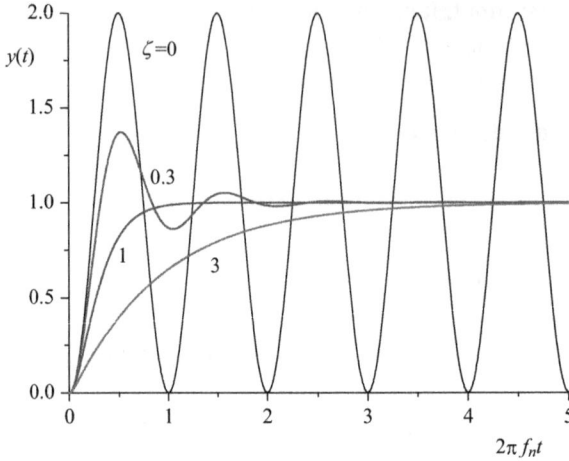

Figure 3.49: The response of the second-order system.

$$y(t) = 1 - \frac{e^{-\left(\zeta - \sqrt{\zeta^2 - 1}\right)2\pi f_n t}}{2\sqrt{\zeta^2 - 1}\left(\zeta - \sqrt{\zeta^2 - 1}\right)} + \frac{e^{-\left(\zeta + \sqrt{\zeta^2 - 1}\right)2\pi f_n t}}{2\sqrt{\zeta^2 - 1}\left(\zeta + \sqrt{\zeta^2 - 1}\right)} \qquad (3.86)$$

The reciprocal of the negative power constant of the exponential term in the error signal is called the time constant. It has been studied that when $\zeta > 1$, the response given to the unit step input of the system does not show the oscillating portion therein, which is called overdamped response. When $\zeta < 1$, the oscillation of the response decays exponentially with a time constant $1/2\pi f_n \zeta$, which is called underdamped response. It can also be studied the case when the damping ratio is 1. In this case, the damping of the response is only controlled by the natural frequency f_n. The actual damping under this condition can be called the critical damping of the response. When the damping ratio is less than 1, the vibration part exists in the response, and when the damping ratio does not exist in the response, it is equal to 1. This is why the damping response is called critical damping when $\zeta = 1$. More accurately, when the damping ratio is equal to 1, the damping is called critical damping and the response is critically damped.

The second-order system has a steady-state error of zero. The response of the second-order system depends on the damping ratio ζ and the natural frequency f_n. The greater the natural frequency, the faster the response. The damping ratio ζ determines the overshoot and oscillation frequency. If the damping ratio is much larger than 1, the second-order system can be seen as a series connection of two first-order systems. For a second-order system with a damping ratio of less than 0.707, the amplitude ratio becomes greater than 1 near $f/f_n = 1$. This is called resonance. At the same time, the phase angle changes quite rapidly around the resonance. Above the frequency ratio of the resonance, the amplitude ratio decreases quite rapidly and the phase angle approaches 180°. As the frequency ratio increases, the amplitude decreases monotonically. The phase angle changes gradually from zero to 180°. The "best" damping ratio of the measuring instrument is $\zeta = 0.707$ because the amplitude ratio remains close to 1 in the maximum frequency range.

3.4.3 Transient response specifications

You can use the transient response of the unit step input function to represent the performance of the measurement system because it can be easily generated. A unit step input signal is given in a second-order measuring system and all initial conditions of the system can be considered as zero. Figure 3.50 shows the time response characteristics of the second-order system at underdamped condition. There are many common terms in transient response characteristics, which are as follows.

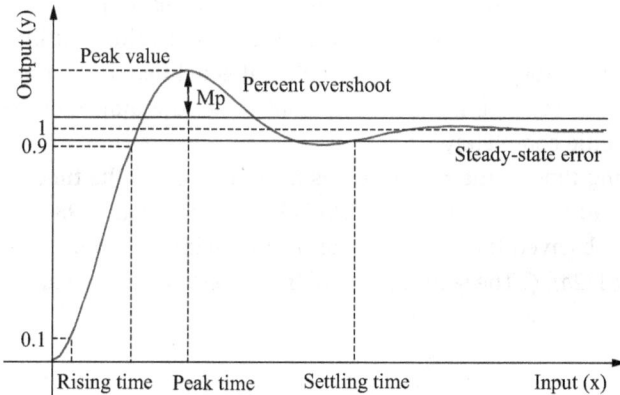

Figure 3.50: The time response characteristics of the second-order system.

(1) Rising time (t_r)
Rising time is the amount of time required to reach the target value from an initial state of zero by an underdamped time response signal during its first cycle of oscillation. When the signal is overdamped, rising time would be defined as the time for the waveform to go from 0.1 to 0.9 of its final value. Some systems never rise to 100% of the final value, and therefore, they will have an infinite rise-time.

From formula (3.83), we know the response of underdamped second-order measuring system with a unit step input. According to the definition of the rising time, the magnitude of the output signal at rising time is 1. Hence

$$y(t_r) = 1 = 1 - \frac{e^{-2\pi f_n \zeta t_r}}{\sqrt{1-\zeta^2}} \sin\left(2\pi f_n \sqrt{1-\zeta^2} t_r + \varphi\right) \tag{3.87}$$

$$\rightarrow \frac{e^{-2\pi f_n \zeta t_r}}{\sqrt{1-\zeta^2}} \sin(2\pi f_n \sqrt{1-\zeta^2} t_r + \varphi) = 0 \tag{3.88}$$

$$\rightarrow \sin\left(2\pi f_n \sqrt{1-\zeta^2} t_r + \varphi\right) = 0 \tag{3.89}$$

$$\rightarrow 2\pi f_n \sqrt{1-\zeta^2} t_r + \varphi = \pi \tag{3.90}$$

$$\rightarrow t_r = \frac{\pi - \varphi}{2\pi f_n \sqrt{1-\zeta^2}} \tag{3.91}$$

(2) Settling time (t_s)

After the initial rising time of the measurement system, some systems will oscillate for a period of time before the system output stabilizes at the final value. Settling time is the time required for the response to become stable. It is defined as the time for the response to reach and stay within 2–5% of its final value. Undamped oscillating systems may never settle completely.

In this book, the settling time of the response has been defined as the time in response to reaching its steady-state condition, which is higher than nearly 98% of its final value. It was also observed that this duration is approximately four times as the signal time constant $1/2\pi f_n \zeta$. The settling time of the second-order measuring system can be given as

$$t_s = \frac{4}{2\pi f_n \zeta} \tag{3.92}$$

(3) Peak time (t_p)

Peak time could be considered as the time required to reach the first or maximum peak (the first overshoot, or peak of the first cycle of oscillation). Based on the definition of the peak time, it can be written as follows:

Let

$$\left. \frac{dy(t)}{dt} \right|_{t=t_p} = 0 \tag{3.93}$$

$$\frac{dy(t)}{dt} = \frac{2\pi f_n}{\sqrt{1-\zeta^2}} e^{-2\pi f_n \zeta t} \sin\left(2\pi f_n \sqrt{1-\zeta^2} t\right) \tag{3.94}$$

$$\frac{dy(t_p)}{dt} = \frac{2\pi f_n}{\sqrt{1-\zeta^2}} e^{-2\pi f_n \zeta t_p} \sin\left(2\pi f_n \sqrt{1-\zeta^2} t_p\right) = 0 \tag{3.95}$$

$$\Rightarrow \sin\left(2\pi f_n \sqrt{1-\zeta^2} t_p\right) = 0 \tag{3.96}$$

$$\Rightarrow 2\pi f_n \sqrt{1-\zeta^2}, \quad t_p = n\pi, \quad n = 0, \pm 1, \pm 2, \dots \tag{3.97}$$

$$t_p = \frac{\pi}{2\pi f_n \sqrt{1-\zeta^2}} \tag{3.98}$$

(4) Percent overshoot (M_p)

Percent overshoot is defined as the amount that the waveform overshoots the steady-state value at the peak time, which is expressed in percentage of the steady-state value of the response. According to the definition of percent overshoot, it can be expressed as

$$M_p = \%\text{F.S.} = \frac{y(t_p) - y(\infty)}{y(\infty)} \times 100\% \tag{3.99}$$

If you put the expression of the peak time in the expression of the output $y(t)$, you will get,

$$y(t_p) = 1 - \frac{e^{-2\pi f_n \zeta t_p}}{\sqrt{1-\zeta^2}} \sin\left(2\pi f_n \sqrt{1-\zeta^2} t_p + \varphi\right)$$

$$= 1 - \frac{e^{-2\pi f_n \zeta \frac{\pi}{2\pi f_n \sqrt{1-\zeta^2}}}}{\sqrt{1-\zeta^2}} \sin\left(2\pi f_n \sqrt{1-\zeta^2} \frac{\pi}{2\pi f_n \sqrt{1-\zeta^2}} + \varphi\right)$$

$$= 1 - \frac{e^{-\zeta \frac{\pi}{\sqrt{1-\zeta^2}}}}{\sqrt{1-\zeta^2}} \sin(\pi + \varphi)$$

$$= 1 + \frac{e^{-\zeta \frac{\pi}{\sqrt{1-\zeta^2}}}}{\sqrt{1-\zeta^2}} \sin(\varphi)$$

$$= 1 + \frac{e^{-\zeta \frac{\pi}{\sqrt{1-\zeta^2}}}}{\sqrt{1-\zeta^2}} \sqrt{1-\zeta^2}$$

$$= 1 + e^{-\zeta \frac{\pi}{\sqrt{1-\zeta^2}}} \tag{3.100}$$

$$M_p = \frac{y(t_p) - y(\infty)}{y(\infty)} = \frac{y(t_p) - 1}{1} = e^{-\zeta\pi/\sqrt{1-\zeta^2}} \qquad (3.101)$$

(5) Steady-state error (e_{ss})

Steady-state error is defined as the difference between the actual output and the desired output at the infinite range of time. According to the definition of steady-state error, it can be expressed as

$$e_{ss} = \lim_{t\to\infty} (y_d(t) - y(t)) \qquad (3.102)$$

(6) Delay time (t_d)

Delay time is defined as the time required to reach at 50% of its steady-state value by a time response signal during its first cycle of oscillation.

3.5 Nondistortion measurement

A noise-free measuring system can be represented by a transfer function, and the output can be written as the function of the input. The relationship between output $y(t)$ and input $x(t)$ shown in Figure 3.51 can be written as follows:

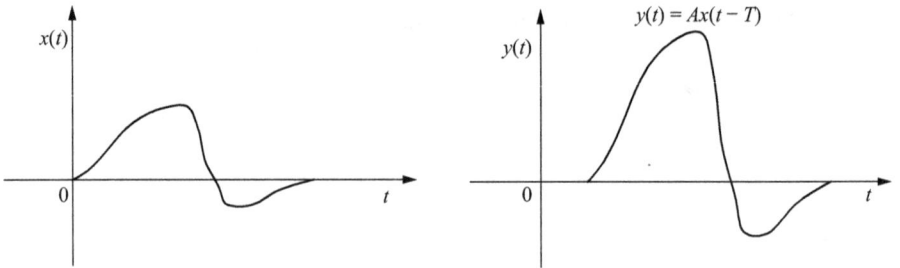

Figure 3.51: The relationship between input and output.

$$y(t) = Ax(t - T) \qquad (3.103)$$

where A is the perfect gain constant and T is the perfect lag time.

If the relationship satisfies formula (3.103), you can say that the output is undistorted. This condition of nondistortion measurement in the time domain can be written as

$$\begin{cases} A = \text{constant} \\ T = \text{constant} \end{cases} \tag{3.104}$$

$$y(t) = Ax(t - T) \Rightarrow Y(f) = Ae^{-j2\pi fT}X(f) \tag{3.105}$$

Similarly, the condition of nondistortion measurement in the frequency domain can be written as

$$\begin{cases} |H(f)| = \left|\frac{Y(f)}{X(f)}\right| = |Ae^{-j2\pi fT}| = A = \text{constant} \\ \phi(f) = -2\pi fT = \text{linear} \end{cases} \tag{3.106}$$

The condition of nondistortion measurement in the frequency domain can be plotted in the logarithmic coordinate system, as shown in Figure 3.52.

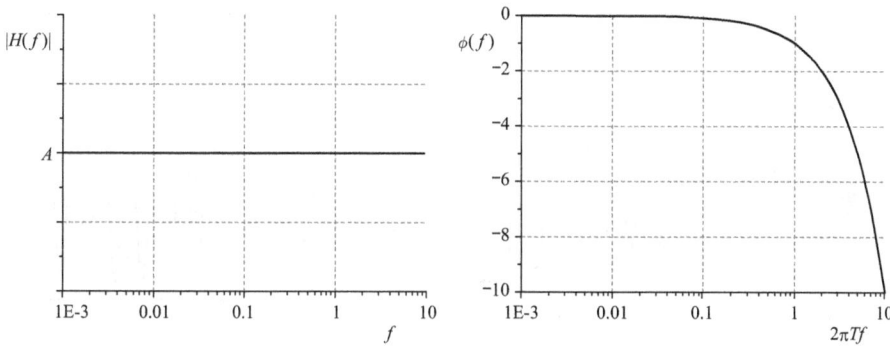

Figure 3.52: The condition of nondistortion measurement in the frequency domain.

A pure sinusoidal signal has no harmonic distortion because it is a signal consisting of a single frequency. A periodic but not purely sinusoidal signal will have a higher frequency component, resulting in harmonic distortion of the signal. In general, the larger the difference between the periodic signal and the sine wave, the stronger the harmonic component and the greater the harmonic distortion. Other conventional periodic signals, such as square waves, generate a large amount of harmonic distortion. In the real world, sinusoidal signals are not completely sinusoidal, and a certain amount of harmonic distortion can occur.

There are usually nonlinear characteristics in the measurement system, which are the main sources of distortion. Figure 3.53 indicates the process that a measured signal $x(t)$ passes through the measuring device:

$$\begin{cases} x(t) = x_1(t) + x_2(t) + x_3(t) + x_4(t) \\ y(t) = y_1(t) + y_2(t) + y_3(t) + y_4(t) \end{cases} \tag{3.107}$$

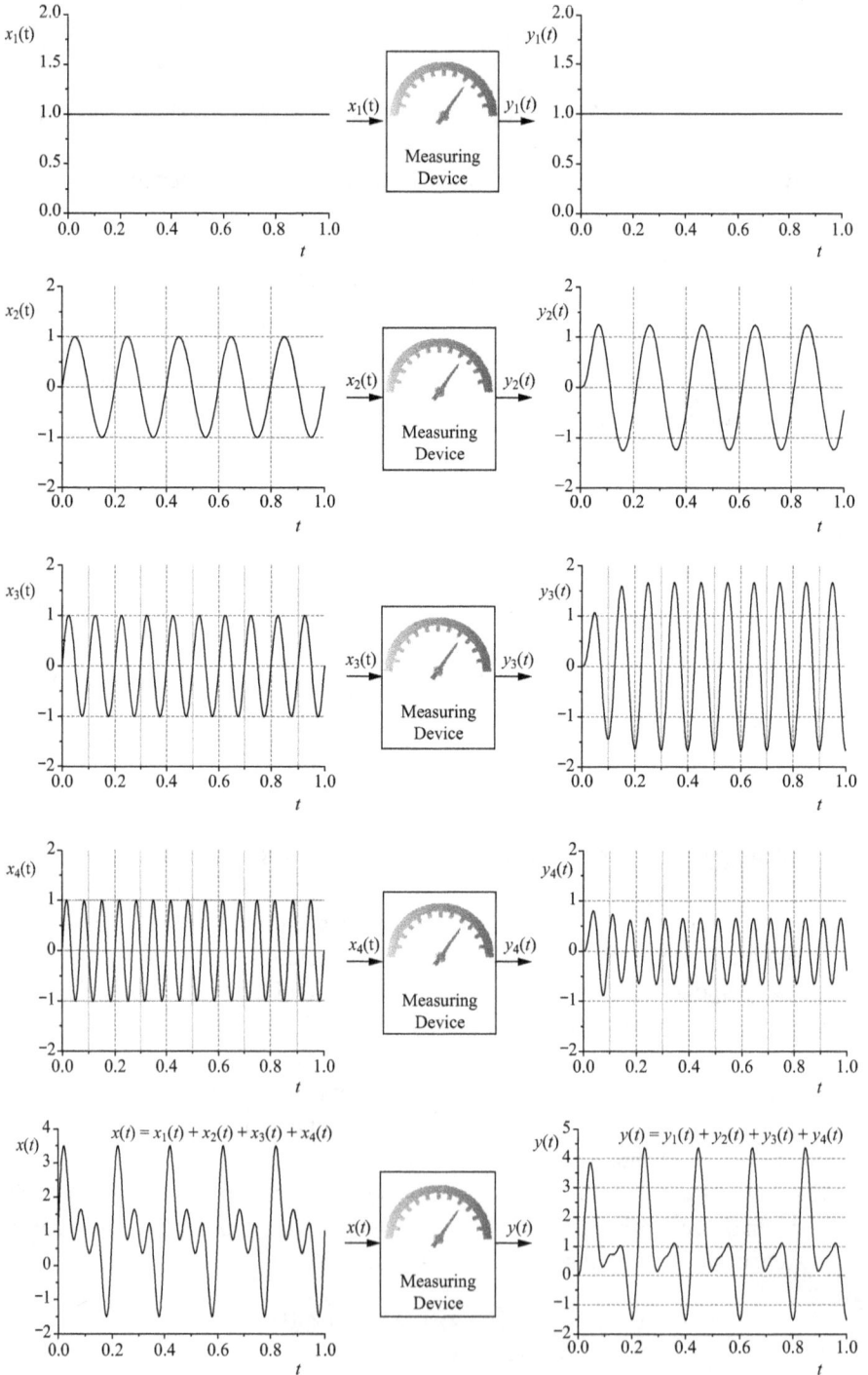

Figure 3.53: The input and the output of the measuring device.

$$\begin{cases} x_1(t) = 1 \\ x_2(t) = \sin(2\pi \cdot 5t) \\ x_3(t) = \sin(2\pi \cdot 10t) \\ x_4(t) = \sin(2\pi \cdot 55t) \end{cases} \qquad \begin{cases} y_1(t) = F(x_1(t)) \\ y_2(t) = F(x_2(t)) \\ y_3(t) = F(x_3(t)) \\ y_4(t) = F(x_4(t)) \end{cases} \qquad (3.108)$$

Figure 3.53 shows that the relationship between the input and the output of the measuring device does not satisfy formula (3.103). It is a distorted measurement.

How to avoid amplitude distortion and phase distortion? For a second-order measurement system, the frequency is close to f_n. The distortion will be worse. There is actually no absolute distortion measurement, as long as the output is within the error range, it can be considered as a nondistortion measurement.

The following steps can effectively reduce the possibility of distortion and achieve nondistortion measurement.

Step 1: Use a suitable measuring device so that the amplitude response and phase response are close to the conditions of the distortion-free measurement.

Step 2: Filter out the noise in the signal.

Step 3: For a first-order system, the smaller the time constant is, the better the performance will be. For a second-order system, it is necessary to design a suitable damping ratio and natural frequency to ensure that nondistortion measurement is achieved in a sufficient frequency range.

3.6 Identification of dynamic characteristics

From Sections 3.3.3.3, 3.3.3.4 and 3.3.3.5, you know that the magnitude response and phase response can be achieved by a swept-sine method, impulse response method and step response method. In this section, you will learn how to identify the dynamic characteristic of the measuring system.

(1) First-order measuring system

For the first-order measuring system, the main parameter is the time constant τ. The value of τ can be determined according to the $M(f)$ response and the $\phi(f)$ response.

According to the magnitude response function of the first-order measuring system, you get

$$M(f) = \frac{1}{\sqrt{1 + (2\pi f \tau)^2}}$$

$$\Rightarrow \tau = \frac{\sqrt{\frac{1}{M(f)^2} - 1}}{2\pi f} \qquad (3.109)$$

According to the phase response function of the first-order measuring system, you get

$$\phi(f) = -\arctg(2\pi f \tau)$$

$$\Rightarrow \tau = \frac{tg(-\phi(f))}{2\pi f} \tag{3.110}$$

If you know point $(M(f), f)$ or point $(\phi(f), f)$, the value of τ can be calculated by formulas (3.109) or (3.110). It is impossible to obtain accurate IRF. Figure 3.48 shows that $y(t)$ is 63.2% of the way to its final value after a time of one time constant τ. So

$$y(\tau) = 63.2\% * y(\infty) \tag{3.111}$$

In this way, time constant τ can also be easily found by the step response method. However, the time constant depends on the single instantaneous value and the whole process is not involved. Thus, the reliability of the measurement results is very poor. If we use the following method to determine the time constant τ, a more reliable result can be obtained.

The step response function of the first-order system

$$y(t) = 1 - e^{-t/\tau} \tag{3.112}$$

Formula (3.112) can be rewritten as

$$1 - y(t) = e^{-t/\tau} \tag{3.113}$$

You can perform a mathematical procedure to transform formula (3.112) into another useful form

$$\begin{cases} \ln[1-y(t)] = -\frac{t}{\tau} \\ y = ax + b \end{cases} \tag{3.114}$$

$$\tau = -\frac{t}{\ln[1-y(t)]} \tag{3.115}$$

The most useful aspect of formula (3.115) is that it can be obtained in the form of a straight line $(y = ax + b)$. Therefore, if we use graphs to represent $(\ln[1-y(t)])$ versus time, it will produce a straight line. The slope of the straight line is very easy to measure so that the time constant τ can be determined. The advantage of this method is that it involves the entire process.

(2) Second-order measuring system
For the second-order measuring system, the main parameters are f_n and ζ. The value of f_n is known, the value of ζ can be determined according to the $M(f)$ response and the $\phi(f)$ response.

According to the magnitude response function of the second-order measuring system, you will get

$$M(f_n) = \frac{1}{\sqrt{\left[1-\left(\frac{f_n}{f_n}\right)^2\right]^2 + 4\zeta^2\left(\frac{f_n}{f_n}\right)^2}}$$

$$= \frac{1}{2\zeta} \tag{3.116}$$

$$\Rightarrow \zeta = \frac{1}{2M(f_n)} \tag{3.117}$$

If you know the point $(M(f_n), f_n)$, the value of ζ can be calculated by formula (3.117). According to the phase response function of the second-order measuring system, you will get

$$\frac{d\phi(f)}{df} = \frac{\dfrac{2\zeta}{\left(2\pi f_n\left(\frac{f}{f_n}\right)^2 - 1\right)} - \dfrac{4(2\pi f)^2\zeta}{(2\pi f_n)^3\left(\left(\frac{f}{f_n}\right)^2 - 1\right)^2}}{1 + \dfrac{4(2\pi f)^2\zeta^2}{(2\pi f_n)^2\left(\left(\frac{f}{f_n}\right)^2 - 1\right)^2}} \tag{3.118}$$

$$\frac{d\phi(f)}{df} = -\frac{2f_n\zeta(f^2+f_n^2)}{2\pi(f^4+f_n^4+f_n^2(4f^2\zeta^2-2f^2))} \tag{3.119}$$

$$\frac{d\phi(f_n)}{df} = -\frac{2f_n\zeta(f_n^2+f_n^2)}{2\pi(f_n^4+f_n^4+f_n^2(4f_n^2\zeta^2-2f_n^2))}$$

$$= -\frac{1}{2\pi f_n\zeta} \tag{3.120}$$

$$\Rightarrow \zeta = -\frac{1}{2\pi f_n\, d\phi\,(f_n)/df} \tag{3.121}$$

$$\phi(f_n) = -90° \tag{3.122}$$

If you know the phase response and find the slope at the point $(\phi(f_n), f_n)$, the value of ζ can be calculated by formula (3.121). The phase is difficult to measure accurately, so it is the best to infer f_n and ζ from the amplitude response curve.

For an underdamped system $(0 < \zeta < 1)$, f_p is the peak frequency in the amplitude response curve. f_p is close to f_n. The relationship between f_p and f_n can be written as

$$f_n = \frac{f_p}{\sqrt{1-2\zeta^2}} \tag{3.123}$$

If ζ is very small, then $f_p \approx f_n$

$$M(f) = \frac{1}{\sqrt{\left[1 - \left(\frac{f}{f_n}\right)^2\right]^2 + 4\zeta^2 \left(\frac{f}{f_n}\right)^2}} \tag{3.124}$$

$$M(f_p) \approx M(f_n) = \frac{1}{2\zeta} \tag{3.125}$$

Let $f_1 = (1 - \zeta)f_p$ and $f_2 = (1 + \zeta)f_n$, you will get

$$M(f_1) \approx \frac{1}{2\sqrt{2}\zeta} = \frac{1}{\sqrt{2}}M(f_p) \approx M(f_2) \tag{3.126}$$

If you draw a horizontal line (with an amplitude equal to $0.707M(f_n)$) in the amplitude response plot, the line will intersect at two points, and the corresponding frequencies should be f_1 and f_2 shown in Figure 3.54. Damping ratio should be

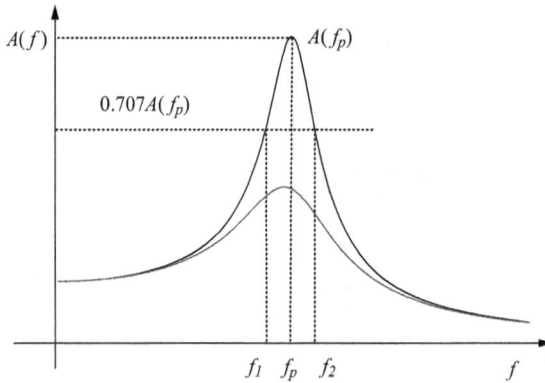

Figure 3.54: The amplitude response of the measuring system.

$$\zeta = \frac{f_2 - f_1}{2f_p} \tag{3.127}$$

Based on formula (3.124), the relationship between M_p and ζ can be written as

$$\zeta = \sqrt{\frac{1}{1 + \left(\frac{\pi}{\ln M_p}\right)^2}} \tag{3.128}$$

The $\zeta - M_p$ curve is drawn based on formula (3.128) shown in Figure 3.55, and it is easy to obtain ζ according to this curve.

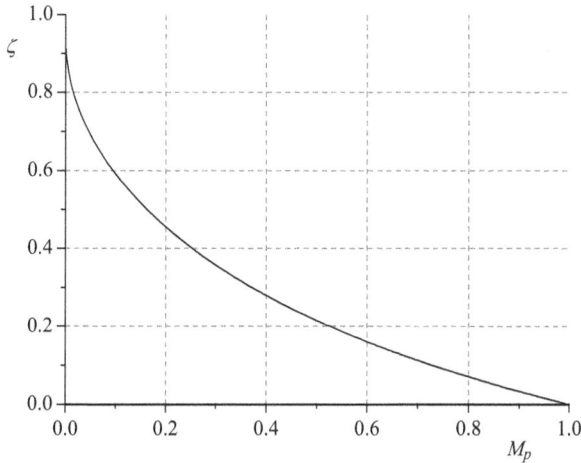

Figure 3.55: The relationship between M_p and ζ.

3.7 Loading effect

Loading effect means that the introduction of a sensing element results in a change in the value of the measured variable. Some loading effects are negligible, but some other loading effects are very serious. The output of the sensor device may deviate from the correct value due to loading effects. For example, the temperature of the IC chip is very high, and the power consumption is very small, about several tens of milliwatts. If a thermometer with a probe is used for measurement, the loading effect of the probe can mislead. For a single degree of freedom system (mass–spring system), if the displacement sensor is mounted on a mass ball, the load effect is caused by a change in mass, resulting in a measurement error. From the perspective of energy, any instrument always absorbs energy from the system, so the system cannot faithfully measure the signal in a form that is undistorted.

The voltmeter is always connected between two points, and the potential difference is measured between these two points, as shown in Figure 3.56. If it is connected through a low resistance Z_S, when the voltmeter resistance Z_m is high, most of the current will pass through the low resistance and produce a voltage drop, which will be a true reading. But if the voltmeter is connected to high resistance Z_S and the two high resistances are connected in parallel, the current will be split into two paths almost equally. Therefore, the meter will record the voltage drop across the high resistance, which will be much lower than the true reading. Therefore, when used in a high resistance circuit, the low sensitivity instrument gives a lower reading than the true reading. This is called the loading effect of the voltmeter. According to the Thevenin equivalent circuit, the reading of voltmeter can be written as

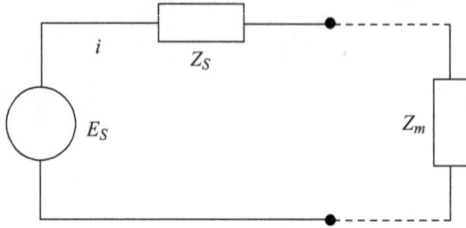

Figure 3.56: The connection of the voltmeter.

$$V_m = i \cdot Z_m = \frac{E_S}{Z_S + Z_m} \cdot Z_m \qquad (3.129)$$

where E_S is voltage source and Z_S is the impedance looking back into the terminal.

Question 2.1:
In the circuit shown in Figure 3.57, the resistor values are given by $R_1 = 2,000\ \Omega$, $R_2 = 2,000\ \Omega$ and $E_{DC} = 24$ V. The voltage across R_2 is measured by a voltmeter whose internal resistance is given by $R_m = 10,000\ \Omega$. Find the reading on the voltmeter and the measurement error due to the loading effect of the voltmeter.

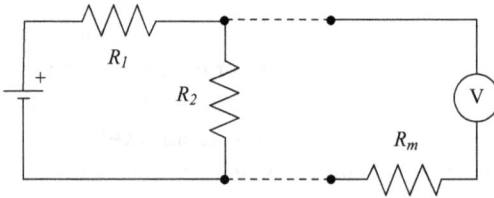

Figure 3.57: The connection of the voltmeter.

Solution:
From formula (3.129), the reading on the voltmeter can be written as

$$V_m = \frac{E_S}{Z_S + Z_m} \cdot Z_m$$

$$= \frac{E_{DC}}{R_1 + \frac{R_m R_2}{R_m + R_2}} \cdot \frac{R_m R_2}{R_m + R_2}$$

$$= \frac{E_{DC} R_m R_2}{R_1(R_m + R_2) + R_m R_2}$$

$$= \frac{24V \cdot 10,000\ \Omega \cdot 2,000\ \Omega}{2,000\ \Omega(10,000\ \Omega + 2,000\ \Omega) + 10,000\ \Omega \cdot 2,000\ \Omega}$$

$$= 10.9\ V \qquad (3.130)$$

By the voltage divider rule, the true voltage of R_2 should be

$$V_{R_2} = \frac{E_{DC}}{R_1 + R_2} \cdot R_2$$

$$= \frac{24\,V \cdot 2,000\,\Omega}{2,000\,\Omega + 2,000\,\Omega}$$

$$= 12\,V \tag{3.131}$$

The measurement error is

$$\text{Measurement error} = \frac{12 - 10.9\,V}{12\,V} 100\%$$

$$= 9.2\% \tag{3.132}$$

Question 2.2:

In the circuit shown in Figure 3.57, the voltage across R_2 is measured by two voltmeters. $R_1 = 2,000\Omega$, $R_2 = 2,000\,\Omega$ and $E_{DC} = 80$ V. The voltmeters used for measuring the voltage across R_2 have a sensitivity of 500 Ω/V or a sensitivity of 2,000 Ω/V. Find out which voltmeter will read more accurately. Both the meters are of the 50 V rang.

Solution:

By the voltage divider rule, the true voltage across R_2 is

$$V_{R_2} = \frac{E_{DC}}{R_1 + R_2} \cdot R_2$$

$$= \frac{80\,V \cdot 2,000\,\Omega}{2,000\,\Omega + 2,000\,\Omega}$$

$$= 40\,V \tag{3.133}$$

For the first voltmeter, the voltmeter resistance will be

$$R_{v1} = SV$$

$$= 500\,\Omega/V \cdot 50\,V$$

$$= 25\,k\Omega \tag{3.134}$$

For the second voltmeter, the voltmeter resistance will be

$$R_{v2} = SV$$

$$= 2,000\,\Omega/V \cdot 50\,V$$

$$= 100\,k\Omega \tag{3.135}$$

The reading on the first voltmeter can be written as

$$V_{m1} = \frac{E_{DC} R_{v1} R_2}{R_1 (R_{v1} + R_2) + R_{v1} R_2}$$

$$= \frac{80\,V \cdot 25,000\,\Omega \cdot 2,000\,\Omega}{2,000\,\Omega(25,000\,\Omega + 2,000\,\Omega) + 25,000\,\Omega \cdot 2,000\,\Omega}$$

$$= 48.1\,V \tag{3.136}$$

The reading on the second voltmeter can be written as

$$V_{m1} = \frac{E_{DC} R_{v1} R_2}{R_1(R_{v1} + R_2) + R_{v1} R_2}$$

$$= \frac{80 \text{ V} \cdot 10,0000 \ \Omega \cdot 2,000 \ \Omega}{2,000 \ \Omega(100,000 \ \Omega + 2,000 \ \Omega) + 100,000 \ \Omega \cdot 2,000 \ \Omega}$$

$$= 49.5 \text{ V} \tag{3.137}$$

Thus the second voltmeter reads more accurately and the high sensitivity voltmeter gives a more accurate reading, though the voltage range for both the meters is the same.

Question 2.2:
Two *RC* integrators are in series connection shown in Figure 3.58. If you have first *RC* integrator with transfer function $H_1(s)$ and second *RC* integrator with transfer function $H_2(s)$, is it right to find the total $H(s)$ just by multiplying $H_1(s)$ and $H_2(s)$?

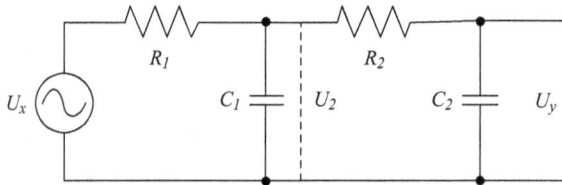

Figure 3.58: Two *RC* integrators are in series connection.

Solution:
The transfer functions of the two RC integrators in Figure 3.58 can be written as

$$H_1(s) = \frac{1}{1 + \tau_1 s}, \quad \tau_1 = R_1 C_1$$

$$H_2(s) = \frac{1}{1 + \tau_2 s}, \quad \tau_2 = R_2 C_2 \tag{3.138}$$

U_2 is the voltage of the C_1, so

$$\frac{U_y(s)}{U_2(s)} = \frac{1}{1 + \tau_2 s} \tag{3.139}$$

The impedance on the right side is

$$Z_2 = R_2 + \frac{1}{C_2 s} = \frac{1 + R_2 C_2 s}{C_2 s} = \frac{1 + \tau_2 s}{C_2 s} \tag{3.140}$$

The impedance of the capacitor C_1 and Z_2 (in parallel connection) is

$$Z = \frac{1}{C_1 s} // Z_2 = \frac{1 + \tau_2 s}{(C_1 + C_2)s + \tau_2 C_1 s^2} \tag{3.141}$$

since

$$\frac{U_x}{Z + R_1} = \frac{U_2}{Z} \tag{3.142}$$

Therefore,

$$\frac{U_2(s)}{U_x(s)} = \frac{Z}{R_1+Z}$$

$$= \frac{1+\tau_2 s}{1+(\tau_1+\tau_2+R_1C_2)s+\tau_1\tau_2 s^2} \quad (3.143)$$

The overall transfer function

$$H(s) = \frac{U_y(s)}{U_x(s)} = \frac{U_2(s)}{U_x(s)}\frac{U_y(s)}{U_2(s)}$$

$$= \frac{1}{1+(\tau_1+\tau_2+R_1C_2)s+\tau_1\tau_2 s^2} \quad (3.144)$$

However,

$$H_1(s)H_2(s) = \frac{1}{1+\tau_1 s}\frac{1}{1+\tau_2 s}$$

$$= \frac{1}{1+(\tau_1+\tau_2)s+\tau_1\tau_2 s^2} \quad (3.145)$$

So

$$H(s) \neq H_1(s)H_2(s) \quad (3.146)$$

If there is a loading effect, the series connection and parallel connection of the LTI system are not applicable. The loading effect is demonstrated by two interacting stages of RC integrators. In order to isolate these circuits from each other, an ideal voltage buffer stage with no reverse transmittance through it can be used, as shown in Figure 3.59.

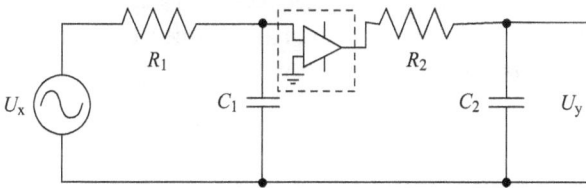

Figure 3.59: Isolated circuit of two RC integrators.

How to minimize the loading effect? The following steps will be taken: (1) use a buffer amplifier to reduce the loading effect; (2) increase the input impedance; (3) use a noncontact measurement; (4) when planning any measurement, the loading effect of the instrument should be considered, and the loading effect should be corrected accurately; (5) appropriate instruments should be used; (6) it is preferable to use a measurement method with negligible load effect or no loading effects.

3.8 Anti-interference technology

Almost all electrical equipment can introduce electrical noise into the measurement circuit. For example, rotating equipment such as motors and generators generate a lot of noise interference. Fixed power sources such as AC power lines and fluorescent lamps also generate considerable noise. When transmitting very high voltages or currents, the noise level of the measurement system is usually higher. In addition, the noise source can be generated by an electric field or a magnetic field. For example, a change charge generated by a varying electric field can be capacitively coupled with a signal line, and a changing current in the wiring can cause a changing magnetic field that is inductively coupled with the signal wiring and causes a varying current to flow. The best remedy for the interference problem is to remove the noise source or to move the instrument to a different location. If the source of noise cannot be removed, other measures can be taken to minimize the interference. Capacitive coupling, such as the capacitance between two adjacent wires or PCB (Printed Circuit Board) traces, is usually unintended. This book will detail the effects of capacitive coupling and inductive coupling.

(1) Capacitive coupling
Capacitive coupling is the displacement current between circuit nodes caused by the electric field. It transfers energy within an electrical network or between distant networks. This coupling has an intentional or unintentional effect. Capacitive coupling is used in a wide range of circuits to pass the required AC signals and leave out the undesired DC signal, as shown in Figure 3.60. Capacitive coupling occurs when a varying electrical field exists (e.g., rapid voltage change, high voltage switching power supply, fluorescent lamp startup process, crosstalk between cables and transforming process between primary and secondary windings) between two adjacent conductors typically less than a wavelength apart inducing a change in voltage across the gap.

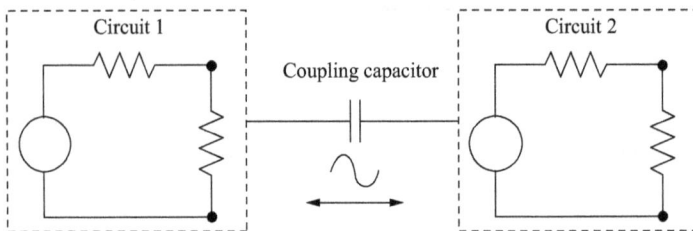

Figure 3.60: Capacitive coupling circuit.

There are several methods for reducing capacitive coupling during measurement: (1) Capacitive coupling with signal wiring can be minimized or eliminated by using shielding. (2) Increase the distance between the interference source and the measuring device. (3) Shield the interference cable. The shield of the measurement cable must be connected and grounded. (4) Increase the signal rising time as much as possible to reduce the voltage variation of the interfering signal. (5) Avoid accidental capacitive coupling between PCB traces or wires close to each other by maintaining a safe separation between the wires and other conductors. (6) the wires are staggered to reduce the effects of capacitive coupling between adjacent signal lines and reduce crosstalk.

(2) Inductive coupling

Inductive coupling refers to the transfer of energy from one circuit component to another through a shared magnetic field. The magnetic field used for inductive coupling is caused by current flowing through the wire. Any time current flows through the wire. It creates a circular magnetic field around the wire. If a second coil is placed in this magnetic field, the coil can induce a current shown in Figure 3.61.

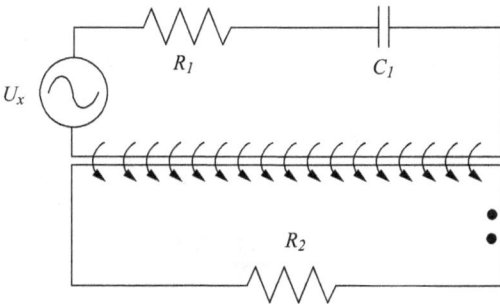

Figure 3.61: Inductive coupling circuit.

Tightly twisted wire can reduce the area of the wire loop in the magnetic field shown in Figure 3.62. The level of induced voltage will be reduced. The tightly twisted wire also minimizes the electric field effect, and the wires are alternately placed at two different positions relative to the electric field. This is equivalent to offsetting the coupling voltage in adjacent wires.

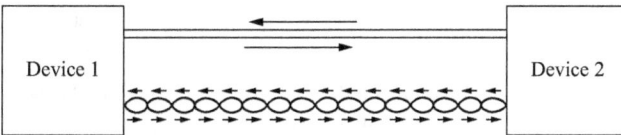

Figure 3.62: The twisted pair.

There are several methods for reducing inductive coupling during measurement: (1) limit the length of the parallel connection cable; (2) increase the distance between the interference cable and the signal cable; (3) grounding a shielded end of a cable reduces the rate of voltage change of the interfering cable and increases the signal rising time; (4) place the cable near the metal surface as much as possible and use the twisted pair; (5) the optical signal output method, such as optical fiber, can be used to completely reduce electromagnetic interference.

(3) Cable shield

In an actual measurement process, the twisted pair is usually combined with a shield, and even a cable with a separate shielded pair can be used as shown in Figure 3.63. These are very useful for sensors such as strain gauges that require a pair of relatively high excitation voltages (AC 10 V) and low-level signal voltages (10 mV). This prevents any noise in the excitation voltage from being coupled with the strain gauge signal.

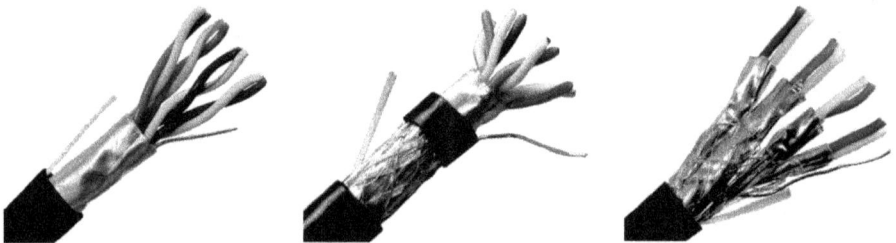

Figure 3.63: The cable shields.

Cable shields are typically made of woven mesh or metal foil of a thin copper wire. The braided shield provides approximately 95% coverage due to the space within the grid, but it does have a very low resistance path. The foil shield provides 100% coverage but has a slightly higher resistance path. The ground cable shield must be properly grounded to eliminate or reduce the possibility of electrical noise coupling with the sensor signal. The shield must be grounded to be effective. If the sensor is equipped with a mating connector and cable, in this case, the shield inside the connector is properly grounded. In an ideal measurement system, all grounding should be terminated at one point so that ground loop current does not flow into the shield. Stranding the active and reference electrode lines minimizes the loop area. The larger the loop area is, the more the capture interference will be.

4 Signal conditioning, processing and recording

4.1 Introduction

Signal conditioning is a technique of sending a signal from a sensor that is suitable for processing by the data acquisition device. Its main purpose is to convert signals that may be difficult to read from conventional instruments to a more readable format. Signal processing involves signal analysis, synthesis and modification. In displacement, force, acceleration and other measurements, signal processing techniques are used to improve signal transmission fidelity and storage efficiency, and to detect the components of interest in the measurement signal. Signal recording is a technique for recording analog/digital signals.

The signals generated by sensors, whether voltages or currents, usually have weak amplitudes (smaller than a few millivolts) that tend toward zero. In order to accurately process these signals, and also to make the system more robust to the effect of noise (contaminated by noise), the signals need to be amplified. The measured weak signal must be amplified so that it is very likely to occupy the dynamic range of the ADC. Therefore, this chapter mainly explains the techniques of signal conditioning, processing and recording.

In this chapter, students will learn (1) bridge circuit; (2) modulation and demodulation; (3) filter; (4) sampling and aliasing; (5) displaying and recording signal; and (6) random signals.

4.2 Bridge circuit

A bridge circuit is a circuit topology where two circuit branches (usually in parallel with each other) are "bridged" along with some intermediate points by a third branch connected between the first two branches. The bridge was originally developed for laboratory measurements. Bridge circuits are often used with transducers for converting physical quantities (temperature, displacement and pressure) into electrical quantities (voltage and current). In addition, the bridge circuit in the power supply design is a structure of a diode to rectify the current, that is, it is converted from an alternating or unknown polarity into a known polarity of direct current.

The most famous bridge circuit is a Wheatstone bridge, which was invented by Samuel Hunter Christie and promoted by Charles Wheatstone to measure resistance. It consists of four resistors, two of which are known values R_1 and R_3, one of which is to be determined as R_x and the other is calibrated and variable R_2. Wheatstone bridges are also widely used for measuring impedance in AC circuits, including resistance, capacitance, inductance and dissipation factor.

https://doi.org/10.1515/9783110624397-004

4.2.1 Classification

In general, a circuit consists of one source and four impedances for measuring various physical quantities. The bridge circuit is very useful in measuring impedance (resistors, capacitors and inductors) and converting the signals of the sensors into relevant voltage or current signals. According to the different excitation voltages, the bridge circuit can be divided into DC bridge circuit and AC bridge circuit. Similarly, depending on the output, the bridge circuit can be divided into an unbalanced bridge circuit and a balanced bridge circuit.

The bridge impedance (Z_1, Z_2, Z_3 or Z_4) shown in Figure 4.1 can be a single impedance (resistor, capacitor or inductor), a combination of impedances or a transducer with varying impedances. For example, a strain gauge is a resistive sensor, whose resistance changes when it is deformed. Another important advantage of the bridge circuit is that it provides higher measurement sensitivity than the sensor. Measurements can be made up with inexpensive precision voltmeters and ammeters. Usually, the voltage form of the signal is the most convenient for information display, control decisions and data storage.

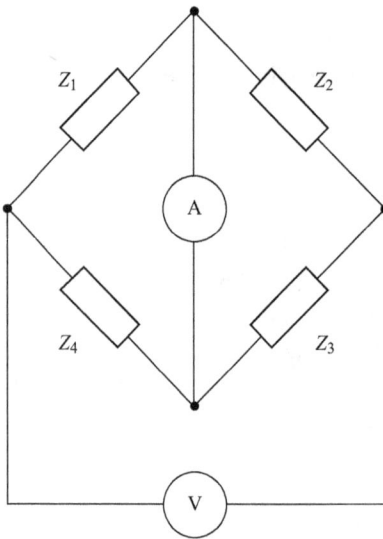

Figure 4.1: The bridge impedance.

4.2.2 Stress and strain

Strain gauges have been used for years and are the basic sensing components for many types of sensors, including load cells, torque sensors, pressure sensors, position sensors. Most of the strain gauges are of foil type and are designed in different shapes and sizes to suit various applications. These consist of a resistive foil pattern mounted on the backing material. When the foil is stressed, they work by changing the resistance of the foil in a defined manner. The stress at certain locations in the material is defined as the applied force per unit of cross-sectional area when a material is loaded with force. The relationship between stress and the applied force can be expressed as follows:

$$\sigma_a = \frac{F}{A} \tag{4.1}$$

where σ_a is an axial stress (axial direction), F is the applied force and A is the original cross-sectional area of the wire.

The strain is defined as the ratio of the length increase to the original length. When a force acts on the wire, its length L is increased by a small increment ΔL, and its cross-sectional area A is decreased. The axial strain is defined as

$$\varepsilon_a = \frac{\Delta L}{L} \tag{4.2}$$

where ε_a is axial strain (axial direction), L is the length of the wire and ΔL is the small increment of wire in the direction of the applied force.

Hooke's law is the law of elasticity, discovered by British scientist Robert Hook in 1660. Hooke's law is a physical principle. This principle states that the force required to stretch or compress the spring at a distance is proportional to the distance. Hooke's law applies only to areas of elastic stress where the load is reversible. According to Hooke's law, for elastic materials, the stress is linearly proportional to the strain:

$$\sigma_a = E\varepsilon_a \tag{4.3}$$

where E is Young's modulus or modulus of elasticity.

The relationship of the electrical resistance R of a wire, the wire length L and cross- sectional area A is given by

$$R = \frac{\rho L}{A} \tag{4.4}$$

where ρ is the resistivity of the wire material.

The wire length L increases as the strain ε_a increases, which increases electrical resistance R. At the same time, the cross-sectional area A of a wire decreases, which also increases the electrical resistance R.

The relationship between electrical resistance R and axial strain ε_a is given by

$$\frac{\Delta R}{R} = S\varepsilon_a \tag{4.5}$$

where S is the strain gauge factor and ΔR is the small change of electrical resistance R.

The electrical resistance R of a commercial strain gauge is typically either 120 or 350 Ω. In strain gauges, the strain gauge factor S of the metal foil used is typically around 2.0. In typical engineering applications with metal beams, axial strain ε_a ranges from 10^{-6} to 10^{-3}. Using formula (4.5), ΔR typically ranges from 0.00024 to 0.24 Ω (S=2.0 and R=120 Ω). This is the main problem with strain gauge: a simple ohmmeter is very small and difficult to measure the change in resistance directly. This is why the bridge circuit is introduced in this section.

The strain gauge is glued to the surface of the beam, and the long portion of the etched metal foil is aligned with the applied axial strain. When the surface is stretched (strained), the strain gauge is also stretched. Therefore, the resistance of the strain gauge increases with the applied strain. The strain gauges are mounted on steel pipeline or rebar to measure shrinkage and monitor strain on structural components such as bridges and buildings, as shown in Figure 4.2.

Figure 4.2: The installation of the strain gauges.

4.2.3 DC (Wheatstone) bridge circuit

The most famous DC (Wheatstone) bridge circuit was invented by Samuel Hunter Christie and was driven by Charles Wheatstone to measure resistance. It consists of four resistors, two of which are known values R_2 and R_4, as shown in Figure 4.3. One of which is determined to be R_3, and the other is variable and calibrated R_1. Two opposing vertices are connected to a battery, and a galvanometer is connected to the other two vertices. Adjust the variable resistor until the ammeter is zero. Then, the ratio between the variable resistor and its neighboring R_2 is equal to the ratio between the unknown resistor and its neighboring R_4. It is possible to calculate the value of the unknown resistor R_3. Wheatstone bridges are also widely used for measuring impedance in AC circuits, resistance, inductance, capacitance and

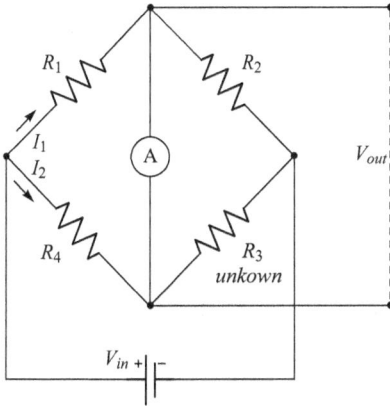

Figure 4.3: Wheatstone bridge circuit.

loss factor, respectively. These variants are called Wien Bridge, Maxwell Bridge and Heaviside Bridge. The output voltage of the Wheatstone bridge is expressed in milli-volts output per volt input.

The output voltage of the bridge circuit can be described as follows:

$$V_{out} = I_1 R_1 - I_2 R_4$$

$$= \frac{V_{in}}{R_1 + R_2} \cdot R_1 - \frac{V_{in}}{R_3 + R_4} \cdot R_4$$

$$= V_{in} \cdot \frac{R_1(R_3 + R_4) - R_4(R_1 + R_2)}{(R_1 + R_2)(R_3 + R_4)}$$

$$= V_{in} \cdot \frac{R_1 R_3 - R_2 R_4}{(R_1 + R_2)(R_3 + R_4)} \tag{4.6}$$

When the bridge is balanced, the output voltage is zero. So the condition for a balanced bridge can be expressed as follows:

$$V_{out} = 0 = V_{in} \cdot \frac{R_1 R_3 - R_2 R_4}{(R_1 + R_2)(R_3 + R_4)} \tag{4.7}$$

$$R_1 R_3 = R_2 R_4 \tag{4.8}$$

Question 4.1:
Determine the value of an unknown resistor R_3 in Figure 4.3 on the condition that current through the galvanometer is zero, given that $R_1 = 35$ kΩ, $R_2 = 45$ kΩ and $R_4 = 28$ kΩ.

Solution:
From Figure 4.3, the product of the resistance in opposite arms of the bridge is in balance, so solving for unknown resistor R_3:

$$R_3 = \frac{R_2 R_4}{R_1}$$

$$= \frac{45\ \mathrm{k\Omega} * 28\ \mathrm{k\Omega}}{35\ \mathrm{k\Omega}} \qquad (4.9)$$

$$= 36\ \mathrm{k\Omega}$$

4.2.3.1 Quarter-bridge strain gauge circuit

The quarter-bridge strain gauge circuit is a simple Wheatstone bridge connected to a stressed strain gauge shown in Figure 4.4 (R_1 is the strain gauge while R_2, R_3 and R_4 are precision resistors). The basic quarter bridge naturally does not have any temperature compensation mechanics, but this can be corrected by the same stress-free strain gauge that is in thermal contact with the active strain gauge. A downward force will bend downward stretching strain gauge when it acts on the free end of the specimen. In the center of the bridge circuit, the voltmeter can detect the imbalance in Figure 4.4. The net force will be zero and the entire circuit will be balanced according to the Wheatstone bridge principle.

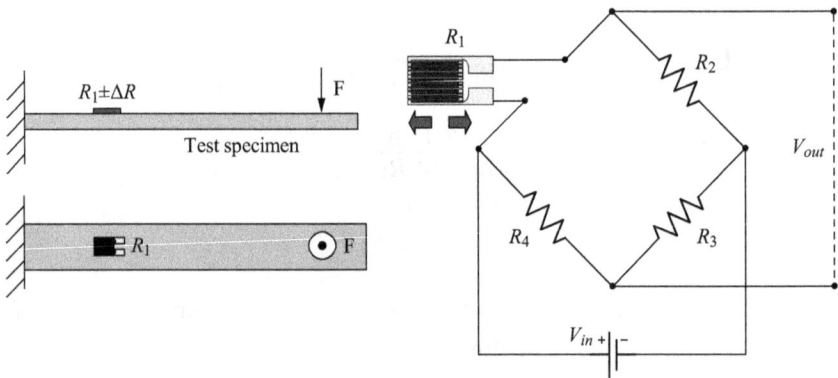

Figure 4.4: Quarter-bridge strain gauge circuit.

According to formula (4.6), the output voltage of the quarter-bridge strain gauge circuit can be expressed as follows:

$$V_{\mathrm{out}} = V_{\mathrm{in}} \cdot \frac{R_1 R_3 - R_2 R_4}{(R_1 + R_2)(R_3 + R_4)} \qquad (4.10)$$

where $R_1 = R_0 + \Delta R$ and $R_2 = R_3 = R_4 = R_0$.

Because $\Delta R \ll R_0$, the output voltage V_{out} can be written as

$$V_{out} = V_{in} \cdot \frac{\Delta R}{4R_0 + 2\Delta R}$$

$$\approx V_{in} \cdot \frac{\Delta R}{4R_0} \qquad (4.11)$$

An unfortunate feature of quarter-bridge strain gauge circuits is that the resistance changes with the temperature. High temperature can affect the internal structure of strain sensing materials such as copper. Temperature can not only influence the characteristics of a strain gauge element but also amend the characteristics of the base material where the strain gauge is attached. Variations in the coefficient of expansion between the gauge and the base material may result in dimensional changes in the sensor elements. They can lead to errors that are very complicated for correction. However, by using a "virtual" strain gauge instead of R_2 in Figure 4.5, the two components of the bridge arm will change the resistance at the same ratio as the temperature changes, thereby counteracting the effects of temperature changes.

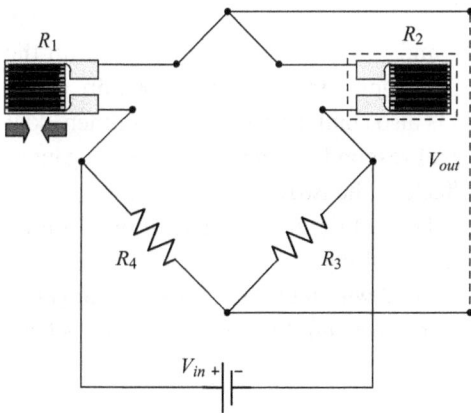

Figure 4.5: Quarter-bridge strain gauge circuit with temperature compensation.

Question 4.2:
In a quarter-bridge circuit, a standard strain gauge is used for measuring the strain of a beam in tension. The strain gauge factor S is 2, and the supply voltage to the Wheatstone bridge is $V_{in} = 10.00$ V. When no load is applied, the bridge is balanced. Assume all resistors are equal when the strain gauge circuit is initially balanced with no load. When the strain gauge circuit is initially balanced with no load, it is assumed that all resistors are equal. For a certain nonzero load, the measured output voltage is $V_{out} = 2.56$ mV. Calculate the axial strain on the beam.

Solution:
From formulas (4.5) and (4.11), we get:

$$V_{out} \approx V_{in} \cdot \frac{\Delta R}{4R_0}$$

$$= \frac{V_{in}}{4} \cdot \frac{\Delta R}{R_0}$$

$$= \frac{V_{in}}{4} \cdot S\varepsilon_a \qquad (4.12)$$

Then

$$\varepsilon_a = \frac{4V_{out}}{SV_{in}}$$

$$= \frac{4 \cdot 2.56 \text{ mV}}{2 \cdot 10 \text{ V}}$$

$$= 0.000512 \qquad (4.13)$$

4.2.3.2 Half-bridge strain gauge circuit

The half-bridge strain gauge circuit uses two strain gauges, as shown in Figure 4.6 (R_1 and R_2 are the strain gauges while R_3 and R_4 are precision resistors), with the same circuit as the quarter-bridge configuration, and the second strain gauge which can be measured by negative strain is also combined with the specimen, so it can respond to mechanical strain. If they all respond in such a way, strain gauges both experience the opposite force. The effects of the two strain gauges are opposite and proportional. This increases the sensitivity while the response to temperature changes is eliminated, thereby reducing the resulting errors. The half-bridge strain gauge circuit has twice the sensitivity compared with a quarter-bridge strain gauge circuit. In the absence of force applied to the specimen, the two strain gauges have

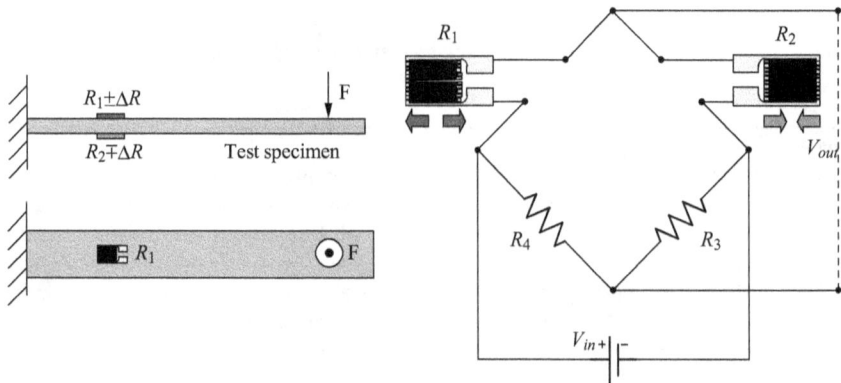

Figure 4.6: Half-bridge strain gauge circuit.

equal electrical resistance, and the bridge circuit is balanced. However, it will bend downward stretching strain gauge R_1 and simultaneously compress strain gauge R_2 when a downward force acts on the free end of the specimen.

Based on formula (4.6), the output voltage of the half-bridge strain gauge circuit can be expressed as follows:

$$V_{out} = V_{in} \cdot \frac{R_1 R_3 - R_2 R_4}{(R_1 + R_2)(R_3 + R_4)} \tag{4.14}$$

where $R_1 = R_0 + \Delta R$, $R_2 = R_0 - \Delta R$ and $R_3 = R_4 = R_0$.

The output voltage V_{out} can be written as

$$V_{out} = V_{in} \cdot \frac{2\Delta R}{4R_0}$$

$$= V_{in} \cdot \frac{\Delta R}{2R_0} \tag{4.15}$$

From formulas (4.15) and (4.11), it can be seen that the sensitivity of a half-bridge strain gauge circuit is twice that of a quarter-bridge strain gauge circuit.

4.2.3.3 Full-bridge strain gauge circuit

The full-bridge strain gauge circuit uses four strain gauges, as shown in Figure 4.7 (R_1, R_2, R_3 and R_4 are all the strain gauges). R_1 and R_3 are used to measure the positive strain, while R_2 and R_4 are used to measure the negative strain. All components of the bridge will be active, which will increase sensitivity. This full-bridge strain gauge circuit is the best bridge strain measuring circuit and provides true linearity based on temperature compensation. Its output voltage is completely proportional to the applied force. But in a quarter-bridge strain gauge circuit, the output voltage is only roughly proportional to the applied force. The only drawback of full-bridge

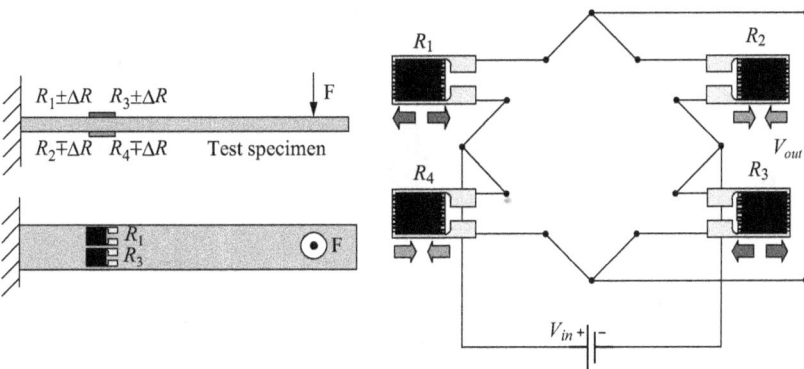

Figure 4.7: Full-bridge strain gauge circuit.

strain gauge circuits is that they are more complex and more expensive than half-bridge gauge circuits and quarter-bridge gauge circuits. A downward force will bend downward stretching strain gauges R_1 and R_3 when it acts on the free end of the specimen while simultaneously compressing strain gauges R_2 and R_4.

Based on formula (4.6), the output voltage of the half-bridge strain gauge circuit can be expressed as follows:

$$V_{out} = V_{in} \cdot \frac{R_1 R_3 - R_2 R_4}{(R_1 + R_2)(R_3 + R_4)} \tag{4.16}$$

where $R_1 = R_0 + \Delta R$, $R_2 = R_0 - \Delta R$, $R_3 = R_0 + \Delta R$ and $R_4 = R_0 - \Delta R$.

The output voltage V_{out} can be written as

$$V_{out} = V_{in} \cdot \frac{4\Delta R}{4R_0}$$

$$= V_{in} \cdot \frac{\Delta R}{R_0} \tag{4.17}$$

From formulas (4.17) and (4.11), it can be seen that the sensitivity of a full-bridge strain gauge circuit is four times that of a quarter-bridge strain gauge circuit.

4.2.3.4 Unbalanced and balanced bridge circuit

Bridge circuits can be divided into two types: unbalanced bridge circuits and balanced bridge circuits. Unbalanced bridge circuits are used to measure many physical quantities, such as strain, temperature and pressure, when V_{out} is not equal to zero. The bridge circuit is balanced at a known point, and then the amount of deviation indicated by the output voltage indicates the amount of change in the measured parameter. Unbalanced bridge circuit shown in Figure 4.8 is utilized in the resistance thermometer temperature detection circuit.

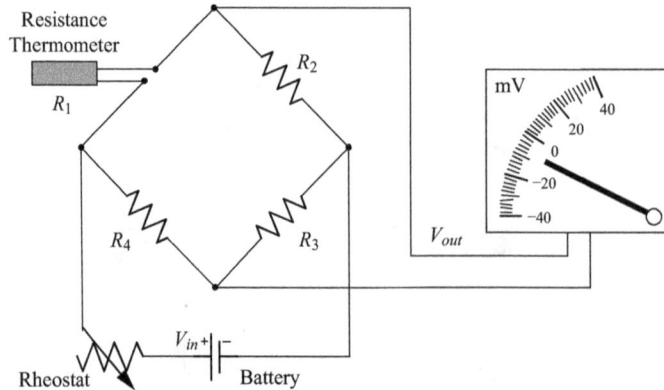

Figure 4.8: Unbalanced bridge circuit.

The unbalanced bridge circuit (Figure 4.8) uses a millivolt voltmeter that is calibrated in temperature units corresponding to the resistance temperature detector. An imbalance occurs when V_{out} is not equal to 0. V_{out} can be converted to a temperature value.

The Wheatstone bridge is in the balanced bridge condition when the output voltage V_{out} is equal to zero. Balanced bridge circuit shown in Figure 4.9 is also utilized in the resistance thermometer temperature detection circuit.

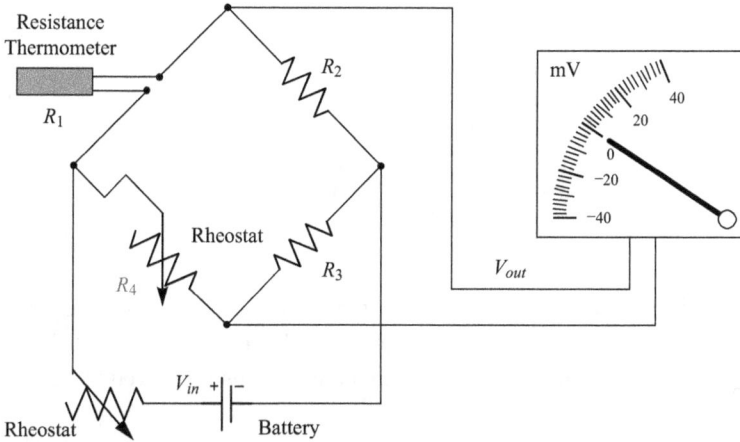

Figure 4.9: Balanced bridge circuit.

As shown in Figure 4.9, a sliding wire resistor (rheostat) is used to balance the bridge arms. As long as the value of the output voltage V_{out} is equal to zero, the circuit will remain balanced. There is a new value for each temperature change. Therefore, the slider must be moved to a new position to balance the circuit. The value of the temperature can be obtained by reading the value of the slip line resistance.

4.2.4 AC bridge circuit

The AC bridge circuit is used to measure the values of unknown resistance, inductance and capacitance, and all AC bridge circuits are based on Wheatstone bridges. The universal AC bridge circuit consists of four impedances, an AC supply source and the balance detector as shown in Figure 4.10. In an AC bridge circuit, the impedance can be pure or complex. Its working principle is same as that of DC bridge circuits that the balance ratio of the impedances will give the balance condition to the circuit. The AC bridge circuit is very convenient and can provide accurate measurement results. AC bridge circuits can be of the "symmetrical" type, where an

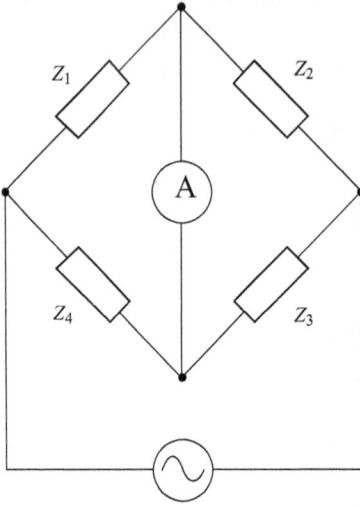

Figure 4.10: AC bridge circuit.

unknown impedance is balanced by a similar type standard impedance on the same side of the bridge. However, balancing the AC bridge circuit is more difficult than the DC bridge circuit because reactive components are also involved in the AC bridge circuit.

When the bridge is in a balanced condition,

$$Z_1 Z_3 = Z_2 Z_4 \tag{4.18}$$

Formula (4.18) states that the product of impedances of one pair of opposite arms must equal the product of impedance of the other pair of opposite arms.

From formula (4.18), we can also obtain

$$|Z_1||Z_3| = |Z_2||Z_4| \tag{4.19}$$

$$\angle Z_1 + \angle Z_3 = \angle Z_2 + \angle Z_4 \tag{4.20}$$

The products of the magnitudes of the opposite arms must be equal, and the sum of the phase angles of the opposite arms must be equal.

Reactance is the imaginary part of the impedance caused by the presence of an inductor or capacitor in the circuit. The reactance is indicated by the symbol X and is in ohms. The following conclusions will be obtained: a resistor's impedance is R and its $X_R = 0$; a capacitor's impedance is $-j(1/\omega C)$ and its $X_C = -(1/\omega C)$; an inductor's impedance is $j\omega L$ and its $X_C = \omega L$.

4.2.4.1 Capacitance bridge circuit
The capacitance bridge circuit has four arms, two have noncapacitive resistance and the other two have capacitances with negligible resistance as shown in Figure 4.11.

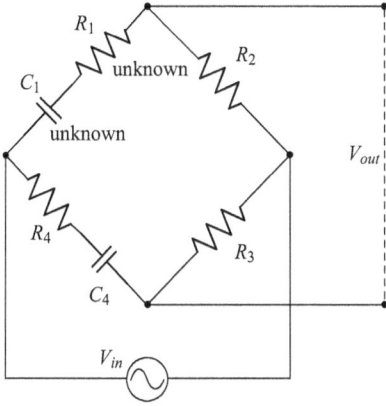

Figure 4.11: Capacitance bridge circuit.

When the bridge is in a balanced condition,

$$Z_1 Z_3 = Z_2 Z_4 \tag{4.21}$$

$$\left(R_1 + \frac{1}{j\omega C_1}\right) R_3 = R_2 \left(R_4 + \frac{1}{j\omega C_4}\right) \tag{4.22}$$

$$R_1 R_3 + \frac{R_3}{j\omega C_1} = R_2 R_4 + \frac{R_2}{j\omega C_4} \tag{4.23}$$

$$R_1 R_3 = R_2 R_4 \;\rightarrow\; R_1 = \frac{R_2 R_4}{R_3} \tag{4.24}$$

$$\frac{R_3}{j\omega C_1} = \frac{R_2}{j\omega C_4} \;\rightarrow\; \frac{R_3}{C_1} = \frac{R_2}{C_4} \;\rightarrow\; C_1 = \frac{R_3 C_4}{R_2} \tag{4.25}$$

The unknown quantities C_1 and R_1 are measured in terms of R_2, R_3, R_4 and C_4.

4.2.4.2 Inductance bridge circuit
The inductance bridge circuit has four arms, two have noninductive resistance and the other two have inductances with negligible resistance as shown in Figure 4.12.
When the bridge is in a balanced condition,

$$Z_1 Z_3 = Z_2 Z_4 \tag{4.26}$$

$$(R_1 + j\omega L_1) R_3 = R_2 (R_4 + j\omega L_4) \tag{4.27}$$

$$R_1 R_3 + j\omega L_1 R_3 = R_2 R_4 + j\omega L_4 R_2 \tag{4.28}$$

$$R_1 R_3 = R_2 R_4 \;\Rightarrow\; R_1 = \frac{R_2 R_4}{R_3} \tag{4.29}$$

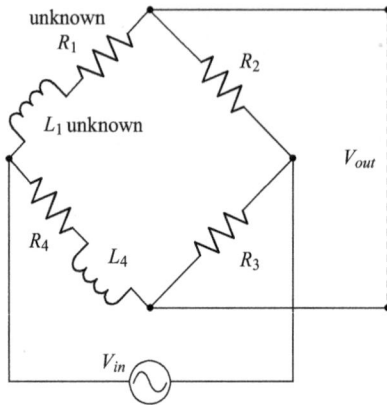

Figure 4.12: Inductance bridge circuit.

$$j\omega L_1 R_3 = j\omega L_4 R_2 \Rightarrow L_1 R_3 = R_2 L_4 \Rightarrow L_1 = \frac{R_2 L_4}{R_3} \tag{4.30}$$

The unknown quantities L_1 and R_1 are measured in terms of R_2, R_3, R_4 and L_4.

4.2.4.3 Typical AC bridge circuits

Wien bridge is a bridge circuit developed by Max Wien in 1891, as is shown in Figure 4.13. The bridge consists of four resistors and two capacitors. The Wien bridge is one of the many typical bridges. Wien's bridges are used to accurately measure capacitance in terms of resistance and frequency. It is also used to measure audio frequencies.

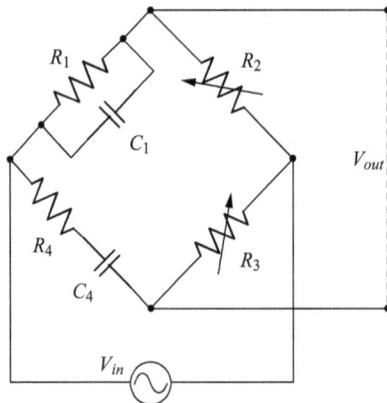

Figure 4.13: Wein bridge circuit.

Maxwell bridge is a bridge circuit developed by James C. Maxwell in 1873, as is shown in Figure 4.14. The Maxwell bridge circuit is an improvement of the Wheatstone bridge for measuring unknown inductance based on calibrated resistance and inductance or resistance and capacitance. It uses the principle that the negative phase angle of the capacitive impedance can compensate the positive phase angle of the inductive impedance when placed in the opposite arm and the circuit is in resonance.

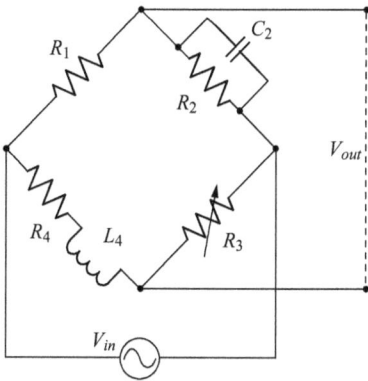

Figure 4.14: Maxwell bridge circuit.

The layout of Owen bridge circuit is shown in Figure 4.15, and is used to measure the large range inductance. In this circuit, the inductance is determined based on the resistance and capacitance, and the balance formula is easily obtained. The advantage of this circuit is that capacitors can be used over a wide range of inductances.

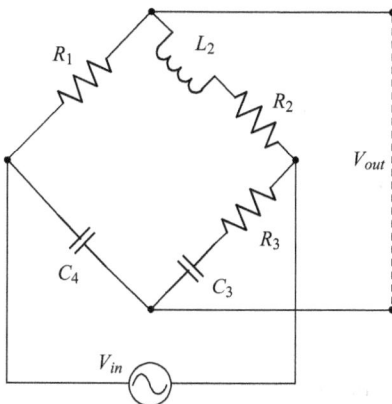

Figure 4.15: Owen bridge circuit.

The Hay bridge is used to determine the self-inductance of the circuit shown in Figure 4.16. In Hay bridge circuit, the capacitor is connected in series with the resistor, and the voltage drop across the capacitor and the resistor varies. The Hay bridge circuit provides a simple representation of the unknown inductance for coils with a mass coefficient greater than 10 Ω.

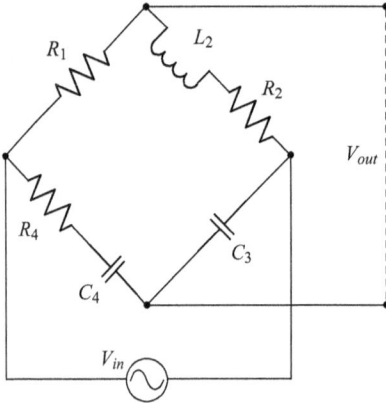

Figure 4.16: Hay bridge circuit.

The Schering bridge is an AC bridge circuit developed by Harald Schering shown in Figure 4.17. The Schering bridge is used to measure capacitor capacitance, dissipation factor, insulator, capacitor bushing, insulating oil and other insulating materials. It is one of the most commonly used AC bridges. The Schering bridge works by balancing the load on its arms. Its advantage is that the equilibrium equation is independent of frequency.

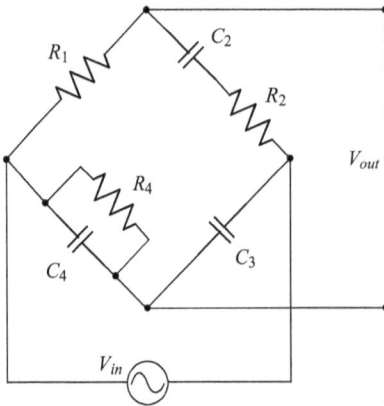

Figure 4.17: Schering bridge circuit.

Question 4.3:
Determine the constants of the unknown arm in Figure 4.18. The impedance of the basic AC bridge is given as follows: $Z_2 = 200\Omega \angle 40°$; $Z_3 = 500\Omega$; $Z_4 = 600\Omega \angle -20°$ and $Z_1 = $ unknown.

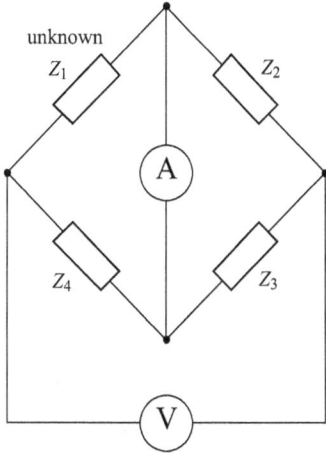

Figure 4.18: Basic AC bridge circuit.

Solution:
From the balance condition of the AC bridge, we get

$$Z_1 Z_3 = Z_2 Z_4 \tag{4.31}$$

$$|Z_1| = |Z_2||Z_4|/|Z_3|$$

$$= \frac{200\Omega * 600\Omega}{500\Omega} = 240\Omega \tag{4.32}$$

$$\angle\theta_1 = \angle\theta_2 + \angle\theta_4 - \angle\theta_3$$

$$= 40° + (-20°) + 0°$$

$$= 20° \tag{4.33}$$

$$Z_1 = 240\Omega \angle 20°$$

$$= 240\cos(20°) + j240\sin(20°)$$

$$= 225.53\Omega + j82.08\Omega \tag{4.34}$$

Question 4.4:

The AC bridge in Figure 4.19 is in balance with the following constants: $R_1 = 400\Omega$; $R_3 = 200\Omega$; $C_3 = 0.4\,\mu F$; $C_4 = 1.0\,\mu F$. The oscillator frequency is 2 kHz. Find the constants of the unknown arm Z_2.

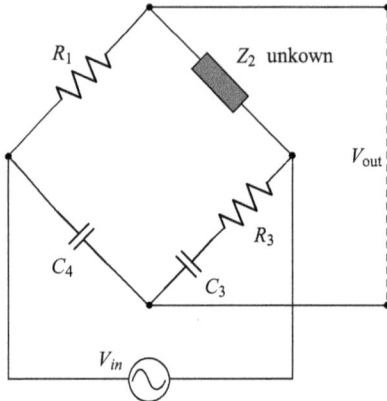

Figure 4.19: AC bridge circuit.

Solution:

From Figure 4.19, we know that

$$Z_1 = R_1$$

$$= 400\Omega\angle 0° \tag{4.35}$$

$$Z_3 = R_3 - j\frac{1}{\omega C_3}$$

$$= 200 - j\frac{1}{2\pi f \times 0.4 \times 10^{-6}}$$

$$= 200 - j\frac{1}{12566.38 \times 0.4 \times 10^{-6}}$$

$$= 200\Omega - j200\Omega$$

$$= 282.8\Omega\angle - 45° \tag{4.36}$$

$$Z_4 = -j\frac{1}{\omega C_4}$$

$$= -j\frac{1}{2\pi f \times 1.0 \times 10^{-6}}$$

$$= -j\frac{1}{12566.38 \times 1.0 \times 10^{-6}}$$

$$= -j79.6\Omega$$

$$= 79.6\Omega\angle - 90° \tag{4.37}$$

From the balance condition of the bridge

$$|Z_2| = |Z_1||Z_3|/|Z_4|$$

$$= \frac{400\Omega * 282.8\Omega}{79.6\Omega}$$

$$= 1,421\Omega \qquad\qquad (4.38)$$

$$\angle\theta_2 = \angle\theta_1 + \angle\theta_3 - \angle\theta_4$$

$$= 0° + (-45°) + 90°$$

$$= 45° \qquad\qquad (4.39)$$

$$Z_2 = 1,421\angle 45°$$

$$= 1004.7\ \Omega + j1004.7\ \Omega \qquad\qquad (4.40)$$

$$R_2 = 1004.7\ \Omega$$
$$\qquad\qquad (4.41)$$
$$\omega L_2 = 1004.7\ \Omega$$

$$L_2 = \frac{1004.7}{\omega}$$

$$= \frac{1004.7}{2\pi f}$$

$$= \frac{1004.7}{12566.38}$$

$$= 0.08H \qquad\qquad (4.42)$$

To sum up, R_2 and L_2 are connected in series.

4.3 Modulation and demodulation

In electronics and telecommunications, modulation is the process of changing one or more attributes of a periodic waveform (referred to as a carrier signal) using a modulated signal that typically contains information to be transmitted. The most common changing characteristics in various forms of modulation include amplitude, frequency and phase. Most radio systems in the twentieth century use frequency modulation (FM) or amplitude modulation (AM) to enable the carrier to carry radio broadcasts. In television transmission, FM is used for sound signals, and AM is used for image signals. Demodulation is a process of extracting information of a signal from a carrier.

4.3.1 Definitions

The amplitude of the high-frequency carrier wave is changed in accordance with the intensity of the signal, which is called AM. There are three fundamental types of modulation: AM, FM and phase modulation. The function of modulation is to put this information on a carrier frequency so that it can be transmitted more readily and with the minimal information loss.

AM wave of a sine signal is shown in Figure 4.20. $x_1(t)$ is the sine signal, $c_1(t)$ is the carrier and $x_{1AM}(t)$ is the AM wave. The carrier signal is usually just a simple, high-frequency sine/cosine wave.

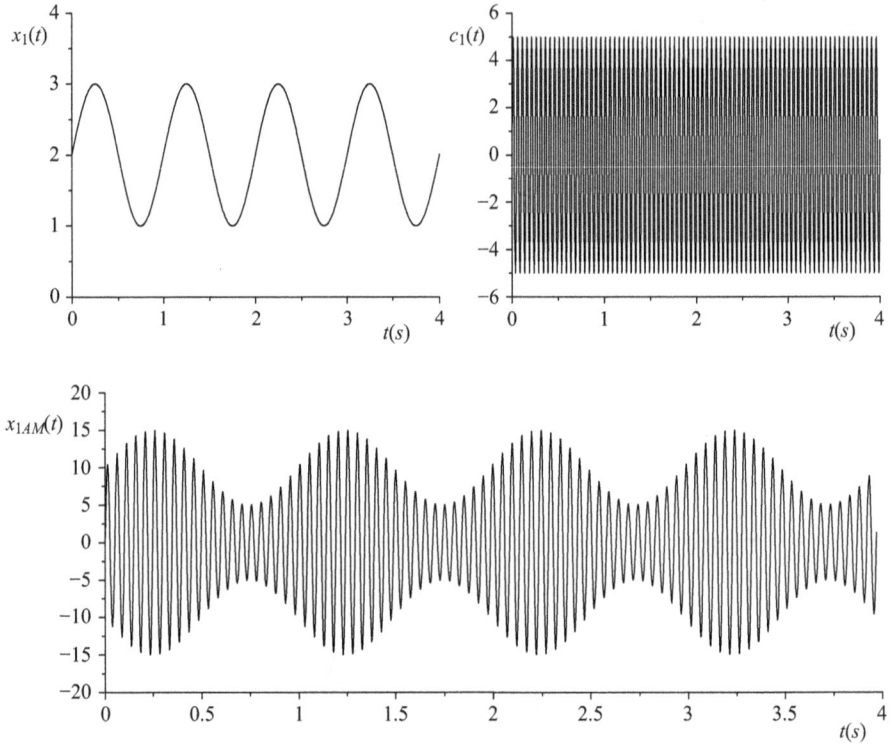

Figure 4.20: Amplitude modulation wave of a sine signal.

AM wave of a square signal is shown in Figure 4.21. $x_2(t)$ is the sine signal, $c_2(t)$ is the carrier and $x_{2AM}(t)$ is the AM wave.

4.3.2 Amplitude modulation process

The modulation process means that the information signal is systematically used to change the amplitude of the carrier signal. The AM wave $x_{AM}(t)$ is obtained by multiplying the carrier $c(t)$ and the signal $x(t)$, as shown in Figure 4.22.

According to the characteristics of the Fourier transform, the following conclusion can be drawn: multiplication in the time domain is equivalent to convolution in the frequency domain. Therefore,

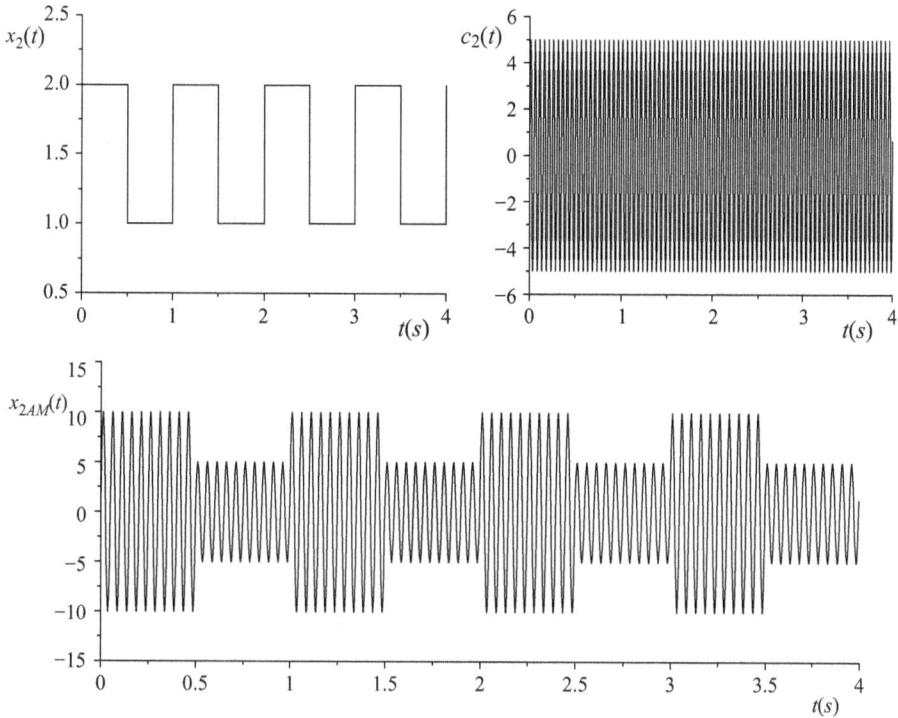

Figure 4.21: Amplitude modulation wave of a square signal.

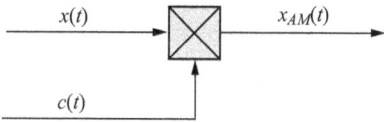

Figure 4.22: Amplitude modulation process.

$$x(t)c(t) \Leftrightarrow X(f)*C(f) \tag{4.43}$$

Description of the modulation process is shown in Figure 4.23. $x(t)$, $c(t)$ and $x_{AM}(t)$ are described in the time domain. $X(f)$, $C(f)$ and $X_{AM}(f)$ are described in the frequency domain. AM results are as follows:

$$x_{AM}(t) = x(t)c(t) \tag{4.44}$$

$$X_{AM}(f) = X(f)*C(f) \tag{4.45}$$

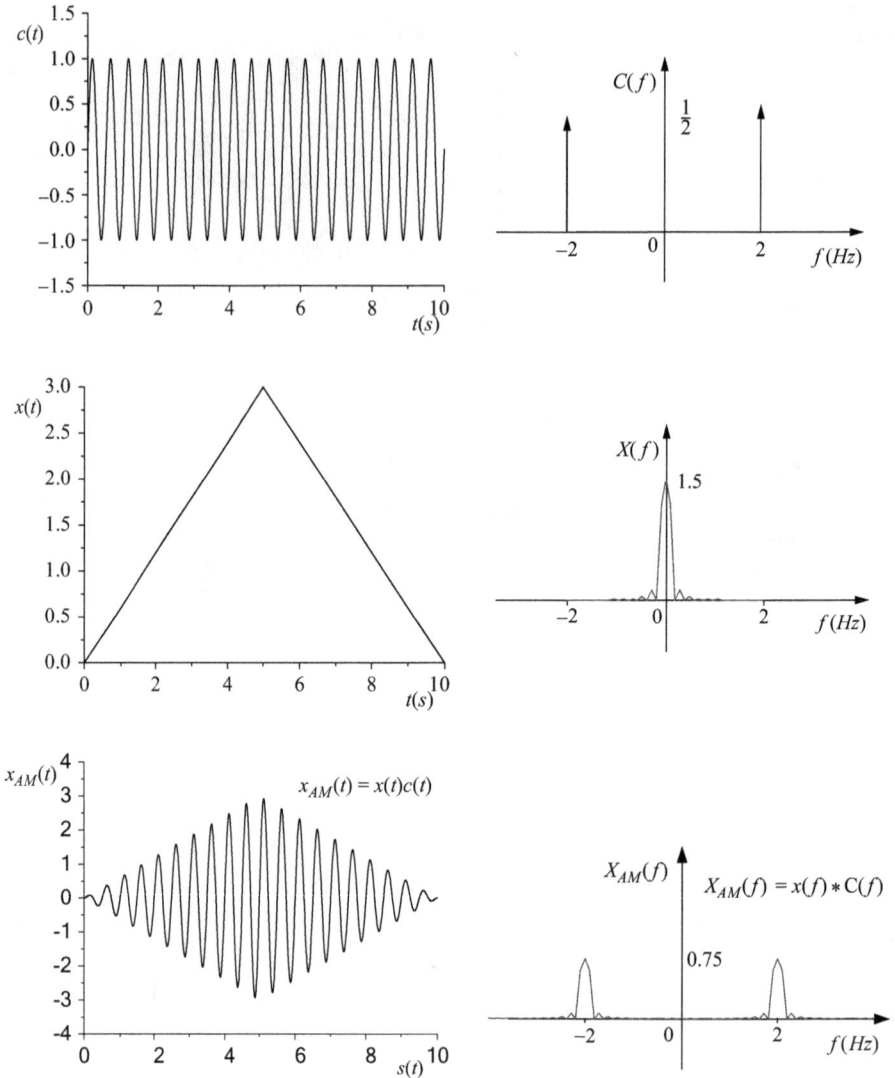

Figure 4.23: Description of amplitude modulation.

4.3.2.1 Synchronous demodulation

Demodulating AM signals typically use synchronous demodulation techniques and provide much lower distortion levels to improve the reception in the presence of selective fading. Synchronous demodulation provides a very low level of distortion with a simple diode rectifier and provides better raw modulation rendering. Synchronous demodulation improves linearity, sensitivity, and signal-to-noise ratio.

In the synchronous demodulation, an AM signal $x_{AM}(t)$ is multiplied with the carrier signal $c(t)$ (exactly the same frequency and same phase), and the product of this multiplication $x_d(t)$ is passed through the low-pass filter, as shown in Figure 4.24.

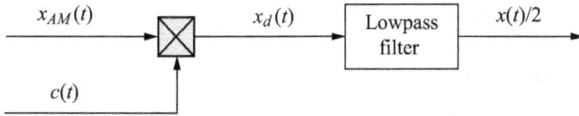

Figure 4.24: Synchronous demodulation process.

Description of synchronous demodulation is shown in Figure 4.25. $x_{AM}(t)$ and $x_d(t)$ are described in the time domain. $X_{AM}(f)$ and $X_d(f)$ are described in the frequency domain.

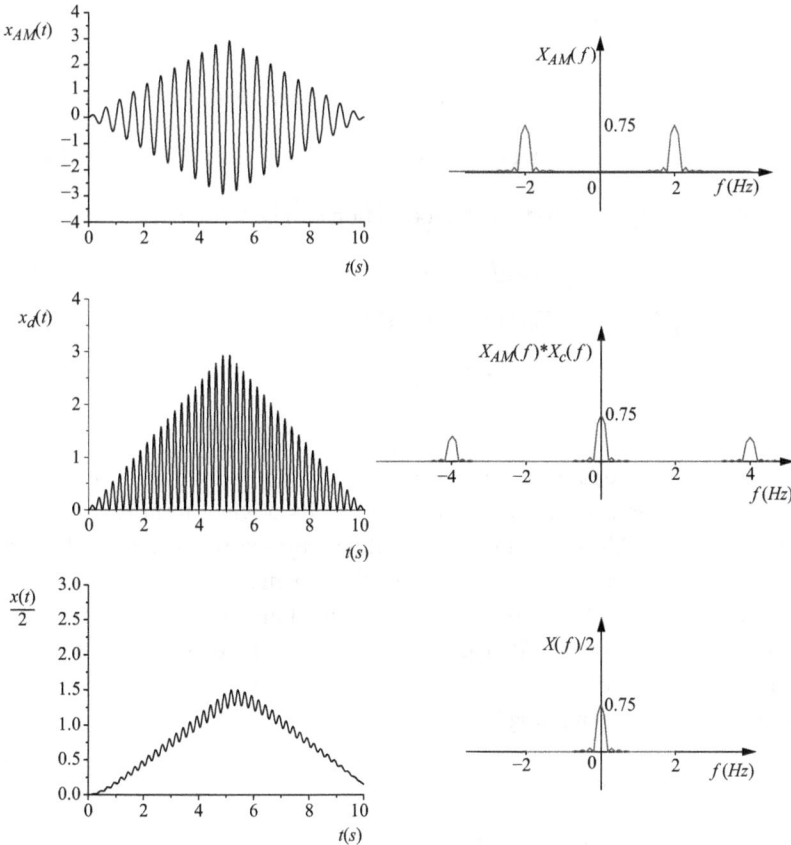

Figure 4.25: Description of synchronous demodulation.

In the time domain, synchronous demodulation can be written as follows:

$$c(t) = \cos 2\pi f_0 t, \quad f_0 = 2 \text{ Hz} \tag{4.46}$$

$$x_{AM}(t) = x(t)c(t)$$

$$= x(t) \cos 2\pi f_0 t \tag{4.47}$$

$$x_d(t) = x_{AM}(t)c(t)$$

$$= x(t)c(t)c(t)$$

$$= x(t) \cos 2\pi f_0 t \cos 2\pi f_0 t$$

$$= \frac{x(t)}{2} + \frac{1}{2}x(t) \cos 4\pi f_0 t \tag{4.48}$$

The inconvenient fraction 1/2 in the output can be eliminated by using a carrier $2 \cos 2\pi f_0 t$ instead of $\cos 2\pi f_0 t$:

$$x_d(t) = x_{AM}(t) \cdot 2 \cdot c(t)$$

$$= x(t)c(t) \cdot 2 \cdot c(t)$$

$$= x(t) \cos 2\pi f_0 t \cdot 2 \cdot \cos 2\pi f_0 t$$

$$= x(t) + x(t) \cos 4\pi f_0 t \tag{4.49}$$

In the frequency domain, synchronous demodulation can be written as

$$X_{AM}(f) = X(f) {}^* C(f) \tag{4.50}$$

$$X_d(f) = X_{AM}(f) {}^* C(f)$$

$$= X(f) {}^* C(f) {}^* C(f) \tag{4.51}$$

4.3.2.2 Conventional demodulation/detection

The process of converting a modulated signal back to its original form is called demodulation/detection. In AM, conventional demodulation/detection can be half-wave or full wave, phase sensitive or nonphase sensitive. In most applications, phase-sensitive detection is required to fully recover the original amplitude and the voltage sign produced by the transducer. After detection, the carrier frequency must be filtered from the signal. This requires a low-pass filter allowing low-frequency signals to pass faithfully but greatly attenuating high frequencies.

The AM wave $x_{AM}(t)$ is obtained by multiplying the carrier $c(t)$ and the signal $x(t)$, as shown in Figure 4.26:

$$x(t) = \sin 2\pi t, \quad 0 \leq t \leq 2 \text{ s} \tag{4.52}$$

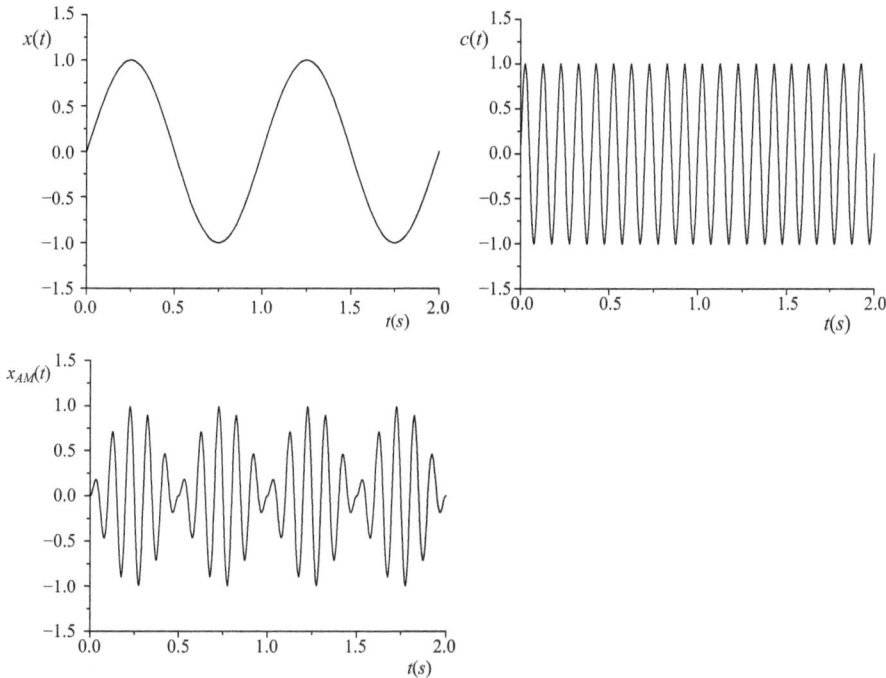

Figure 4.26: Description of the modulated signal.

$$c(t) = \sin 2\pi f_0 t, \quad f_0 = 10 \text{ Hz} \tag{4.53}$$

$$x_{AM}(t) = x(t)c(t) \tag{4.54}$$

The demodulation/detection signals are described in Figure 4.27, including half-wave nonphase-sensitive detection, half-wave phase-sensitive detection, full-wave nonphase-sensitive detection and full-wave phase-sensitive detection. Nonphase-sensitive detection has limited applications and cannot be used to recover signal polarity. For example, the output of the linear variable differential transformer is a sine wave whose amplitude is proportional to the displacement of the core. Only phase-sensitive detection is used for this kind of measurement. In many scientific and engineering fields, phase-sensitive detection is a powerful mathematical tool that recovers signals at low noise/signal ratios. It began in the 1960s and has become a ubiquitous experimental technique.

The half-wave nonphase-sensitive detection generates a half-wave-rectified output $x_d(t)$, which is positive when the $x_{AM}(t)$ is positive and zero when the $x_{AM}(t)$ is negative. The half-wave phase-sensitive detection generates a half-wave-rectified output $x_d(t)$ which is positive when the input $x(t)$ is in phase with the carrier $c(t)$ and is negative when the input $x(t)$ is 180° out of phase with $c(t)$. The full-wave nonphase-sensitive detection generates a full-wave-rectified output $x_d(t)$, which is

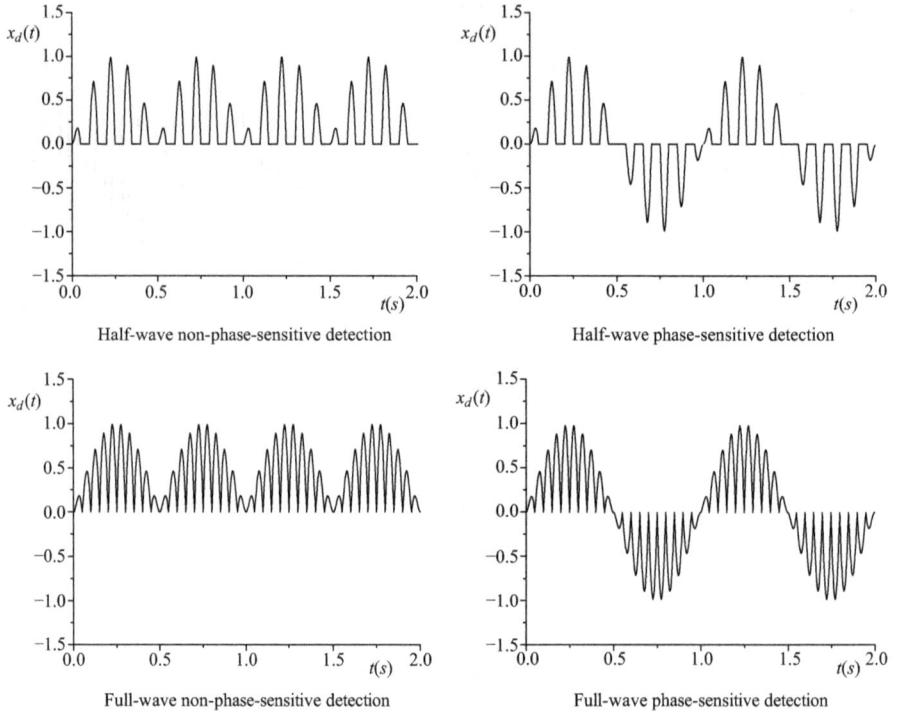

Figure 4.27: Description of demodulation/detection signals.

always positive. The phase-sensitive detector generates a full-wave-rectified output $x_d(t)$, which is positive when the input $x(t)$ is in phase with the carrier $c(t)$ and is negative when the input $x(t)$ is 180° out of phase with $c(t)$.

In electronics, ring modulation is a signal processing function that is implemented by mixing two signals, one of which is usually a sine wave, and the other is the signal to be modulated. The ring modulator is an electronic device for ring modulation, as is shown in Figure 4.28. It consists of four diodes and two transformers. The operation of the ring modulator is explained by assuming that the diode acts as a perfect switch and is turned on and off by the carrier signal. Without the modulating signal and with the modulating signal, the operation can be divided into different modes. In this way, full-wave phase-sensitive detection can be achieved.

As shown in Figure 4.28, the full-wave phase-sensitive output also passes through the RC low-pass filter to eliminate noise. The filtered results are depicted in Figure 4.29 using RC low-pass filters with different time constants (0.2, 0.1 and 0.05 s). From the output results, it can be found that the smaller the time constant is, the cleaner the high-frequency noise will be. Comparing the other graphs, it can also be found that the small time constant also causes some reduction in the amplitude of the output signal.

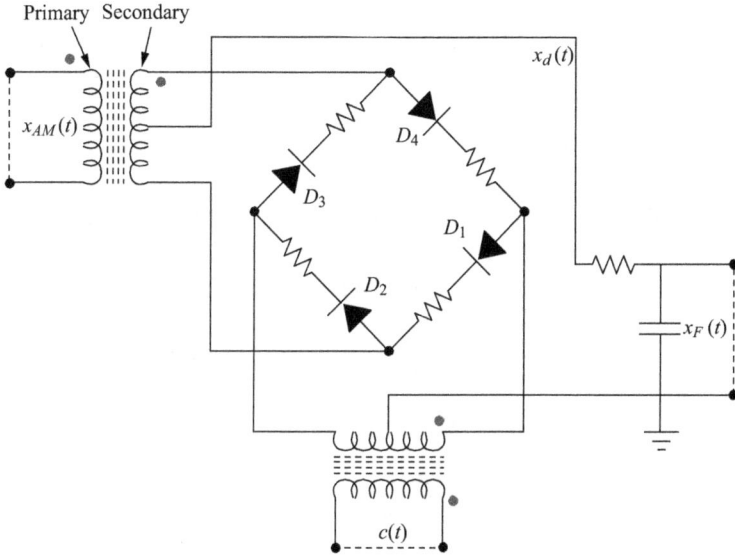

Figure 4.28: A ring demodulator.

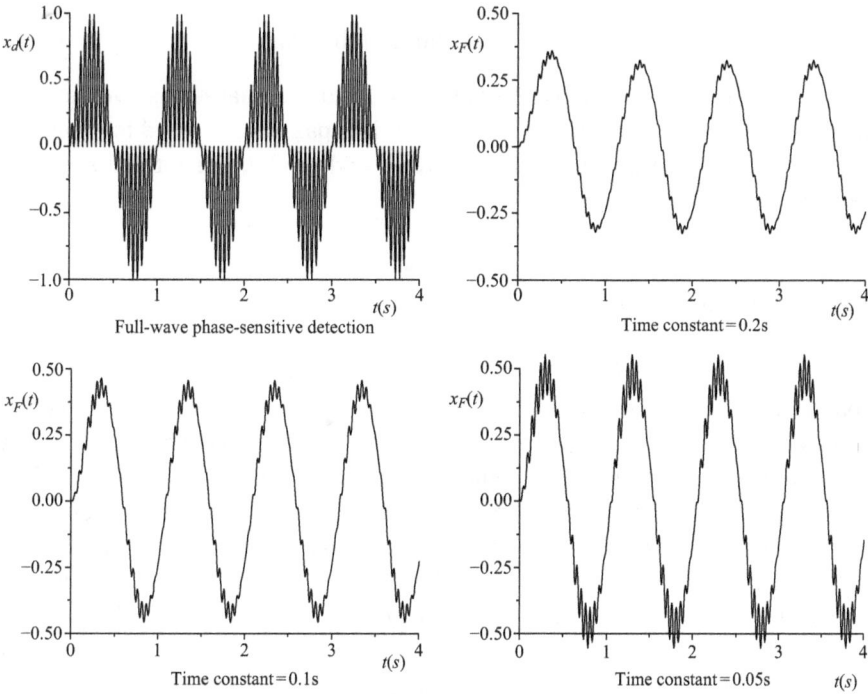

Figure 4.29: Description of filtering results.

4.3.2.3 Dynamic strainmeter

The strain that changes with time is called dynamic strain and is measured using a dynamic strain gauge. The dynamic behavior of the structure can be drawn in waveform by measuring the dynamic strain caused by vibration. The analog signal is amplified, output and recorded by an external recorder. The block diagram of the dynamic strain gauge is shown in Figure 4.30.

The AM process is illustrated using a quarter bridge strain gauge circuit as an example (R_1 is the strain gauge while R_2, R_3 and R_4 are precision resistors). According to the principle of the Wheatstone bridge, the entire circuit will be balanced. The following measurement signal $x(t)$, the carrier $c(t)$ and AM $x_{AM}(t)$ can be written as follows:

$$\Delta R(t) = 4R_0 \sin(2\pi t) \tag{4.55}$$

$$c(t) = \sin(20\pi t) \tag{4.56}$$

$$x(t) = \frac{\Delta R(t)}{4R_0 + 2\Delta R(t)}$$

$$\approx \frac{\Delta R(t)}{4R_0}$$

$$= \sin(2\pi t) \tag{4.57}$$

$$x_{AM}(t) = x(t) \cdot c(t)$$

$$= \sin(2\pi t)\sin(20\pi t) \tag{4.58}$$

The amplifier is used to amplify the AM $x_{AM}(t)$, and the phase-sensitive detector generates a full-wave-rectified output $x_d(t)$. RC low-pass filter is used to eliminate noise in the full-wave phase-sensitive output, and finally, the filtered signal is obtained.

4.3.3 Frequency modulation

FM has been widely used for many years. FM is still a very important form of modulation, though there are many other forms of data transmission. In fact, in the early days of wireless, it was thought that a narrower bandwidth was needed to reduce noise and interference, and AM is dominant. Since FM was introduced for the first time, the status of FM has greatly increased. Broadband FM is still considered to be a high-quality transmission medium for high-quality broadcasting and is also widely used in communications. FM uses the instantaneous frequency of the modulated signal (voice, music, data, etc.) to directly change the frequency of the carrier signal.

The frequency of carrier wave $c(t)$ is changed in accordance with the intensity of the measuring signal $x(t)$, and the FM process can be written as

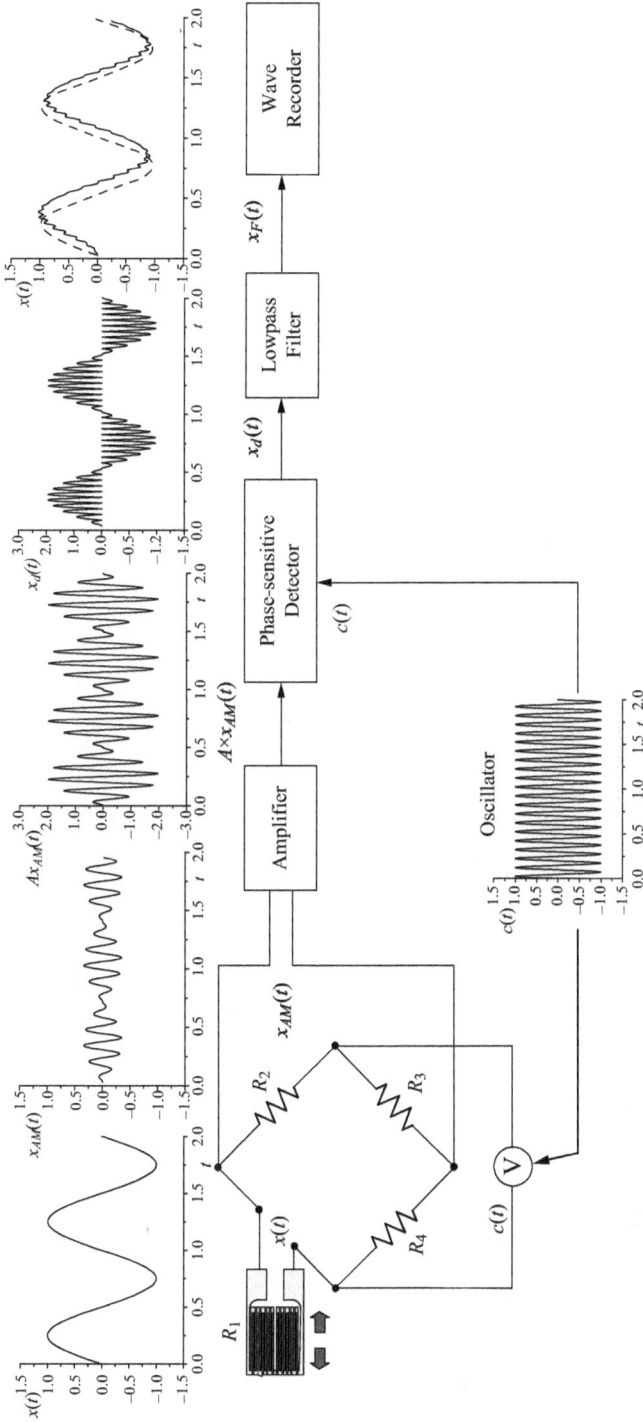

Figure 4.30: Block diagram of the dynamic strainmeter.

$$c(t) = A \sin(\omega_c t) \qquad (4.59)$$

$$x(t) = B \sin(\omega_0 t) \qquad (4.60)$$

$$x_{FM}(t) = A \sin[(\omega_c + kx(t))t] \qquad (4.61)$$

Figure 4.31 shows the results after FM. The measured signal on the left is the sine wave, and the measured signal on the right is a ramp. From the FM result, it can be seen that the frequency of the carrier changes with the amplitude of the measured signal.

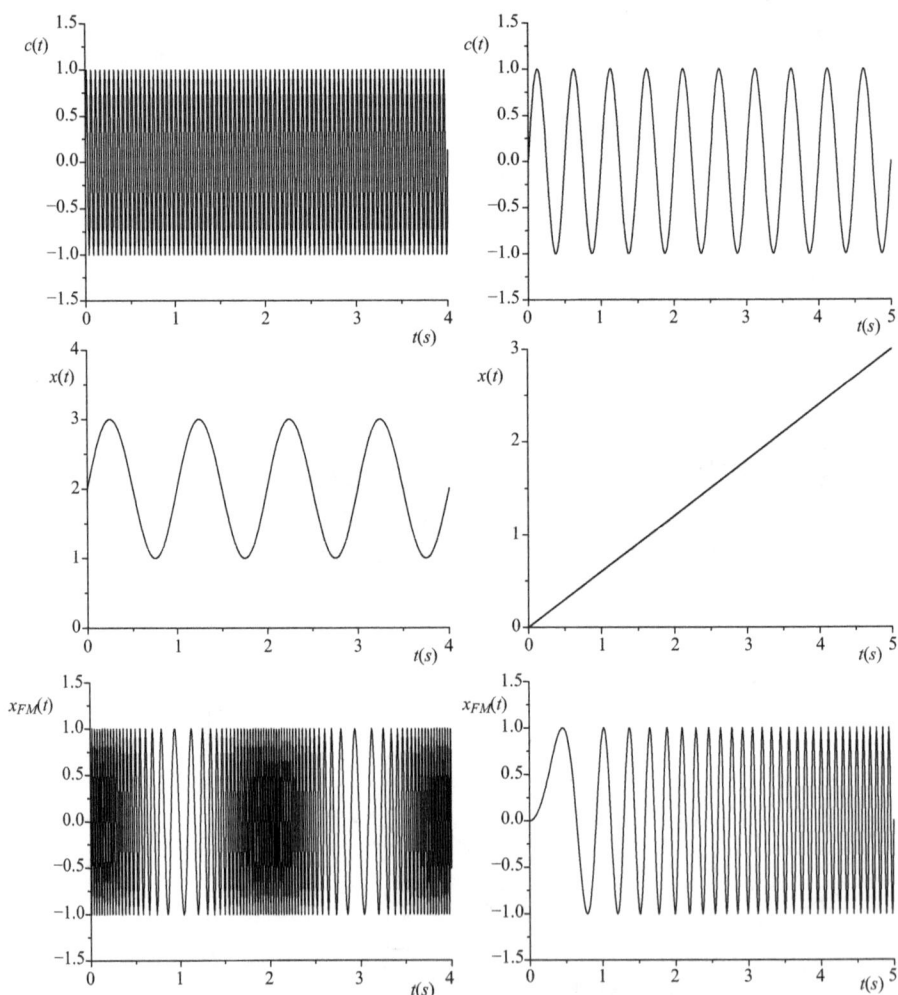

Figure 4.31: Description of frequency modulation.

A discriminator is a device that converts the properties of an input signal (such as frequency) into amplitude changes depending on the difference between the signal and the standard or reference signal. The frequency identification process is shown in Figure 4.32.

$$v_{out} = k_d (f_{in} - f_d)$$ (4.62)

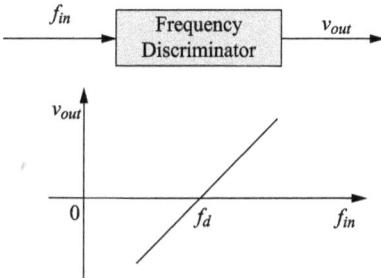

Figure 4.32: Frequency discrimination process.

In order to be able to convert frequency changes into voltage changes, the demodulator must be frequency dependent. The ideal response is a completely linear voltage–frequency characteristic. There are many circuits available for demodulating FM. Each type has its own advantages. Foster Seeley is a common type of FM detector circuit invented by Dudley E. Foster and Stuart William Seeley in 1936, primarily for radios built using discrete components. The Foster Seeley circuit is characterized by the use of transformers, chokes and diodes in the circuit to form the basis of its operation.

The FM ratio detector provided a better level of AM. FM ratio detector circuit is shown in Figure 4.33. As most noise is amplitude noise, it enabled the circuit to provide a greater level of noise immunity. It also enabled the FM detector to operate more effectively even with lower levels of limiting and offer a good level of performance and reasonable linearity. When a stable carrier is applied to the circuit, the diode is used to generate a stable voltage across the resistor, with the result that the capacitor is charged. The transformer enables the circuit to detect changes in the frequency of the input signal. It has three winding primaries and the third winding is untuned. The coupling between the primary winding and the third winding is very tight.

4.3.4 Typical applications

AM is the earliest modulation technique used to transmit speech over the air. This type of modulation technique is used for electronic communication. In this modulation, the amplitude of the carrier signal varies according to the message signal, and other factors such as phase and frequency remain constant. This modulation

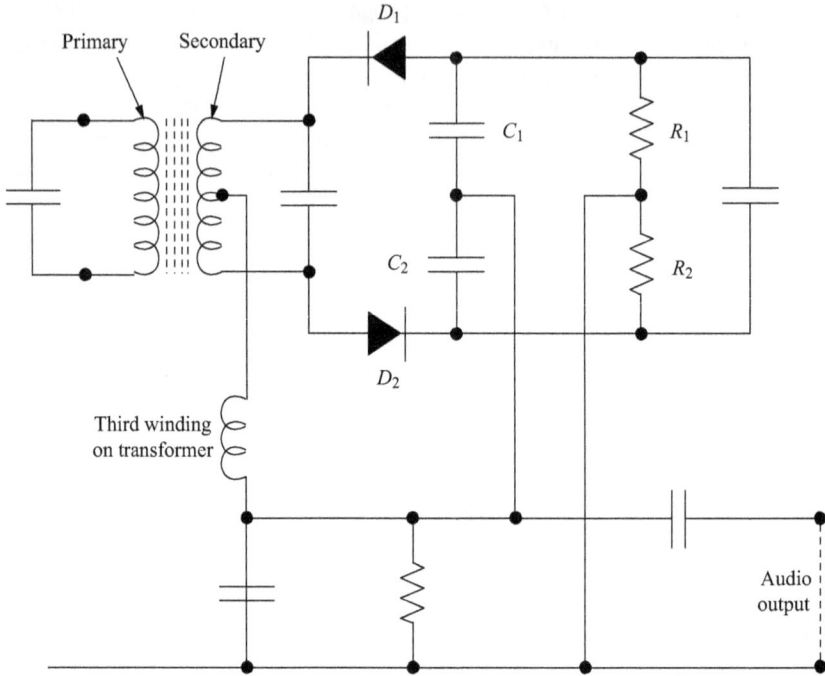

Figure 4.33: FM ratio detector circuit.

requires more power and more bandwidth, and filtering is very difficult. AM is used for VHF aircraft radios, computer modems and portable two-way radios. In this book, torque measurement is introduced by using AM.

In FM, the frequency of the carrier signal varies with the change of the message signal, while other parameters such as amplitude and phase remain unchanged. FM can be used for different applications such as radio and telemetry, radar, seismic exploration and monitoring of neonatal seizures by electroencephalography. One of the main advantages of FM is to reduce noise. However, FM requires a more complex demodulator. In this book, railway signal transmission is introduced by using FM.

4.3.4.1 Torque measurement

The torque sensor consists of a specially designed structure that is executed in a predictable and repeatable manner when torque is applied. This torque is converted into a signal voltage by the change in resistance of the strain gauge, which is organized in the circuit. The change in resistance represents the degree of deflection, which in turn represents the structural torque. The circuit consists of strain gauges, usually connected in a four-arm (Wheatstone bridge) configuration.

A reaction torque sensor is used to measure the rotational force or moment applied to the stationary portion of the device. Therefore, it cannot rotate 360° without cable winding, and this kind of sensor is typically used to measure the torque of back-and-forth agitated motion. A rotary torque sensor can rotate 360°, so this sensor must be able to transmit signals from the rotating element to the stationary surface. This can be done in the following ways: slip ring and rotary transformer. A slip ring, resolver and telemetry have been used in this measurement as shown in Figure 4.34. The strain gauge bridge is connected to four slip rings mounted on a rotating shaft. A graphite brush rubs on these slip rings and provides an electrical path for the AC bridge excitation and output signals.

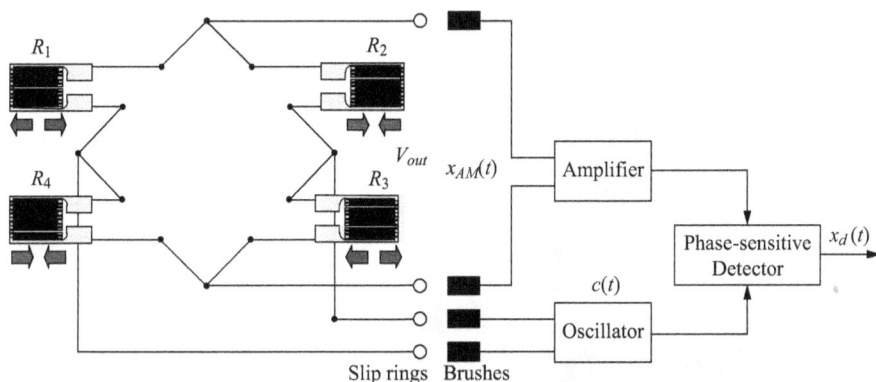

Figure 4.34: Schematic diagram of rotational torque measurement using slip rings.

One disadvantage of slip rings is that slip rings are easily wearing. Alternatively, a rotary transformer can be used to connect the strain gauges for power and returned signal as shown in Figure 4.35. In addition to the primary or secondary coil rotation, the rotary transformer works like any conventional transformer. The rotary transformer is simple and easy to use. When the transducer is mounted in line with the rotating shaft, the rotary torque transformer is best suited for measuring torque. The real photo of the rotary transformer is shown in Figure 4.36, and it is designed and manufactured by Forsentek Co., Limited, China.

4.3.4.2 Railway signal transmission
The railway signal is the most important factor in the train control system. It carries commands and information for controlling the train line, time interval and speed, and displays the current train line and equipment status to ensure the safety of high-speed train traffic. Different types of railway signals and train control systems are used in the world. The AC count signal is one of the railway signal standards,

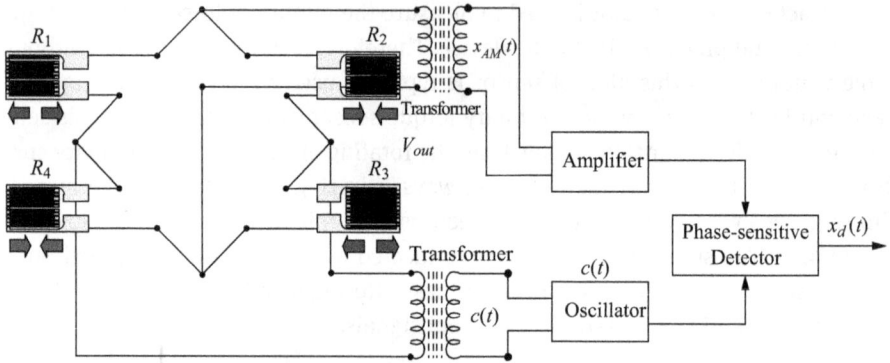

Figure 4.35: Schematic diagram of rotational torque measurement using rotary transformers.

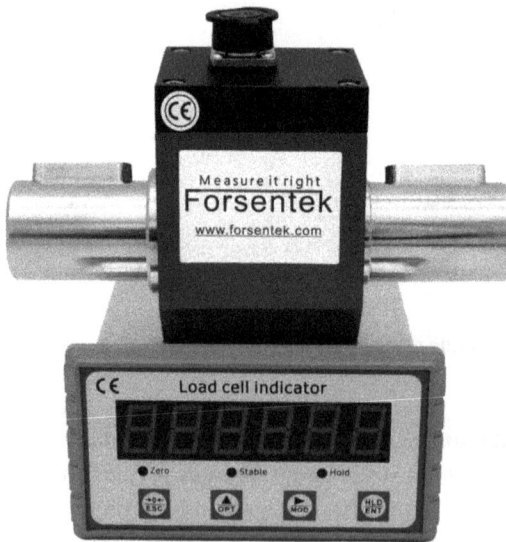

Figure 4.36: Torque sensor using rotary transformers (Forsentek, China).

and the AC count signal comes from the Soviet Union Russia. The characteristic points of this signal are the carrier frequency and the duration of the codes "0" and "1." The AC count signal is periodic with a period from 1,600 to 1,900 ms. In one cycle, the codes "0" and "1" alternate between different durations, and the carrier frequency is 25 or 50 Hz.

Radio wave communication can be done by AM or FM. A more popular communication method is FM because it is less susceptible to external interference and noise than AM. This noise usually affects the amplitude of the radio wave, but it does not affect its frequency, so the FM signal remains the same. FM can also transmit stereo better than AM.

4.4 Filter

In signal processing, a filter is a device or process that removes unwanted components or features from a signal. Filtering is a type of signal processing that is characterized by complete or partial suppression of certain aspects of the signal. In most cases, the filter needs to remove some frequencies or bands. Filters are widely used in recording, control systems, television, electronics and telecommunications, radio, image processing, music synthesis and computer graphics.

4.4.1 Classification

There are many different ways to classify filters. Depending on the frequency band removed, the filters can be divided into (1) low-pass filter (low frequencies are passed, high frequencies are attenuated); (2) high-pass filter (high frequencies are passed, low frequencies are attenuated); (3) band-pass filter (a certain band of frequencies is passed); and (4) band-stop filter (a certain band of frequencies is attenuated). Different filters are shown in Figure 4.37.

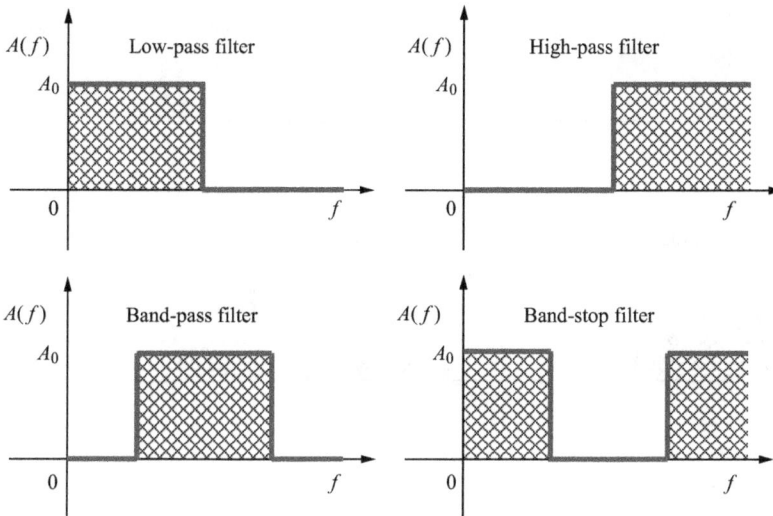

Figure 4.37: Different filters.

Filters can also be classified as digital filters and analog filters. Digital filters are implemented using digital computers or dedicated digital hardware. Analog filters can be classified as passive or active, usually with R, L and C components. Digital filters began with the advent of digital computers and dedicated signal processing boards. Digital filters are just implementations of equations in computer software, so there is no R, L and C components. Digital filter can also be built directly into a dedicated computer in hardware, but it still executes in software.

4.4.2 Real filter

Filters in Figure 4.37 are called ideal filters, and the ideal filters are not realizable in practical applications. A real filter cannot fulfill all the criteria of an ideal filter. The frequency range from the passband edge frequency to the stopband edge frequency is a transition band with unspecified frequency response. In fact, the actual filter always has a finite transition band between the passband and the stopband, as shown in Figure 4.38. From Figure 4.38, it can be seen that the low-pass and high-pass filters have one passband and one stopband, the band-pass filter has one passband and two stopbands, and the band-stop filter has two passbands and one stopband.

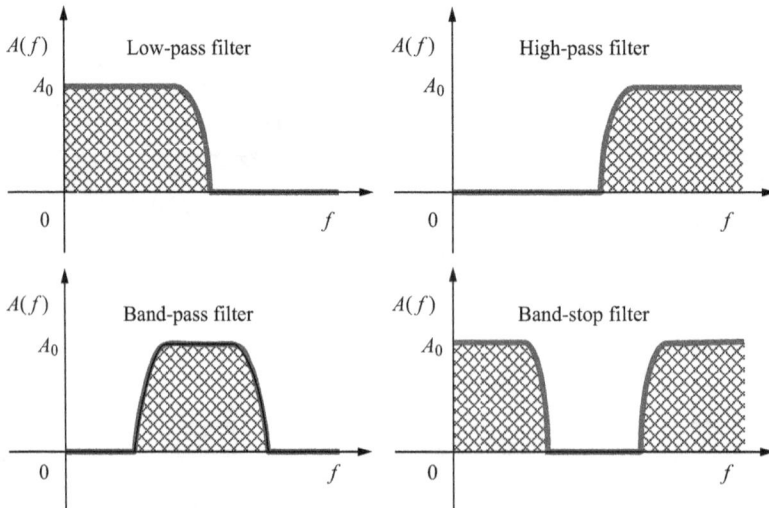

Figure 4.38: Real filters.

If a low-pass filter and a high-pass filter are connected in series, a band-pass filter can be obtained. Similarly, if a low-pass filter and a high-pass filter are connected in parallel, a band-stop filter can be obtained. The combination processes of a low-pass filter and a high-pass filter are shown in Figure 4.39.

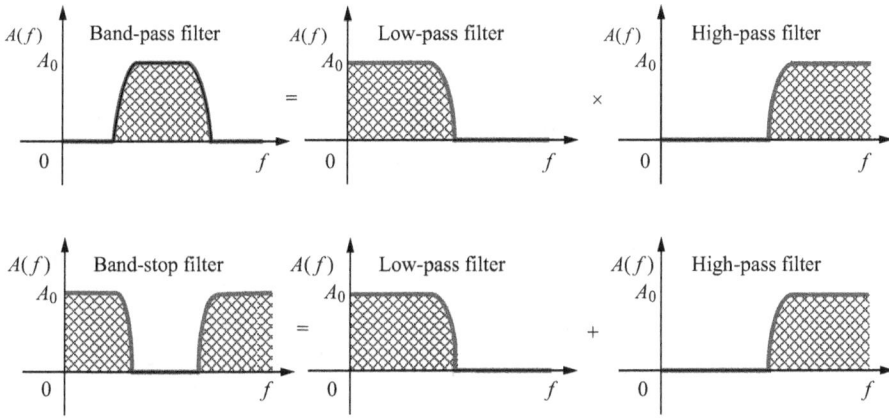

Figure 4.39: Combination processes of a low-pass filter and a high-pass filter.

The following Figure 4.40 illustrates the magnitude frequency response of a band-pass filter, which allows a certain band of frequencies to pass. In the transition band, the gain of the filter changes gradually from A_0 in the passband to zero ($-\infty$ dB) in the stopband. Frequency characteristics of the real band-pass filter are introduced in this section.

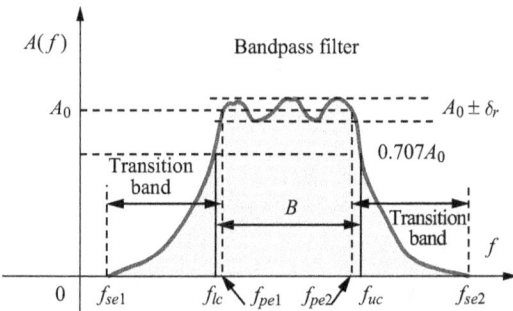

Figure 4.40: Real band-pass filter.

(1) Cutoff frequency

In electrical engineering, the cutoff frequency is a boundary in a system's frequency response, where energy flowing through the system begins to be attenuated rather than passing through. The cutoff frequency is characteristic of filtering devices, and it is defined as the frequency at which the ratio of the output/input has a magnitude of 0.707 ($\sqrt{2}/2$). This magnitude that converted into decibels using the following equation is equal to −3 dB, often referred to as the −3 dB down point:

$$\text{Magnitude} = 20 \log_{10} \left(\frac{\text{output}}{\text{input}} \right) \tag{4.63}$$

Beyond the upper −3 dB cutoff frequency f_{uc} or within the lower −3 dB cutoff frequency f_{lc}, the filter will attenuate all other frequencies in Figure 4.40.

(2) Ripple

The filter band-pass and band-stop contain oscillations, which are known as ripples. A typical example appears in Figure 4.40. δ_r is the magnitude of the passband ripple, which equals the maximum deviation of the A_0. The gain of the filter in the transition band (from f_{se1} to f_{pe1} or from f_{pe2} to f_{se2}) is unspecified. The gain usually changes gradually through the transition band from A_0 ($A_0 = 1$, 0 dB) in the passband to 0 (−∞ dB) in the stopband. The band-pass ripple can be measured in decibels, as shown in the following formulas:

$$\text{Band} - \text{pass ripple} = -20 \log_{10}(1 - \delta_r) \tag{4.64}$$

For example,
  If band-pass ripple equals 0.2 dB, that is

$$\text{Band} - \text{pass ripple} = -20 \log_{10}(1 - \delta_r)$$
$$0.2 = -20 \log_{10}(1 - \delta_r) \tag{4.65}$$

then

$$\delta_r = 0.02276$$

(3) Bandwidth

Bandwidth B is the difference between the upper −3 dB cutoff frequency f_{uc} and the lower −3 dB cutoff frequency f_{lc} in a continuous band of frequencies. It is typically measured in Hertz:

$$B = f_{uc} - f_{lc} \tag{4.66}$$

For a low-pass filter, the bandwidth equals the upper frequency.

(4) Q (Quality) factor

The Q factor is a dimensionless parameter that describes the underdamping of a filter and characterizes the bandwidth of the filter relative to its center frequency (f_0).

A higher Q indicates a lower energy loss rate relative to the stored energy of the filter. The definition is the frequency-to-bandwidth ratio of the filter:

$$Q = \frac{f_0}{B} \tag{4.67}$$

4.4.3 Low-pass filter

A low-pass filter is a filter that passes a signal having a frequency below a cutoff frequency and attenuates a signal having a frequency higher than the cutoff frequency. There are two basic circuits to accomplish this objective: the inductive low-pass (LR) filter and the capacitive low-pass (RC) filter in Figure 4.41. You should remember that inductors pass low frequencies and block high frequencies, while capacitors do the opposite. High frequencies in the input signal are attenuated in an electronic low-pass RC/LR filter for voltage signals, but it has little attenuation below the cutoff frequency determined by its time constant.

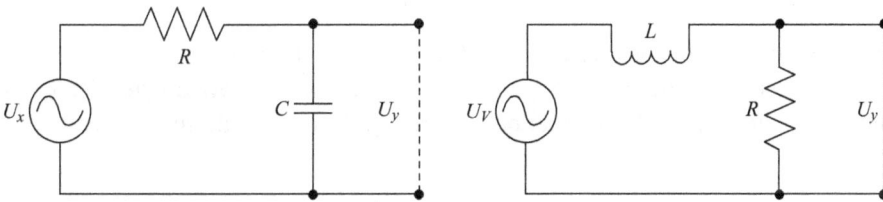

Figure 4.41: Low-pass filter circuits.

Low-pass RC filter:

$$RC\frac{dU_y}{dt} + U_y = U_x \tag{4.68}$$

$$RC = \tau \tag{4.69}$$

$$H(s) = \frac{U_y(s)}{U_x(s)} = \frac{1}{\tau s + 1} \tag{4.70}$$

"$j\omega$" is substituted for "s" where ω is frequency in rad/s, and frequency response function can be written as

$$\frac{U_y(j\omega)}{U_x(j\omega)} = \frac{1}{1 + j\omega RC} = \frac{1}{1 + j\omega\tau} \tag{4.71}$$

Low-pass LR filter:

$$\frac{U_y}{U_x} = \frac{Z_R}{Z_R + Z_L} = \frac{R}{R + j\omega L} \tag{4.72}$$

$$\frac{L}{R} = \tau \tag{4.73}$$

$$\frac{U_y(j\omega)}{U_x(j\omega)} = \frac{1}{1 + \frac{j\omega L}{R}} = \frac{1}{1 + j\omega\tau} \tag{4.74}$$

The magnitude of formula (4.71) or (4.74) is

$$M(\omega) = \sqrt{\frac{1}{1 + \omega^2\tau^2}} \tag{4.75}$$

$$M(f) = \frac{1}{\sqrt{1 + (2\pi f\tau)^2}} \tag{4.76}$$

The phase of formula (4.71) or (4.74) is

$$\phi(\omega) = -\arctan\omega\tau \tag{4.77}$$

$$\phi(f) = -\arctan 2\pi f\tau \tag{4.78}$$

The magnitude and phase responses of the low-pass filter are given in Figure 4.42. It can be seen that the voltage gain is approximately 1 at low frequencies (ω is small) and the magnitude attenuation is more significant at high frequencies.

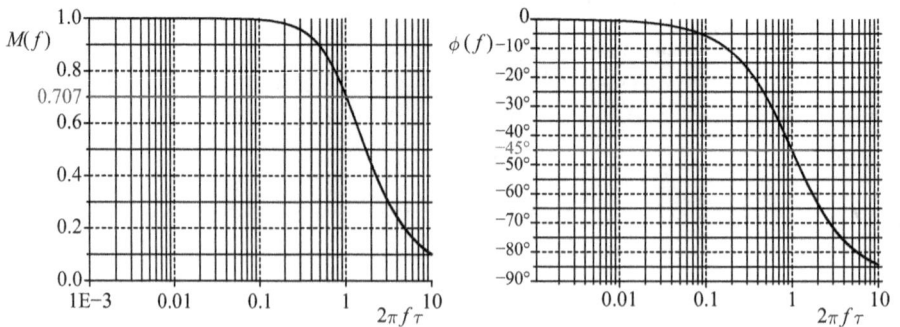

Figure 4.42: Magnitude and phase responses for the low-pass filter.

The ideal low-pass filter passes the frequencies below the cutoff frequency and eliminates all those above this value completely. A brick-wall filter's frequency response is a rectangular function. An ideal filter does not have the transition region in practical filters. An ideal low-pass filter can be mathematically realized by multiplying a signal and the rectangular function in the frequency domain.

The cutoff frequency occurs at $M(\omega) = 0.707$, and this can be substituted for

$$M(\omega) = \sqrt{\frac{1}{1+\omega^2\tau^2}} = 0.707 = \sqrt{\frac{1}{2}} \qquad (4.79)$$

The upper cutoff frequency f_{uc} is given by

$$f_{uc} = \frac{1}{2\pi\tau} \qquad (4.80)$$

4.4.4 High-pass filter

A high-pass filter is a filter that passes a signal having a frequency beyond a cutoff frequency and attenuates a signal having a frequency lower than the cutoff frequency. There are two basic circuits to accomplish this objective: the inductive high-pass (RL) filter and the capacitive high-pass (CR) filter in Figure 4.43. In an electronic high-pass CR/RL filter for voltage signals, low frequencies in the input signal are attenuated.

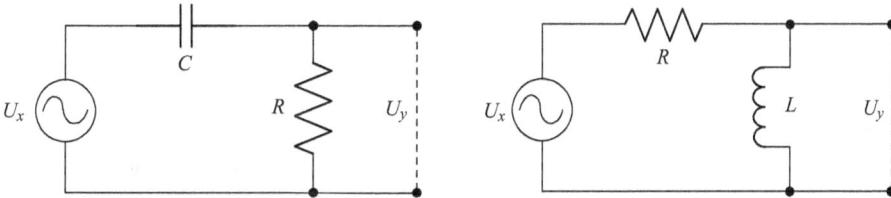

Figure 4.43: High-pass filter circuits.

High-pass CR filter:

$$u_y + \frac{1}{RC}\int u_y dt = u_x \qquad (4.81)$$

$$H(s) = \frac{\tau s}{\tau s + 1} \qquad (4.82)$$

"$j\omega$" is substituted for "s," where ω is the frequency in rad/s, and frequency response function can be written as

$$H(\omega) = \frac{j\omega\tau}{1+j\omega\tau} \qquad (4.83)$$

High-pass RL filter

$$\frac{U_y}{U_x} = \frac{Z_L}{Z_R + Z_L} = \frac{j\omega L}{R + j\omega L} = \frac{j\omega L/R}{1 + j\omega L/R} \tag{4.84}$$

$$\frac{L}{R} = \tau \tag{4.85}$$

$$\frac{U_y(j\omega)}{U_x(j\omega)} = \frac{\frac{j\omega L}{R}}{1 + \frac{j\omega L}{R}} = \frac{j\omega\tau}{1 + j\omega\tau} \tag{4.86}$$

The magnitude of formula (4.84) or (4.86) is

$$M(\omega) = \frac{\omega\tau}{\sqrt{1 + (\omega\tau)^2}} \tag{4.87}$$

$$M(f) = \frac{2\pi f\tau}{\sqrt{1 + (2\pi f\tau)^2}} \tag{4.88}$$

The phase of formula (4.84) or (4.86) is

$$\phi(\omega) = \operatorname{arctg}\frac{1}{\omega\tau} \tag{4.89}$$

$$\phi(f) = \operatorname{arctg}\frac{1}{2\pi f\tau} \tag{4.90}$$

The magnitude and phase responses of the high-pass filter are given in Figure 4.44. It can be seen that the voltage gain is approximately 1 at high frequencies (ω is large) and the magnitude attenuation is more significant at low frequencies.

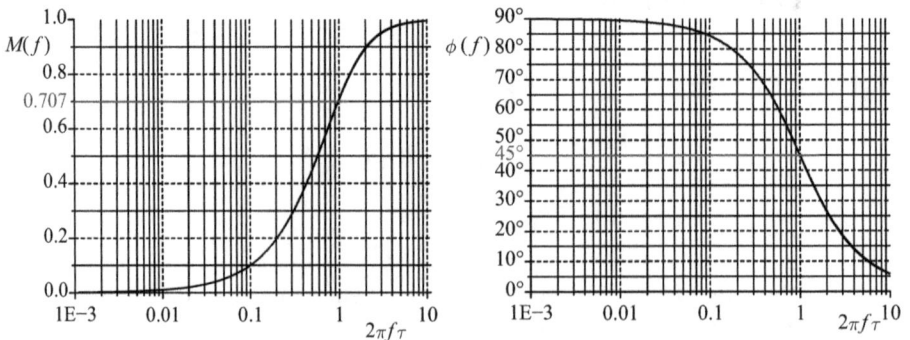

Figure 4.44: Magnitude and phase responses for the high-pass filter.

The cutoff frequency occurs at $M(\omega) = 0.707$, and this can be substituted for

$$M(\omega) = \sqrt{\frac{\omega^2\tau^2}{1+\omega^2\tau^2}} = 0.707 = \sqrt{\frac{1}{2}} \tag{4.91}$$

The lower cutoff frequency f_{lc} is given by

$$f_{lc} = \frac{1}{2\pi\tau} \tag{4.92}$$

4.4.5 Band-pass filter

A band-pass filter is a device that allows a particular band to pass through while shielding other bands. If a low-pass filter and a high-pass filter are connected in series, a band-pass filter can be obtained. Two basic circuits (*RL* low-pass filter and *CR* high-pass filter) are used to build a band-pass filter as shown in Figure 4.45.

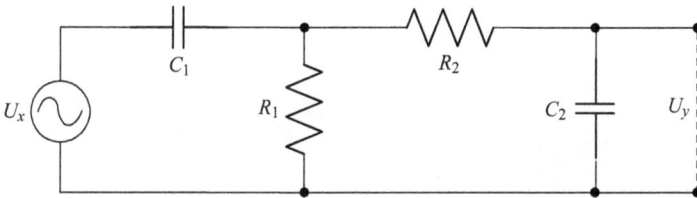

Figure 4.45: Band-pass filter circuit.

Similarly, the magnitude of the filter is

$$M(f) = \frac{2\pi f\tau_1}{\sqrt{1+(2\pi f\tau_1)^2}} \times \frac{1}{\sqrt{1+(2\pi f\tau_2)^2}} \tag{4.93}$$

The phase of the filter is

$$\phi(f) = \text{arctg}\left(\frac{1}{2\pi f\tau_1}\right) - \text{arctg}(2\pi f\tau_2) \tag{4.94}$$

The magnitude and phase responses of the band-pass filter are given in Figure 4.46. It can be seen that the voltage gain is approximately 1 within a certain range, and the magnitude attenuation is more significant outside that range.

The cutoff frequency occurs at $M(f) = 0.707$, and this can be substituted into

$$M(f) = \frac{2\pi f\tau_1}{\sqrt{1+(2\pi f\tau_1)^2}} \times \frac{1}{\sqrt{1+(2\pi f\tau_2)^2}} = 0.707 \tag{4.95}$$

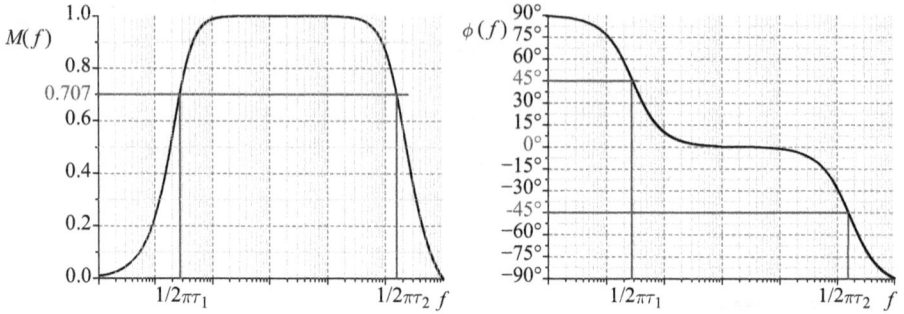

Figure 4.46: Magnitude and phase responses for the band-pass filter.

The lower cutoff frequency f_{lc} and the upper cutoff frequency f_{uc} are given as follows:

$$f_{lc} = \frac{1}{2\pi\tau_1} \tag{4.96}$$

$$f_{uc} = \frac{1}{2\pi\tau_2} \tag{4.97}$$

4.4.6 Filter bank

A filter bank is a band-pass filter array used to separate an input signal into several components. Each component carries a single frequency sub-band of the original signal. The separating process is shown in Figure 4.47. The decomposition process performed by the filter bank is called analysis. The reconstruction process is called

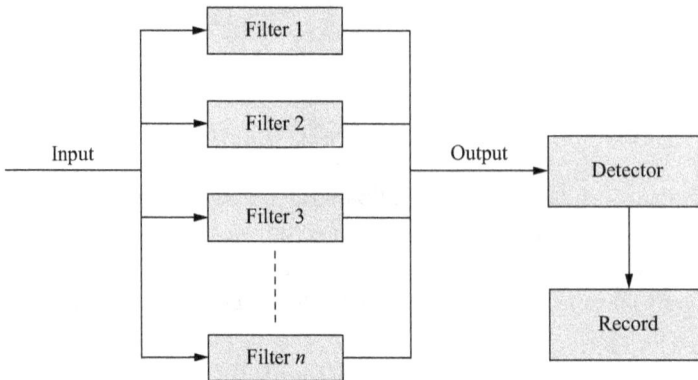

Figure 4.47: Multidimensional analysis filter banks.

synthesis, which means reconstructing the complete signal generated by the filtering process. The most commonly used method of observing signals is to display them in time domain with an oscilloscope. It is also useful to display signals in the frequency domain. The apparatus used to provide this frequency domain view is the spectrum analyzer. Spectrum analyzers use a parallel filter bank to analyze frequency.

The spectrum analyzers may use a single band-pass filter, and its center frequency is adjustable. Due to the limitations of the value adjustment of the R or C of a band-pass filter, the adjustable range is usually limited. It is also possible to use a series of band-pass filters with different center frequencies and a fixed frequency range. The filter bank is a good choice for analyzing the signal. The filter bank utilizes a set of filters to cover the frequency range. The central frequencies and bandwidth of these filters are chosen to overlap each other, as shown in Figure 4.48.

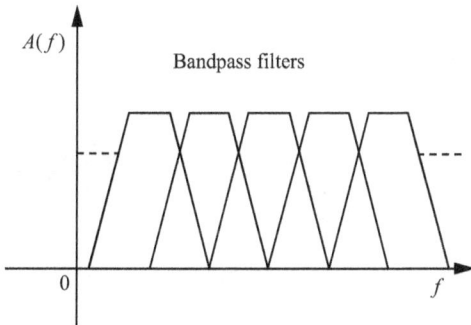

Figure 4.48: Overlapping filter bank.

There are usually two different filter banks: equal resolution filter bank and proportional resolution filter bank as shown in Figures 4.49 and 4.50. An equal resolution filter bank is a filters group with constant bandwidth. This means that the frequency resolution is uniform at all frequencies. Unlike an equal resolution filter bank, a proportional resolution filter bank has a proportional resolution scale that is proportional to the frequency. The higher resolution can be achieved at lower frequencies.

To show the frequency characteristics of the signal, the octave analysis passes the signal to a set of band-pass filters. The octave filter has a proportional bandwidth. If the lower cutoff frequency and the upper cutoff frequency of the band-pass filter are f_{lc} and f_{uc}, its relationship can be represented by the following formula:

$$f_{uc} = 2^n f_{lc} \tag{4.98}$$

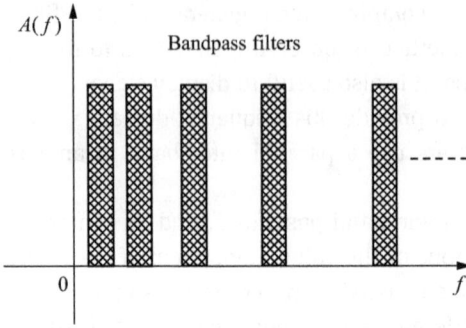

Figure 4.49: Equal resolution filter bank.

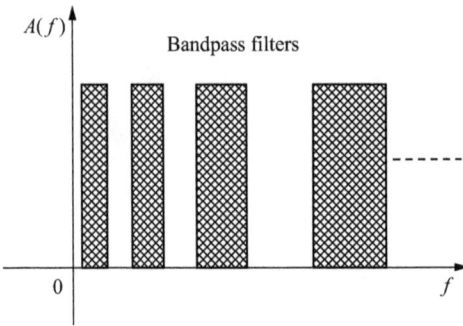

Figure 4.50: Proportional resolution filter bank.

where n is the octave number. If $n = 1$, it is called octave filter bank; if $n = 1/3$, it is called the third octave filter bank.

The center frequency f_0 is approximately equal to

$$f_0 = \sqrt{f_{lc} f_{uc}} \tag{4.99}$$

Then

$$f_{lc} = 2^{-n/2} f_0 \tag{4.100}$$

$$f_{uc} = 2^{n/2} f_0 \tag{4.101}$$

$$f_{uc} - f_{lc} = B = \frac{f_0}{Q} \tag{4.102}$$

$$\frac{1}{Q} = \frac{B}{f_0} = 2^{n/2} - 2^{-n/2} \tag{4.103}$$

If $n = 1$, $Q = 1.41$; if $n = 1/3$, $Q = 4.38$; if $n = 1/5$, $Q = 7.2$.

For the octave filter bank ($n = 1$), the center frequency of each filter is twice that of the previous filter. Each filter covers twice that of the bandwidth of the previous band and half that of the width of the next band.

4.4.7 Typical applications

(1) Programmable filter

The LTC1564 is an independent, continuous-time, variable gain, high-order analog low-pass filter, designed by Linear Technology Corporation (USA) as shown in Figure 4.51, and it is a new type of continuous time filter for anti-aliasing, reconstruction and other band-limiting applications. The shape of the low-pass response is fixed while the cutoff frequency (f_c) and gain are programmable. For the signal frequency component below the cut-off frequency f_c, the gain between the input and output pins is approximately constant, while for the above frequency, the gain decreases rapidly. The f_c ranges from 10 to 150 kHz in 10 kHz steps. This response rolls off approximately by 100 dB from f_c to $2.5f_c$. It can be used in DSP systems, communications systems, scientific instruments, programmable data rates, processing signals hidden in noise and so on.

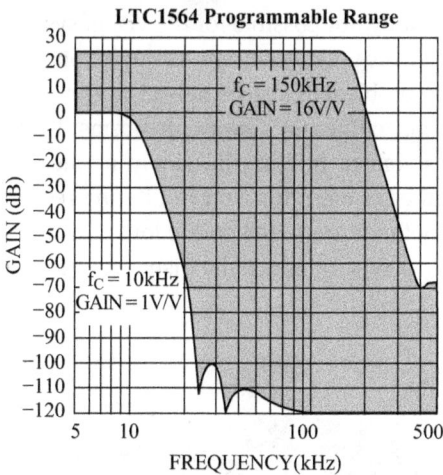

Figure 4.51: LTC1564 programmable range (Linear Technology Corporation).

(2) Signal noise filtering

This application shows how to eliminate high-frequency noise in the measurement data with a low-pass filter shown in Figure 4.52. The low-pass filter is designed to pass lower frequency components through and block higher frequency components

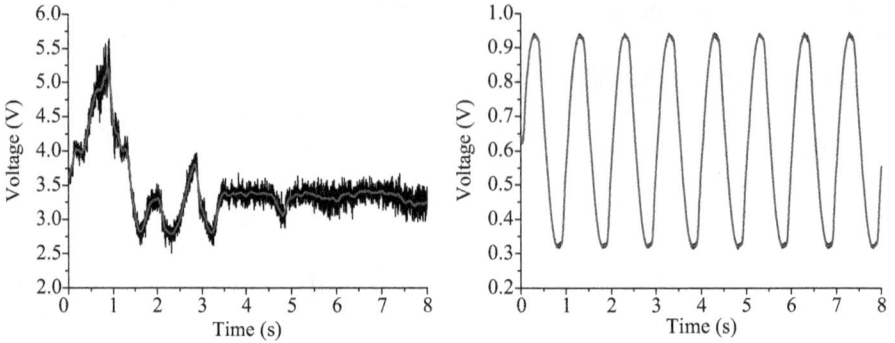

Figure 4.52: Signal noise filtering.

of the signal. High-frequency noise is due to the fact that the component of the signal changes faster than the signal of interest. A cutoff frequency with 5–15 Hz can be used without affecting the data of interest. Eliminating high-frequency noise can make the signal of interest more compact and smoother, and enable more accurate analysis.

(3) Image filtering

When the camera captures an image, it is usually not possible for the vision system directly using it. Random variations in intensity, variations in illumination and poor contrast can corrupt the image. These problems must be solved in the early stages of vision processing. The aim of the image filter for image enhancement is to eliminate these undesirable characteristics. Image filtering is widely applied in smoothing, sharpening, removing noise and edge detection. Image filtering results are shown in Figure 4.53. Most smoothing methods are based on the low-pass filter. The image is smoothed by nearby averaging pixels to reduce the disparity between

Original   Lowpass filtering   Highpass filtering

Figure 4.53: Image filtering.

pixel values. The use of low-pass filters tends to preserve low-frequency information within the image while reducing high-frequency information. High-pass filters
tend to preserve high-frequency information within the image while reducing low-
frequency information. The high-pass filter is the basis for most sharpening methods. When the contrast between adjacent areas increases and the brightness or
darkness changes very little, the image becomes sharp.

4.5 Sampling and aliasing

4.5.1 Sampling

In signal processing, sampling is a process of converting a continuous-time signal
into a discrete-time signal (a numeric sequence). The sampling process is shown in
Figure 4.54. This section describes how signal sampling affects its spectrum. $x(t)$ is
a continuous signal for sampling, which measures the value of the continuous function every T_s seconds. T_s is called the sampling interval or the sampling period.

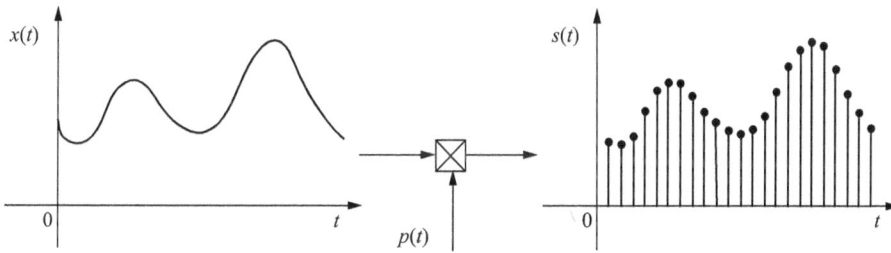

Figure 4.54: Sampling process.

The sampling signal $p(t)$ is an infinite "impulse train"

$$p^{(t)} = \sum_{n=-\infty}^{\infty} \delta(t - nT_s), \quad T_s - \text{period} \quad n = 0, \pm 1, \pm 2, \ldots \tag{4.104}$$

Fourier transform of $p(t)$ is

$$P(f) = \frac{1}{T_s} \sum_{k=-\infty}^{\infty} \delta(f - kf_s) = \frac{1}{T_s} \sum_{k=-\infty}^{\infty} \delta\left(f - \frac{k}{T_s}\right) \tag{4.105}$$

The frequency spectrum of the sampled signal is still an impulse sequence. If the
period is T_s in the time domain, the period in the frequency is $1/T_s$ in the frequency
domain. The pulse intensity is 1 in the time domain, which is $1/T_s$. $p(t)$ and $P(f)$ are
described in Figure 4.55.

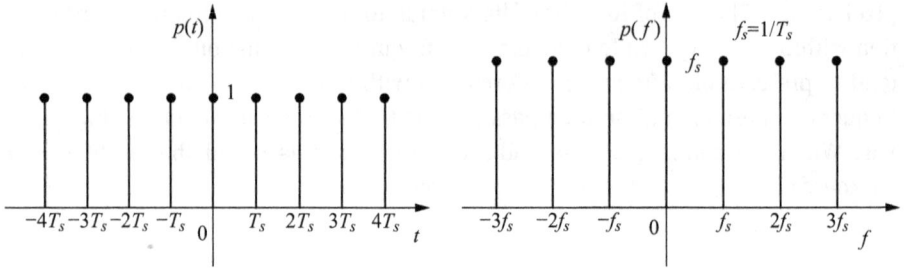

Figure 4.55: The sampling signal.

In the time domain, $s(t)$ is

$$s(t) = x(t)p(t) \qquad (4.106)$$

In the frequency domain, $S(f)$ is

$$S(f) = X(f)^*P(f) \qquad (4.107)$$

Convolution process in the frequency domain is described in Figure 4.56. Convolution with the delta function should result in the same function.

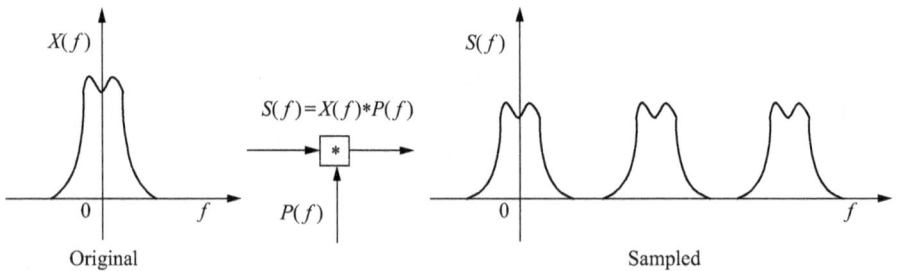

Figure 4.56: Convolution process in the frequency domain.

Surprisingly, the sampling process does not damage the spectrum in the original "baseband." More notably, the output can be recovered without loss by using an ideal low-pass filter to filter the output, as shown in Figure 4.57.

4.5.2 Nyquist–Shannon sampling theorem

The name of the Nyquist–Shannon sampling theorem, also known as the Shannon sampling theorem, is to commemorate Harry Nyquist and Claude Shannon. The sampling theorem was described by Nyquist in 1928 and proven by Shannon in 1949.

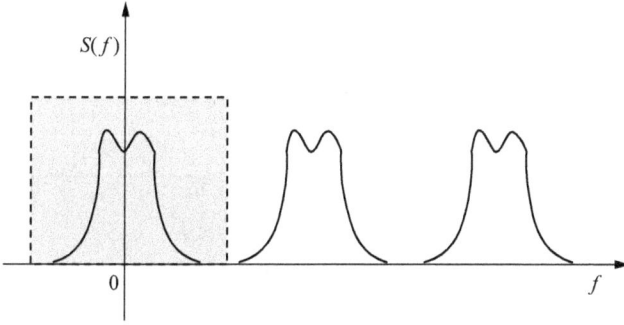

Figure 4.57: Signal filtering.

When the bandwidth of the signal is limited, and the sampling frequency is more than twice that of the bandwidth, the continuous time signal can be accurately reconstructed from its sample. The sampling theorem indicates that the equally spaced sampling frequency f_s must be more than twice that of the maximum frequency for a finite bandwidth (band-limited) signal with a maximum frequency f_{max} (Nyquist limit):

$$f_s > 2f_{max} \tag{4.108}$$

Under the above condition, the signal can be reconstructed uniquely without aliasing. The frequency $2f_{max}$ is referred to as the Nyquist sampling rate and half of this value f_{max} is called the Nyquist frequency.

4.5.3 Aliasing

In signal processing, aliasing is an effect that makes different signals indistinguishable when sampling. It also refers to distortion when the signal reconstructed from the sample is different from the original continuous signal. Figure 4.58 shows the aliasing phenomenon. No aliasing occurs on the left, and the result on the right is aliased. If the Nyquist–Shannon sampling theorem is not satisfied, the higher frequency components will enter the reconstructed signal and create signal distortion. This type of signal distortion (the overlapping) is called aliasing.

From the diagram, it can be seen that this condition occurs if

$$f_s < 2f_{max} \tag{4.109}$$

When the sampling rate is lower than or equal to the Nyquist rate, it is impossible to reconstruct the original signal according to the sampling theorem. Figure 4.59 shows that the lower the sampling rate, the farther the reconstructed signal constructed by the sampled data is different from the original signal.

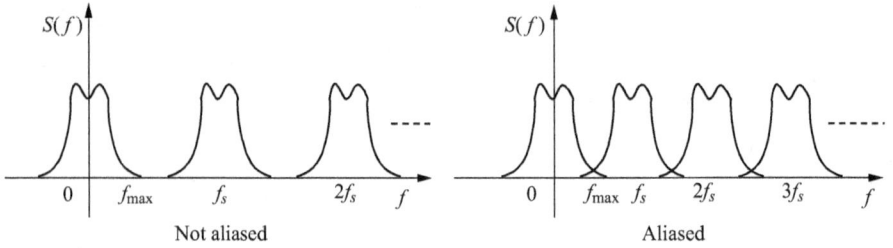

Figure 4.58: Aliasing phenomenon in the frequency domain.

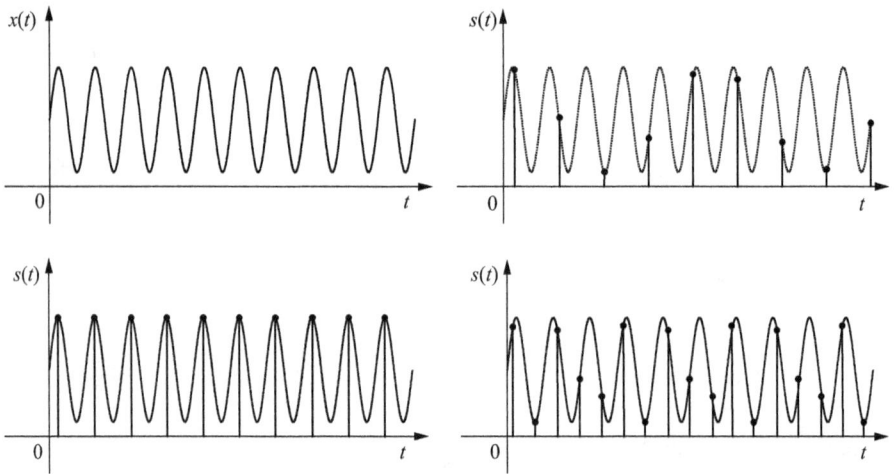

Figure 4.59: Different sampling rates.

How to avoid aliasing? The actual solution is to use an anti-aliasing filter to limit the signal bandwidth to less than half of the sampling rate, which is applied to the input channel in front of the A/D converter. Because once the digitization is done and the aliasing signal has been created, the original signal cannot be retrieved. The specific steps of anti-aliasing shown in Figure 4.60 are as follows: (1) according to the sampling theorem, sampling must be performed at a rate that is at least twice that of the highest frequency component of interest within the input signal; (2) any frequency component above half the sampling rate must be eliminated by the anti-aliasing filter prior to sampling; and (3) the ideal low-pass filter is used for recovering the original signal.

The anti-aliasing filter is implemented on each input channel in front of the A/D card to eliminate the unwanted high-frequency noise and the interference introduced before sampling. This reduces system cost by allowing lower sample rates, collecting storage requirements and analysis time. For example, if 600 Hz is

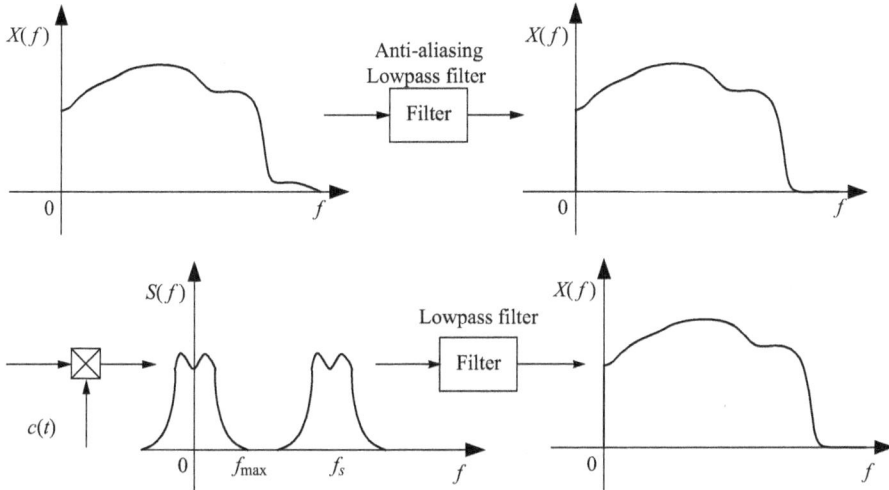

Figure 4.60: Steps to avoid aliasing.

assumed to be an interfering signal due to unwanted mechanical vibration, the 600 Hz frequency must be removed by an anti-aliasing filter. If the cutoff point is set around 400 Hz, a filter with a steep roll-off slope will eliminate the 600 Hz frequency while the input frequency of interest below the filter cutoff frequency (400 Hz) will still pass through the system.

The following describes a typical anti-aliasing filter. The 6211LP/HP (Microwave Filter Company, Inc., USA) are a set of low-pass and high-pass filters. They are factory-assembled networks, consisting of channel deletion filters and custom-designed low-pass and high-pass filters. The typical frequency responses of 6211LP and 6211HP are shown in Figure 4.61. The 6211LP/HP are designed with such a steep cutoff slope that they resemble a "brick wall." The "brick wall" filter eliminates the channel block while preserving the relevant associated channels. These types of filters have often been used as anti-aliasing filters for A/D converters.

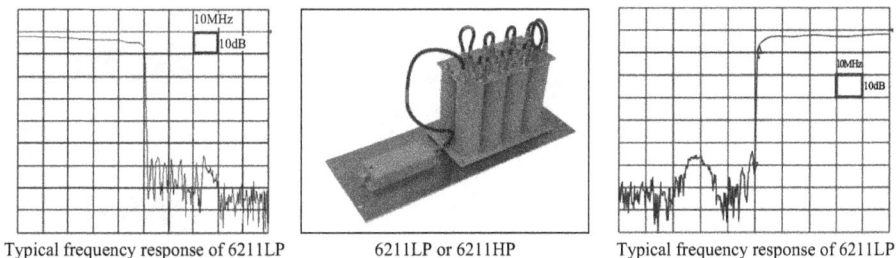

Typical frequency response of 6211LP   6211LP or 6211HP   Typical frequency response of 6211LP

Figure 4.61: 6211LP/HP filters (Microwave Filter Company, Inc., USA).

4.6 Displaying and recording signal

This section focuses on how to display and record the data after obtaining undistorted data easily. This section introduces the techniques that can be used to display data. In the field of mechanical engineering, paper or display screens can be used as display media to present data in graphical or tabular form.

4.6.1 Analog and digital display

For analog signals, the measured signal can be displayed by an oscilloscope or a voltage or ammeter as shown in Figure 4.62 (left picture, Wenzhou Deshen Electric Co., Ltd, China). However, if the signal is converted into the digital, it can be displayed using an light emitting diode (LED) digital display or a computer monitor as shown in Figure 4.62 (middle and right pictures, Bell & Gossett, USA). LED digital display uses two common formats: seven-segment display and 5×7 dot-matrix display. Both formats have the advantage of being able to display alphanumeric information. Typically, digital displays accept serial or parallel digital input signals in binary-coded decimal or ASCII format. Techniques for the various components in the display are LEDs or liquid crystal elements. Computer monitors provide an excellent mechanism for displaying and storing information, in addition to displaying letters and numbers. Other information such as factory floor plans and process flow layouts can also be displayed. When the data triggers an alarm, it can be easily determined by its position on the factory schematic. In order to realize human–computer interaction, a touch screen can also be used for display, which realizes the input of commands without using a keyboard. Moreover, in addition to displaying measured values in real time, it is often necessary to continuously record the measured values for later analysis. These records can help you quickly and easily find the cause of the failure when a failure occurs in the system. Devices for recording data include chart recorders, digital oscilloscopes, digital recorders and printers. The following is a brief introduction of some displaying and recording devices commonly used in the field of mechanical engineering.

Figure 4.62: Analog and digital display (Bell & Gossett, USA).

4.6.2 Pen strip chart recorder

A pen strip chart recorder is an electromechanical device that records the trend of an electronic or mechanical input onto a piece of paper (the chart). The working principle of the pen strip chart recorder is to record data on a continuous chart paper moving at a constant speed. The logger records data for one or more variables that change with time. The pen strip chart recorder includes a marked pen, drive system, long scroll chart paper, chart paper drive mechanism and a speed selection switch (as shown in Figure 4.63).

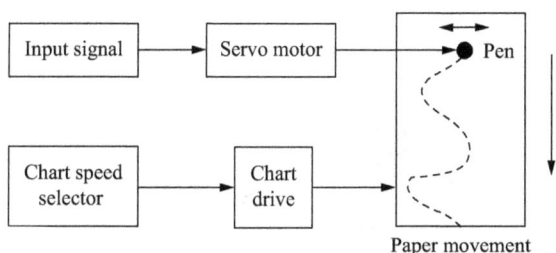

Figure 4.63: Structure of pen strip chart recorder.

A multipoint strip chart recorder is an improvement of the pen chart recorder that uses a dot-matrix printhead instead of a pen to strike the ink ribbon (Figure 4.64). A typical model utilizes a six-color ribbon to record up to 24 different signal inputs simultaneously. A photograph of the pen strip chart recorder by Yokogawa Electric Corporation is shown in Figure 4.65. The strip chart recorders trend electrical or mechanical signals onto paper. The pen servo unit utilizes an ultra-small, rack-and-pinion stepping motor. The servo unit is usually smaller and consumes less power. The recorder can be used as a monitoring device and a quality control instrument in many application fields such as process temperature monitoring, pollution, construction and medical diagnosis.

4.6.3 Circular chart recorder

The circular chart recorder consists of a servo-driven pen assembly that records the measured signal on a rotating circular paper chart. The main advantage of the circular chart recorders over other forms is that it is compact. A photograph of the circular chart recorder by Dickson, USA, is shown in Figure 4.65. The chart recorder uses a single or double pen or "hot pen" to generate a chart image. It accepts directly connected process signals, enabling it to process data and create charts

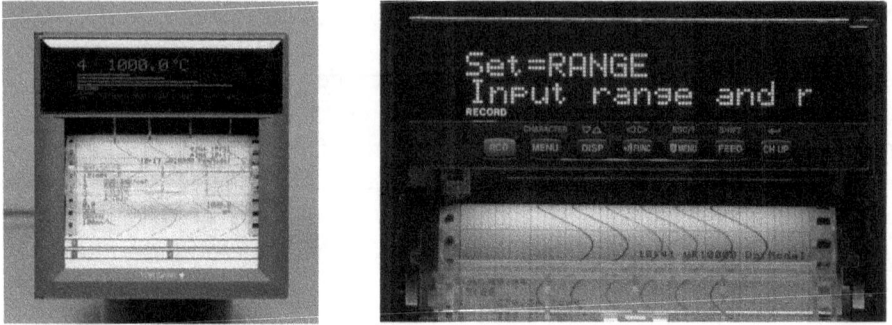

Figure 4.64: Pen strip chart recorder (Yokogawa, Japan).

Figure 4.65: Circular chart recorder (Dickson, USA).

under real-world environmental conditions. It can be used in food processing, pharmaceuticals, heat treatment, manufactured goods and wastewater treatment.

4.6.4 Paperless chart recorder

The paperless chart recorder uses a black/white or color matrix liquid crystal display to electronically display the measured signal. The device can avoid regular replacement of chart papers and ink cartridges associated with other forms of paper recorders. Reliability is also improved compared to paper recorders. In addition to displaying measurement signals in real time on the screen, it is also convenient to store history records. This recording method has the advantages of the low maintenance cost, a large amount of recorded data, fast dynamic response and accurate data recording. Figure 4.66 shows some paperless chart recorders produced by ABB (Switzerland), Honeywell (USA) and Yokogawa (Japan).

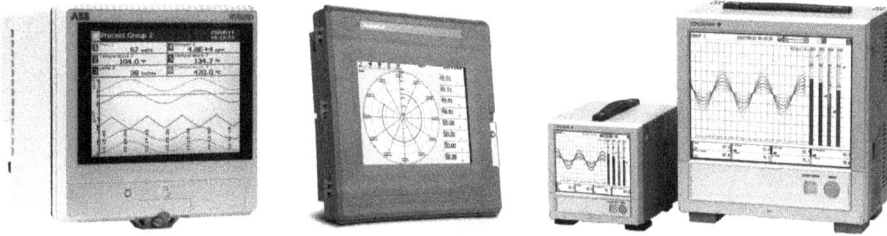

Figure 4.66: Paperless chart recorder.

Digital recorders can input both analog and digital signals. The allowed analog input signals include direct voltage, direct current, alternating voltage and alternating current. The digital input can typically be a switch from switch closure or relay operation. The device can record some standard output signals from sensors such as accelerometers, thermocouples, thermistors, resistance thermometers and differential transformers. Digital signals such as encoders, counters, timers, tachometers and clocks can also be recorded. High-resolution AD converters in digital recorders provide better accuracy, and higher sampling frequencies enable higher speed data sampling. In addition, inkjet and laser printers are standard output devices. The data can be output to a printer to store the data of the measurement system in paper form.

5 Correlation measurement

5.1 Introduction

The word correlation consists of co- (meaning "together") and relation. When the two sets of data are closely linked, it can be said that they have a high correlation. The degree of correlation is useful for describing the nature of the association between two variables. If the two variables change in the same direction, the correlation is called positive correlation. Similarly, if the values of variables change in the opposite direction, the correlation is called negative correlation. Correlation analysis is a very important technology that is used widely in vibration testing and analysis, radar ranging, acoustic emission testing and so on. This section discusses the application of correlation technology in mechanical engineering measurement.

In this chapter, students will learn (1) correlation analysis, (2) spectral analysis and (3) digital image correlation measurement.

5.2 Correlation analysis

Relevance is a statistical tool that measures and analyzes the relational degree between two or more variables. If two related variables are interdependent, there must be a causal relationship between the two phenomena. If this relationship does not exist, the two phenomena cannot be related. For example, if the two variables X and Y are positively correlated, it means that as X increases, Y increases, or as X decreases, Y also decreases. A negative correlation between the two variables X and Y means that Y decreases as X increases, or Y increases as X decreases. There are some examples of positive correlations, such as the more the homework is completed, the better the grades will be. Examples of negative correlations are as follows: the higher the price of the commodity is, the less the sales will be.

In signal processing, if the statistical characteristics can be derived from a single, sufficiently long, random sample of the process, then the stochastic process is called ergodic process. The mean of the sample function is referred to as the sample mean of the ergodic process and is defined via

$$\mu_x = \lim_{T\to\infty} \frac{1}{T} \int_{-(T/2)}^{T/2} x(t)dt \tag{5.1}$$

where $x(t)$ is the sample function and T is the observing time.

https://doi.org/10.1515/9783110624397-005

The sample variance of the random process is defined similarly via

$$\sigma_x^2 = \lim_{T \to \infty} \frac{1}{T} \int_{-(T/2)}^{T/2} [x(t) - \mu_x]^2 dt \qquad (5.2)$$

The standard deviation σ_x is a statistic that measures the dispersion of a data set relative to its mean and is calculated as the square root of the variance. By determining the change between each data point relative to the average, it is calculated as the square root of the variance. If the data points are further from the average, the deviation within the data set is greater. Therefore, the more dispersed the data is, the larger the standard deviation will be.

The standard deviation measures the typical distance between each data point and the mean. The formula for standard deviation depends on whether the data is its own population or as a sample of a larger population. If the data itself is considered as a population, the following formula is divided by the number of data points, N:

$$\sigma_x^2 = \frac{1}{N} \sum_{i=1}^{N} (x_i - \mu)^2 \qquad (5.3)$$

$$\mu = \frac{1}{N} \sum_{i=1}^{N} x_i \qquad (5.4)$$

where σ_x is the population standard deviation and μ is the population mean.

If the data is a sample of a larger population, the following formula is divided by $N-1$ of data points in the sample.

$$s^2 = \frac{1}{N-1} \sum_{i=1}^{N} (x_i - \bar{x})^2 \qquad (5.5)$$

$$\bar{x} = \frac{1}{N} \sum_{i=1}^{N} x_i \qquad (5.6)$$

where s is the sample standard deviation and $\bar{x}$ is the sample mean.

The calculation process is illustrated by taking the population standard deviation as an example:

Step 1: Calculate the average of the data using formula (5.4).

Step 2: Find the deviation $x_i - \mu$. Subtract the mean from each data point. The deviation is obtained: the data smaller than the average value is negative deviation, and the data higher than the average value is positive deviation.

Step 3: You can square each deviation and let it be always positive $(x_i - \mu)^2$.

Step 4: You can add the squared deviations together $\left(\sum_{i=1}^{N} (x_i - \bar{x})^2 \right)$.

Step 5: You can divide the sum by the number of data points in the population, and you will get the result variance $\left(1/N \left(\sum_{i=1}^{N} (x_i - \mu)^2 \right) \right)$.

Step 6: Take the square root of the variance to get the population standard deviation (σ_x).

5.2.1 Correlation coefficient

A correlation method that is used to characterize the relationship between two variables is called the correlation coefficient. Correlation coefficients are usually represented by symbols and range from −1 to +1. When the correlation coefficient is very close to 0, whether it is positive or negative, there is little or no relationship between the two variables. A correlation coefficient close to positive 1 means that there is a positive correlation between the two variables, that is, if one variable increases, the other variable also increases. A correlation coefficient close to −1 indicates a negative correlation between the two variables, with one variable increasing and the other decreasing. Different correlation coefficients are described in Figure 5.1.

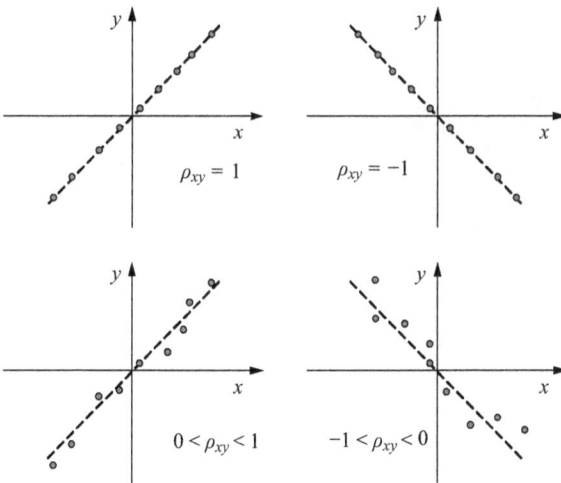

Figure 5.1: Different correlation coefficients.

The population correlation coefficient can be expressed as follows:

$$\rho_{xy} = \frac{\sigma_{xy}}{\sigma_x \sigma_y}$$

$$= \frac{E\left[(x - \mu_x)(y - \mu_y)\right]}{\sigma_x \sigma_y}$$

(5.7)

where E is the mathematical expectation; μ_x is the mean value of the random variable x; μ_y is the mean value of the random variable y; σ_x is the standard deviation of x; σ_y is the standard deviation of y:

$$\mu_x = E[x] \tag{5.8}$$

$$\mu_y = E[y] \tag{5.9}$$

$$\sigma_x^2 = E\left[(x - \mu_x)^2\right] \tag{5.10}$$

$$\sigma_y^2 = E\left[(x - \mu_y)^2\right] \tag{5.11}$$

The sample correlation coefficient can be expressed as follows:

$$r_{xy} = \frac{S_{xy}}{S_x S_y}$$

$$= \frac{N\sum xy - \sum x \sum y}{\sqrt{N\sum x^2 - (\sum x)^2}\sqrt{N\sum y^2 - (\sum y)^2}} \tag{5.12}$$

Question 5.1:

Calculate the population correlation coefficient of the random variable x and the random variable y.

No.	x	y	$x - \mu_x$	$x - \mu_y$	$(x - \mu_x)(x - \mu_y)$	$(x - \mu_x)^2$	$(x - \mu_y)^2$
1	1	2	-3	-2	6	9	4
2	3	3	-1	-1	1	1	1
3	5	3	1	-1	1	1	1
4	5	6	1	2	2	1	4
5	6	6	2	2	4	4	4
Total	20	20			14	16	14

Solution:

Step 1: Find the mean value μ_x and μ_y

$$\mu_x = E[x] = \frac{20}{5} = 4$$

$$\mu_y = E[y] = \frac{20}{5} = 4$$

Step 2: Subtract the mean from each variable. The results of $x - \mu_x$ and $x - \mu_y$ are shown in the above table.

Step 3: Calculate $(x - \mu_x)(x - \mu_y)$.

Step 4: Square each deviation from each variable. The results of $(x - \mu_x)^2$ and $(x - \mu_y)^2$ are shown in the above table.

Step 5: Find σ_{xy}, σ_x and σ_y

$$\sigma_{xy} = E\left[(x - \mu_x)(y - \mu_y)\right]$$

$$= \frac{14}{5} = 2.8$$

$$\sigma_x = \sqrt{E\left[(x - \mu_x)^2\right]}$$

$$= \sqrt{\frac{16}{5}} = 1.789$$

$$\sigma_y = \sqrt{E\left[(y - \mu_y)^2\right]}$$

$$= \sqrt{\frac{14}{5}} = 1.673$$

Step 6: Find the population correlation coefficient

$$\rho_{xy} = \frac{\sigma_{xy}}{\sigma_x \sigma_y}$$

$$= \frac{2.8}{1.789 \cdot 1.673} = 0.936$$

5.2.2 Autocorrelation function

Autocorrelation is the correlation between a signal and its own with a delayed copy as a delay function. Informally, it is the similarity between them. Autocorrelation analysis is a mathematical tool for finding repetitive patterns, such as the presence of periodic signals that are obscured by noise, or the identification of missing fundamental frequencies in signals that are implied by their harmonic frequencies.

The value of ρ_{xy} can be from −1 to +1 and is independent of the unit of measurement. Based on formula (5.7), the following conclusion can be obtained:

$$E\left[(x - \mu_x)(y - \mu_y)\right]^2 \leq E\left[(x - \mu_x)^2\right] E\left[(y - \mu_y)^2\right] \tag{5.13}$$

This formula is the Cauchy–Schwarz inequality, which has been encountered in many different settings, such as linear algebra, analysis, probability theory, vector algebra and other fields. The autocorrelation function can be used for detecting nonrandomness in data. And if the data are not random, it can also be used for identifying an appropriate time series model.

Given a signal $x(t)$, the continuous autocorrelation $R_x(\tau)$ is most often defined as the continuous cross-correlation integral of $x(t)$ with itself, at a lag τ:

$$R_x(\tau) = \lim_{T \to \infty} \frac{1}{T} \int_{-(T/2)}^{T/2} x(t)x(t+\tau)dt \tag{5.14}$$

$x(t)$ and $x(t+\tau)$ have the same average value μ_x and standard deviation σ_x. The autocorrelation of a random process describes the correlation between the values of the process at different times.

Formula (5.14) is founded on a power signal $x(t)$. For an energy signal, the definition of autocorrelation can be written as

$$R_x(\tau) = \int_{t_1}^{t_2} x(t)x(t+\tau)dt \tag{5.15}$$

where the signal exists only in the interval $t_1 \le t \le t_2$.

For a periodic signal, the definition of autocorrelation can be written as follows:

$$R_x(\tau) = \frac{1}{T} \int_{t_0}^{t_0+T} x(t)x(t+\tau)dt \tag{5.16}$$

where T is the period.

The following formula can be deduced based on formulas (5.7) and (5.14):

$$\rho_{x(t)x(t+\tau)} = \rho_x(\tau)$$
$$= \frac{\lim_{T \to \infty} \frac{1}{T} \int_{-(T/2)}^{T/2} [x(t) - \mu_x][x(t+\tau) - \mu_x]dt}{\sigma_x^2}$$
$$= \frac{\lim_{T \to \infty} \frac{1}{T} \int_{-(T/2)}^{T/2} [x(t)x(t+\tau) - x(t)\mu_x - \mu_x x(t+\tau) + \mu_x \mu_x]dt}{\sigma_x^2}$$
$$= \frac{\lim_{T \to \infty} \frac{1}{T} \int_{-(T/2)}^{T/2} [x(t)x(t+\tau)]dt - \mu_x^2}{\sigma_x^2}$$
$$= \frac{R_x(\tau) - \mu_x^2}{\sigma_x^2} \tag{5.17}$$

Question 5.2:
Find the autocorrelation function of the square pulse of amplitude $x(0)$ and duration t_0 shown in Figure 5.2.

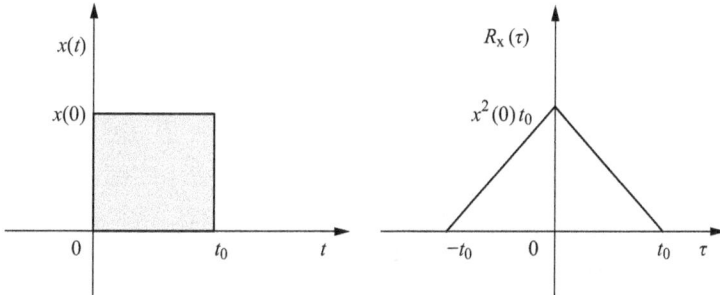

Figure 5.2: Square pulse and its autocorrelation.

Solution:
The signal is an energy signal, and the autocorrelation function is

$$R_x(\tau) = \int_0^{t_0-\tau} x(t)x(t+\tau)dt$$

$$= x^2(0)(t_0 - |\tau|) \tag{5.18}$$

Question 5.3:
Find the autocorrelation function of $x(t) = x_0 \sin(\omega_0 t + \phi)$, where ϕ is a random variable.

Solution:
Since $x(t)$ is periodic, the autocorrelation function is defined as follows:

$$R_x(\tau) = \frac{1}{T_0}\int_{t_0}^{t_0+T_0} x(t)x(t+\tau)dt$$

$$= \frac{1}{T_0}\int_0^{T_0} x_0^2 \sin(\omega_0 t + \phi)\sin[\omega_0(t+\tau)+\phi]dt$$

$$= \frac{x_0^2}{2\pi}\int_0^{2\pi} \sin(\omega_0 t + \phi)\sin(\omega_0 t + \phi + \omega_0\tau)d\theta$$

$$= \frac{x_0^2}{2}\cos\omega_0\tau \tag{5.19}$$

where $T_0 = 2\pi/\omega_0$; $t_0 = 0$.
It can be seen that $R_x(\tau)$ is periodic with period $2\pi/\omega_0$ and is independent of phase ϕ.

5.2.2.1 Properties of autocorrelation function

$$\rho_x(\tau) = \frac{R_x(\tau) - \mu_x^2}{\sigma_x^2} \qquad (5.20)$$

(1) $R_x(\tau)$ is an even function, that is, $R_x(\tau) = R_x(-\tau)$;
(2) If $\tau \to \infty$, $\rho_x(\infty) \to 0$:

$$\rho_x(\infty) = \frac{R_x(\infty) - \mu_x^2}{\sigma_x^2} \to 0 \qquad (5.21)$$

$$\rho_x(\infty) = \frac{R_x(\infty) - \mu_x^2}{\sigma_x^2} \to 0 \qquad (5.22)$$

(3) A maximum value of $R_x(\tau)$ occurs at delay $\tau = 0$:

$$\rho_x(0) = \frac{R_x(0) - \mu_x^2}{\sigma_x^2} = 1 \qquad (5.23)$$

$$\Rightarrow R_x(0) = \mu_X^2 + \sigma_X^2 \qquad (5.24)$$

(4) If $x(t)$ is periodic with period T, $R_x(\tau)$ is also periodic with period T.
(5) $R_x(\tau)$ ranges from $\mu_x^2 - \sigma_x^2$ to $\mu_x^2 + \sigma_x^2$.

5.2.2.2 Extracting useful signal

Autocorrelation (correlating the signal to itself) can be used to extract the signal from the noise. Figure 5.3 shows the extraction of useful signal (periodic information) from noise using autocorrelation. Details in Figure 5.3 are as follows: (1) useful sine wave is $10\sin(2\pi t)$; (2) amplitude of noise is less than 2.0; (3) $x(t) = 10\sin(2\pi t) + \text{Noise}$; (4) the autocorrelation function $R_x(\tau)$. The autocorrelation function of the periodic signal itself is a periodic signal whose period is the same as the period of the original signal. Autocorrelation can be used to separate the signal from noise because the autocorrelation function of the noise is easily distinguished from the autocorrelation function of the signal.

5.2.3 Cross-correlation function

The cross-correlation function is a function that indicates to what extent the two arbitrary functions are similar or to what extent the two functions are shifted. It can be used for pattern recognition, single particle analysis, averaging, electron tomography, cryptanalysis and neurophysiology. The cross-correlation function is a method to measure self-similarity between two waveforms $x(t)$ and $y(t)$.

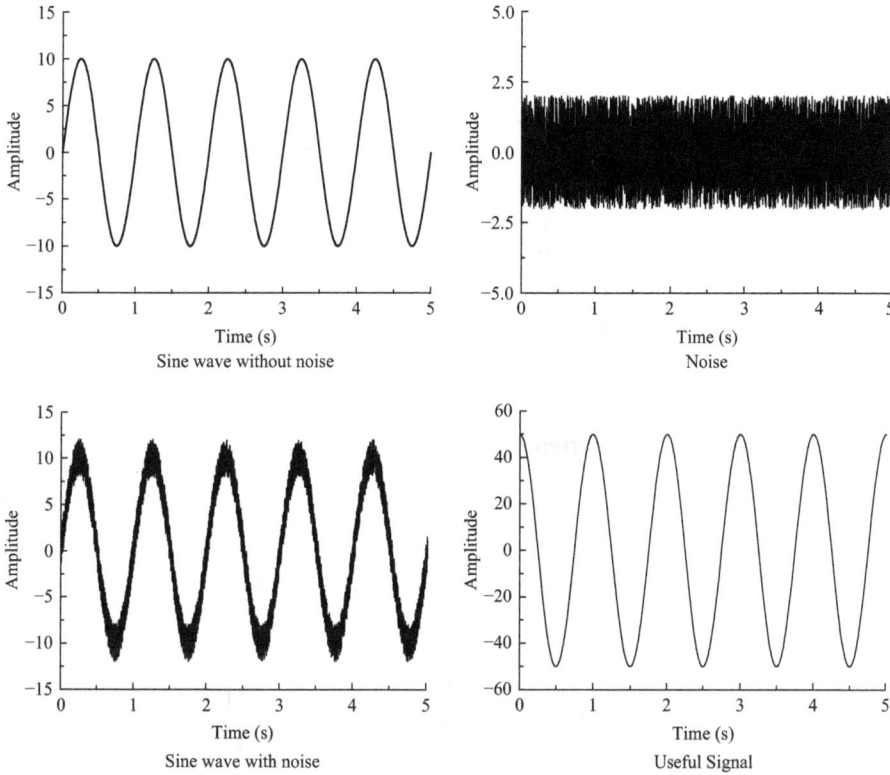

Figure 5.3: Extracting useful signal.

The cross-correlation functions in the case of infinite duration waveforms can be defined as

$$R_{xy}(\tau) = \lim_{T \to \infty} \frac{1}{T} \int_{-(T/2)}^{T/2} x(t)y(t+\tau)dt \qquad (5.25)$$

The cross-correlation functions in the case of finite duration waveforms can be defined as

$$R_{xy}(\tau) = \int_{-\infty}^{\infty} x(t)y(t+\tau)dt \qquad (5.26)$$

Question 5.4:

Find the cross-correlation function of the $x(t)$ and $y(t)$. $x(t) = x_0 \sin(\omega_0 t + \theta)$, $y(t) = y_0 \sin(\omega_0 t + \theta - \phi)$ and ϕ is a random variable.

Solution:
The cross-correlation function is defined by

$$R_{xy}(\tau) = \lim_{T\to\infty} \frac{1}{T}\int_0^T x(t)y(t+\tau)dt$$

$$= \frac{1}{T_0}\int_0^{T_0} x_0 \sin(\omega_0 t + \theta)y_0 \sin[\omega_0(t+\tau)+\theta - \phi]dt$$

$$= \frac{1}{2}x_0 y_0 \cos(\omega_0\tau - \phi) \qquad (5.27)$$

where $T_0 = 2\pi/\omega_0$.

It can be seen that $R_{xy}(\tau)$ is periodic with period $2\pi/\omega_0$ and is independent of the phase θ.

5.2.3.1 Properties of cross-correlation function

$$\rho_{xy}(\tau) = \frac{R_{xy}(\tau) - \mu_x\mu_y}{\sigma_x\sigma_y} \qquad (5.28)$$

(1) $R_{xy}(\tau) = R_{yx}(-\tau)$, $R_{xy}(\tau)$ is not necessarily an even function.
(2) If $\tau \to \infty$, $\rho_{xy}(\infty) \to 0$

$$\rho_{xy}(\infty) = \frac{R_{xy}(\infty) - \mu_x\mu_y}{\sigma_x\sigma_y} \to 0 \qquad (5.29)$$

$$\Rightarrow R_{xy}(\infty) \to \mu_x\mu_y \qquad (5.30)$$

(3) A maximum value of $R_{xy}(\tau)$ occurs at $\tau = \tau_0$, where τ_0 is the time lag between $x(t)$ and $y(t)$.
(4) If $R_{xy}(\tau) = 0$ for all τ, then $x(t)$ and $y(t)$ are said to be uncorrelated.
(5) $R_{xy}(\tau)$ ranges from $\mu_x\mu_y - \sigma_x\sigma_y$ to $\mu_x\mu_y + \sigma_x\sigma_y$.
(6) Two periodic signals with the same frequency are correlated, and two periodic signals with the different frequencies are not correlated.

5.2.3.2 Determining time delays

Cross-correlation is useful for determining the time delay between two signals. After the cross-correlation between the two signals is calculated, the maximum value of the cross-correlation function (or minimum if the signal is the negative correlation) indicates the point in time at which the signal is optimally aligned. Cross-correlation is often used for the best estimates of delays, such as echolocation (radar, sonar) and GPS receivers.

Radar is the abbreviation of radio detection and ranging. It is a system that was developed during the Second World War to measure the distance and speed of objects using radio waves as shown in Figure 5.4.

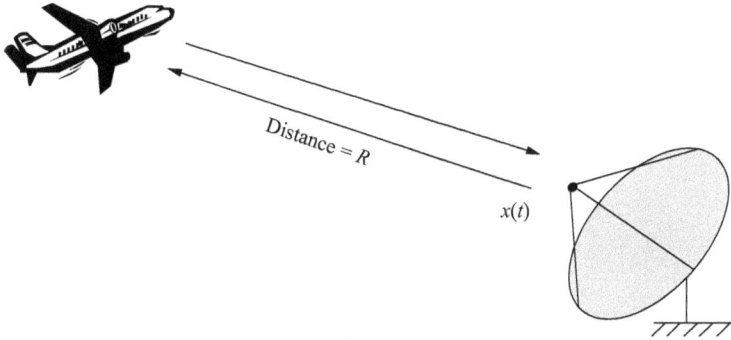

Figure 5.4: Distance and speed measurements.

In the above radar system, a transmitted waveform $x(t)$ is reflected off an airplane at a distance R and received a time T_0 later. The received signal $y(t) = A_0x(t - T_0) + z(t)$ is attenuated by a factor A_0 and is contaminated by additive noise $z(t)$:

$$R_{xy}(\tau) = \int_{-\infty}^{\infty} x(t)y(t+\tau)dt$$

$$= \int_{-\infty}^{\infty} x(t)[A_0x(t+\tau-T_0)+z(t+\tau)]dt$$

$$= \int_{-\infty}^{\infty} x(t)z(t+\tau)dt + A_0\int_{-\infty}^{\infty} x(t)x(t+\tau-T_0)dt$$

$$\qquad\qquad\qquad\qquad\qquad\qquad\qquad (5.31)$$

$$= R_{xz}(\tau) + A_0R_x(\tau-T_0)$$

The transmitted waveform $x(t)$ and additive noise $z(t)$ are not correlated, that is, $R_{xz}(\tau) = 0$. Then

$$R_{xy}(\tau) = A_0R_x(\tau-T_0) \qquad\qquad (5.32)$$

A scaled and shifted version of the autocorrelation function of the transmitted waveform will have its peak value at $\tau = T_0$, which may be used for measuring the distance and speed of objects.

5.3 Spectral analysis

Consider the transient case

$$S_x(f) = \int_{-\infty}^{\infty} R_x(\tau)e^{-j2\pi f\tau}d\tau \qquad\qquad (5.33)$$

For an energy signal, formula (5.33) can be written as

$$S_x(f) = \int_{-\infty}^{\infty} \left(\int_{-\infty}^{\infty} x(t)x(t+\tau)dt \right) e^{-j2\pi f\tau} d\tau \qquad (5.34)$$

$$S_x(f) = \int_{-\infty}^{\infty} x(t)e^{j2\pi ft} dt \int_{-\infty}^{\infty} x(t+\tau)e^{-j2\pi f(t+\tau)} d(t+\tau)$$

$$= X(-f)X(f)$$

$$= |X(-f)^2| \qquad (5.35)$$

where $S_x(f)$ is the energy density spectrum of the transient waveform $x(t)$. Similarly, the Fourier transform of the power-based autocorrelation function can be written as

$$S_x(f) = \int_{-\infty}^{\infty} \lim_{T\to\infty} \frac{1}{T} \int_{-(T/2)}^{T/2} x(t)x(t+\tau)dt e^{-j2\pi f\tau} d\tau \qquad (5.36)$$

$S_x(f)$ in formula (5.36) is the power density spectrum of an infinite duration waveform. The power density spectrum shows the intensity of the change (energy) as a function of frequency. In other words, it shows which frequencies change very strongly and which frequencies change very weakly.

Spectral analysis is used in many different areas. In vibration monitoring, the spectral content of the measured signal gives information on the wear and other characteristics of the mechanical components in the study. In radar and sonar systems, the spectral content of the received signal provides information about the location of the source (or target) located in the field of view. In medicine, spectral analysis of various signals measured from patients, such as electroencephalography or electrocardiogram signals, can provide useful information for diagnosis. In seismology, spectral analysis of signals recorded before and during seismic events gives useful information about the ground motion associated with these events. Seismic spectral analysis is also used for predicting subsurface geological structures in natural gas and oil exploration.

5.4 Digital image correlation measurement

5.4.1 Introduction

In general, the measurement methods, which use traditional strain gauges, gain precise data only when detected strain is small and uniformly distributed. However, it is common to come across local and high-degree strain gradient deformation in practice. Under such circumstance, noncontact full-field optical measurements show irreplaceable advantages. After the development of the last several decades, the optical metering methods, such as holographic interference, electric speckle pattern, speckle photography and digital image correlation (DIC) method, come into being. Among these methods, DIC method, invented in the early 1980s, detects the displacements and deformations of an object by calculating the correlation of speckle patterns captured before and after deformation. DIC has advantages in the following aspects: (1) The simpleness of measuring system and devices: the specimen of speckle pattern can be acquired by either artificial speckle manufacture technology or collecting the natural speckles on objects' surface; (2) low requirement for experimental environment and easy to use in engineering application: DIC usually uses white light as the light source, and interference fringes processing is not required; (3) high automation in measuring and data collecting process; (4) the application of microinstrument enables the multiscale measurement from microscale to macroscale.

After being proposed, DIC has been applied in fracture mechanics, elastic mechanics and damage detection, to detect stress and strain in complicated environments. Besides, to overcome the limitation, the measured displacements must lay on the measured plane of the object. The latest developed 2.5D-DIC measuring system and stereo-DIC measuring system adopt multiple cameras or single camera with the specific optical system. Nevertheless, these systems are generally expensive and can only detect the relatively small region of interest (ROI).

Since two-dimensional (2D)-DIC measuring system requires only a single camera, the budget of equipment is consequently saved. Meanwhile, measurement accuracy remains relatively high. Therefore, 2D-DIC measurement system has a wide range of applications.

For the last two decades, the 2D-DIC, a kind of noncontact measuring method based on image, has been developed and become mature. This literature mainly focuses on summarizing research results of the last two decades by expounding the improved 2D-DIC algorithms on its increased arithmetic efficiency and measurement accuracy. Furthermore, it systematically summarized the research results on error analysis of 2D-DIC measuring system to provide a reference for further study on the related field.

5.4.2 Basic principle of two-dimensional DIC

Before analyzing and employing the specimen images, it is crucially important to make adequate preparation for the experimental setup of the standard 2D-DIC system to gain a good test effect. It primarily includes the specimen preparation, light source preparation and image acquisition hardware preparation. Figure 5.5 shows the basic structures of the 2D-DIC experimental system. The surface of a specimen records the full deformation information. In addition, a good quality pattern should be with the characteristic of highly contrasted, stochastic and isotropic. Depending on the surface material and specimen size, several methods for creating typically artificial speckles have been put forward. Spraying black and white paints with

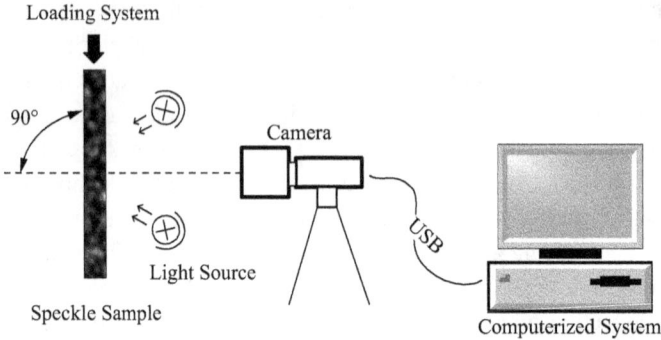

Figure 5.5: A schematic diagram of the experimental device for standard 2D-DIC system.

airbrushes has become the most common technique for large specimen sizes, due to its usability and practicability. A mathematical model of the airbrush gun and the main influential factors on generating the speckle size was proposed by Lionello and Cristofolini, which demonstrates how to use the airbrush to produce a desired speckle size. For instance, when measuring biological soft tissue, aluminum for thermal properties and other materials, it is important to notice that covered speckle should not change the material properties and should not be peeled off when deformation takes place. If the specimen size is at micrometer grade, various approaches will be used, including etching the counterpart of the specimen to fabricate speckle patterns, and spinning an epoxy resin with powder to form speckle patterns. For nanoscale speckle sample, in order to obtain a good quality of speckle surfaces, approaches including deposited fluorescent nanoparticle tracers, beam lithography techniques, focused ion beam and other technologies are used.

In addition, there are also some particularly noteworthy matters in 2D-DIC measurements. The measured surface should be a flat plane, and the optical axis of the optical imaging lens should be perpendicular to the measured flat plane. During the loading process, the stable and mild light source should be provided, and the loading system should have good accuracy. For complex loading systems (e.g., multiaxial loading, thermal loading and dynamic loading), the destruction of speckles on specimen's surface caused by loading should be noticed to avoid serious loss of gray information of speckles. High-quality cameras should be adopted to guarantee the accuracy of DIC measurements.

Basically, the idea of the subset-based 2D-DIC technique is to extract the displacement and the strain information within an ROI by tracking a one-to-one correspondence position between matching subsets in the reference image and the deformed image. The standard DIC leverages specifically match algorithms to solve initial guesses with integer-pixel estimation, and then employ iterative nonlinear optimization schemes and certain interpolation to achieve the subpixel accuracy. Figure 5.6 shows the basic principles of the standard subset-based DIC method.

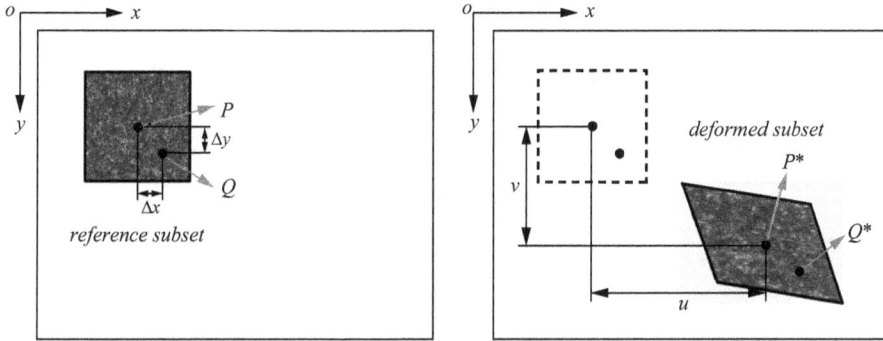

Figure 5.6: Schematic illustration of the reference square subset and the corresponding deformed subset in 2D-DIC.

To be more specific, a square subset of $(2M+1) \times (2M+1)$pixels contained a center calculation reference point $P$from the initial undeformed image. This point is used as an input, and then finds its corresponding location in the corresponding deformed image. The first-order shape function is commonly used for describing the linear transformation, as expressed in the form of:

$$x^* = u + \frac{\delta u}{\delta x}\Delta x + \frac{\delta u}{\delta y}\Delta y$$

$$y^* = v + \frac{\delta v}{\delta x}\Delta x + \frac{\delta v}{\delta y}\Delta y$$

(5.37)

where x^* and y^* are the final displacements of the reference subset point along x axial and y axial; u and v are the displacement components of the subset center P in the x axial and y axial directions; Δx and Δy are the initial distances between an arbitrary subset point Q and subset center point P; $\delta u/\delta x$, $\delta u/\delta y$, $\delta v/\delta x$, $\delta v/\delta y$ are the displacement gradient components of the reference subset. The basic form of the displacement of P is $(u, v, \delta u/\delta x, \delta u/\delta y, \delta v/\delta x, \delta v/\delta y)^T$. Figure 5.7 shows six different forms of linear transformation for the reference subset.

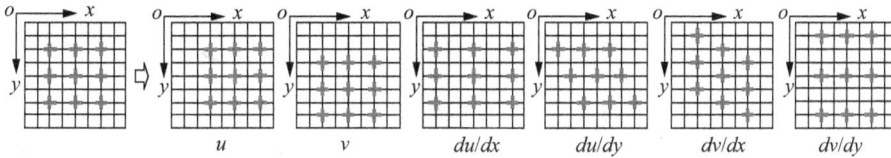

Figure 5.7: The six different forms of linear transformation for the reference subset.

In order to evaluate the similarity of intensity between the reference subset and the current subset, the justified correlation criteria method is carried out. Both the zero-mean normalized sum-of-square difference criterion (ZNSSD) and zero-mean normalized cross-correlation criterion (ZNCC) have been widely used in DIC analysis, which have been proved to be robust correlation criteria since the ZNSSD and ZNCC correlation criteria are unaffected by the changes of brightness and image contrast. It is worth pointing out that the relationship between the ZNCC and the ZNSSD is listed as the following formula:

$$C_{\text{ZNCC}}(p) = 1 - 0.5 \times C_{\text{ZNSSD}}(p) \tag{5.38}$$

ZNSSD and ZNCC are defined as follows:

$$C_{\text{ZNSSD}}(p) = \sum_{i=-M}^{M} \sum_{j=-M}^{M}$$

$$\left[\frac{f(x,y) - f_m}{\sqrt{\sum_{i=-M}^{M} \sum_{j=-M}^{M} [f(x,y) - f_m]^2}} - \frac{g(x^*,y^*) - g_m}{\sqrt{\sum_{i=-M}^{M} \sum_{j=-M}^{M} [g(x,y) - g_m]^2}} \right]^2 \tag{5.39}$$

$$C_{\text{ZNCC}}(p) = \sum_{i=-M}^{M} \sum_{j=-M}^{M}$$

$$\left[\frac{[f(x,y) - f_m] \times [g(x^*,y^*) - g_m]}{\sqrt{\sum_{i=-M}^{M} \sum_{j=-M}^{M} [f(x,y) - f_m]^2} \times \sqrt{\sum_{i=-M}^{M} \sum_{j=-M}^{M} [g(x^*,y^*) - g_m]^2}} \right]^2 \tag{5.40}$$

where $f(x,y)$ is the gray-level function at coordinates (x,y) in the reference subsets of the reference image; $g(x^*,y^*)$ is the gray-level function at coordinates (x^*,y^*) in the current subsets of the current image; f_m and g_m are the mean grayscale values of the reference and current subsets, respectively, as shown below:

$$f_m = \frac{1}{(2M+1)^2} \sum_{i=-M}^{M} \sum_{j=-M}^{M} [f(x,y)]$$

$$g_m = \frac{1}{(2M+1)^2} \sum_{i=-M}^{M} \sum_{j=-M}^{M} [g(x^*,y^*)] \tag{5.41}$$

The main step of implementation DIC method is to use nonlinear optimization scheme to achieve high subpixel accuracy with a better match, and a specific interpolation is undertaken subsequently. The specific implementation process can be found in the literature. It should be stressed that the finite element-based global DIC has been developed into a new global measuring method, which is distinguished from the traditional subset-based local DIC. Pan et al. analyzed the experimental and theoretical errors of both of these methods. Although the basic principle of DIC above gives a brief description, numerous papers introduce other methods to increase measurement accuracy and efficiency to meet different measurement demands. In the following chapters, the main details of the development of 2D-DIC will be expatiated.

5.4.3 Applications of two-dimensional DIC

Over the years, the application of 2D-DIC technique gets extensively applied and rapidly developed. Various mechanical constants of material properties and several descriptive parameters of stress field were analyzed via DIC such as the coefficient of thermal expansion (CTE), Young's modulus (E), Poisson's ratio, stress intensity factor and J-integral. In addition, this method can also obtain the properties of the material in the dynamic test. Strain localization of dual-phase high-strength steel was characterized by Tarigopula et al. in tensile tests under dynamic loading at a strain-rate between 150 and 600/s.

From the viewpoint of the sort of materials, the typical applications of the main materials are listed. The research employed in various classes of metallic materials such as the damage mechanisms of duplex stainless steels using this technique under low-cycle fatigue test is analyzed by Bartali et al. The method was applied in the polymer analysis, and the microscopic strain localization of natural wood-reinforced polypropylene matrix composites was investigated by Godara et al. The strain behavior of glassy polymers corresponds to uniaxial compression, which was investigated in the literature. Regarding biological materials, that is, cell, studies on the mechanical strain about the bone have been reported in the literature.

With the long-time integration of technology, the improved DIC method can handle many complicated environments reliably. Under high-temperature conditions, Grant et al. applied the DIC technique combined with the application of filters and blue illumination to measure CTE and Young's modulus of a nickel base up to 1,000 °C. A noncontact high-temperature deformation measuring system by means of RG-DIC technique was designed by Pan et al., which can accurately reflect the thermal deformation of metals and alloys in the environments with the temperature under 550 °C. According to Guo et al., a stable speckle pattern by spraying tungsten powder at 2,600 °C was introduced into measurement of the thermomechanical properties of carbon fibers.

The DIC method is also a versatile and flexible technique that can achieve multiple length measurement depending on the magnification of the observation tools. In addition to traditional photogrammetry for macroscopic detection, more quantitative subtle deformation information at the micro/nanoscales and the requirement in measurement of miniature specimen can be satisfied through DIC with optical microscope (OM), scanning electron microscope (SEM), atomic force microscope

(AFM) and scanning tunneling microscope (STM). For example, the use of an OM in DIC method to acquire the Young's modulus of commercial aluminum, bovine pericardium and Cu-Al-Be shape memory alloy was employed by Sánchez-Arévalo and Pulos. The DIC system with an attached microscopic lens to characterize the strain deformation information of single crystal during a micro-tensile test was investigated in the article. In terms of the combination of DIC method with SEM, Stinville et al. succeeded in using DIC with SEM to measure the heterogeneous strain fields of the nickel-based superalloy at the subgrain level. An analytical method via 2D-DIC for analyzing local damage mechanisms of porous carbonate at different scales by observation integrates both light-based observations and SEM was introduced by Dautriat et al. Alternatively, the deformation data of microstructure can also be extracted from STM or AFM, and its related application will not be enumerated one by one here. The DIC technique is successfully developed into a powerful deflection methodology, and it has vast application in prospect for its feasibility.

6 Sensors

6.1 Introduction

A sensor is a device that detects and responds to certain inputs in the physical environment. The measurands can be force, torque, temperature, displacement, pressure, velocity, acceleration, flow, level and so on. The sensor can convert a signal from one form of energy to another form of energy. The primary sensing element generates the output signal and maintains the natural characteristics of the sensing technology. The transducer is also a sensor, but the transducer always converts the nonelectric pressure signal into an electrical signal. Therefore, all transducers are sensors, but a sensor may not be a transducer. Different types of sensors are suitable for various industrial applications, from the automotive, medical, home appliance and aerospace and defense, to commercial transportation.

A good sensor is designed to have a linear relationship with some simple mathematical functions of the input measurand, which obeys the following rules: first, it should be sensitive to the measured property (measurand). Second, it should be insensitive to any other property that encountered in the measurement. Finally, it should not affect the measured properties (without or low load effect). For the last rule, using modern technologies like MEMS (microelectromechanical system), more and more sensors can be manufactured on a microscopic scale. In this way, the sensor has little effect on the measurement results. The conditioned output signal can be measured with voltage (0 to 10 V, 0 to 5 V, –5 to 5 V or –10 to 10 V) or current (0–20 or 4–20 mA). An analog sensor signal processed in a digital device should be converted to a digital signal, using an AD (analog-to-digital) converter. In general, the sensors can be electromagnetic (radio waves, microwaves and X-rays), electromechanical (strain gauge, piezoelectric actuator and quartz), electro-optical (photodiode and solar cell), electrostatic (electrometer), electrochemical (pH electrode), electroacoustic (microphone and loudspeaker) and thermoelectric (thermoresistor and thermocouple). Classification of sensors in the textbooks is based on the type of detection. There are temperature sensors, pressure sensors, flow sensors, displacement sensors, velocity sensors, acceleration sensors, force/torque sensors, level sensors, image sensors, proximity/presence sensors, color, contrast and luminescence sensors, environmental sensor and so on.

6.2 Temperature sensors

A temperature sensor is a device that can detect and measure hotness and coolness from a specific source and convert the data into an understandable electrical signal. Temperature measuring sensors are different, but they all measure temperature by

https://doi.org/10.1515/9783110624397-006

sensing some changes in physical properties. Temperature sensors play an important role in many applications. They are installed in equipment for accurately and effectively measuring the temperature of a medium in a given set of requirements, such as monitoring industrial heating and cooling systems, and maintaining a certain temperature. Different types of temperature measuring sensors discussed in this chapter include thermocouples, resistor temperature detectors, thermistors, integrated circuit (IC) temperature sensors, infrared (IR) temperature sensors and bimetallic temperature devices.

6.2.1 Thermocouple

The thermocouple is the most common type of temperature sensors for measuring the temperature of solids, liquids and gases, which is abbreviated as TC. A thermocouple consists of two different metal wires. The wires are welded together at one end to form a measuring (hot) junction. The other end, called the reference (cold) junction, is connected by an electronic measuring device (voltmeter). These metal wires work on the principle of thermoelectric effect or Seebeck effect (the German physicist Thomas Johann Seebeck discovered that in 1821, when the two ends are joined with different metals, and there is a temperature difference between the joints, the magnetic field is observed due to thermoelectric current). The voltage will be generated at a single junction of two different types of wire. Its magnitude depends on the type of wire used, which is proportional to the temperature changes. When the temperature goes down/up, the output voltage also decreases/increases. The schematic diagram of a thermocouple is shown in Figure 6.1.

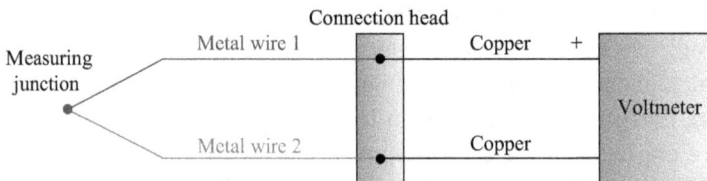

Figure 6.1: The schematic diagram of a thermocouple.

The thermocouple is usually housed in a metal or ceramic shield that protects it from different environments. Type J, type K, type T and type E are the most common types of thermocouples, while type R, type S and type B thermocouples are used for high-temperature applications. The thermocouple is typically selected for its low cost, high-temperature limits, wide-temperature range and durability, and it can be used for long distance. Thermocouples can be found in almost all industrial, scientific and OEM (Original Equipment Manufacturer) applications. A picture of type K mineral insulated probe (maximum tip temperature 1,100 °C, 1 m plain long lead, and the sensor in the photography was made by RS Components Ltd., UK) is shown in Figure 6.2. It is a very rugged and flexible mineral insulated thermocouple suitable for high pressure and vacuum applications, and it can withstand high levels of vibration.

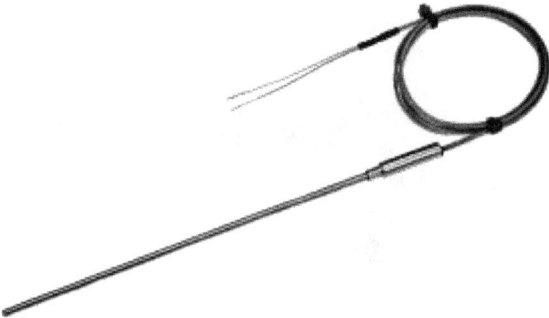

Figure 6.2: Type K mineral insulated probe.

Advantages of the thermocouple include the following:
The thermocouple is of low cost and has small size, fast response, good robustness, wide operating range, large temperature variation and stable accuracy.

Disadvantages of the thermocouple include the following:
The thermocouple has very weak output, nonlinear characteristic, complex transitions from electromotive force to temperature and limited accuracy for small variations. It is also sensitive to electrical noise.

Typical applications of the thermocouple include the following:
The thermocouples are used where high temperatures are involved: metallurgy, power generation, ceramics, oil/gas, pharmaceutical, biotechnology, cement, paper and pulp and so on. Thermocouples are also used in household appliances such as stoves, ovens and toasters.

6.2.2 Resistance temperature detector

Resistance temperature detector (RTD) can accurately detect temperature with excellent repeatability and component interchangeability, which is also known as a resistance thermometer. The RTD uses a variable resistor whose resistance value changes in proportion to the temperature. As the temperature goes up, resistance increases in an accurate, repeatable and nearly linear manner and vice versa. When in use, a small current flows through the RTD and a voltage is generated across the RTD. By measuring the voltage, its resistance and the corresponding temperature can be determined. The photo of the RTD is shown in Figure 6.3, which consists of the connecting head, fixing nut, protecting tube, sensing element and so on.

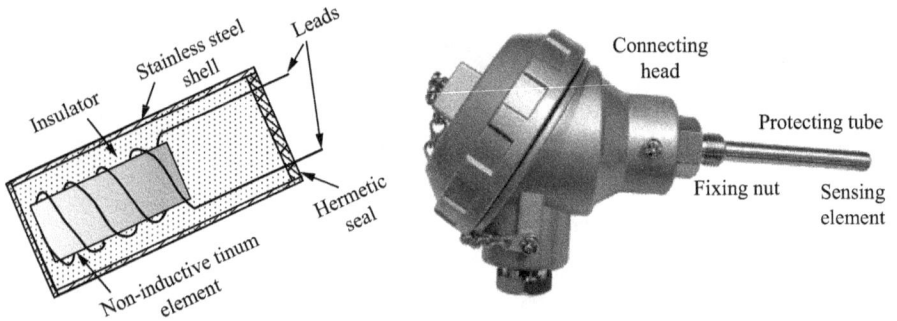

Figure 6.3: Resistance temperature detector (RTD).

RTD is made from platinum, nickel and copper metals, which has a wide range of temperature measurement capabilities. Platinum is an ideal material for precision measurement (±0.1 °C) because of its perfect linearity in its purity and temperature coefficient of resistance. Platinum RTD with a thin platinum film on a plastic film typically has a resistance of 100 Ω at 0 °C. RTD can also be manufactured cheaply from copper and nickel, but the latter has a restricted range due to nonlinearity and wire oxidation problem of copper. Summary of the basic characteristics of nickel, copper and platinum materials is shown in Table 6.1.

Table 6.1: The basic characteristics of nickel, copper and platinum materials.

Element material	Temperature range	Resistance	Advantages
Platinum	−260 to 850 °C	100 Ω at 0 °C	Good linearity and stability, high purity and precision
Copper	−100 to 260 °C	100 Ω at 0 °C	Best linearity
Nickel	−100 to 260 °C	120 Ω at 0 °C	High sensitivity and low cost

There are three types of techniques for measuring temperatures through RTD, called two-wired, three-wired and four-wired method, as shown in Figure 6.4. In the two-wired method, the current goes through the RTD to measure the changing voltage, which is a very simple connection method. However, all wires and connections

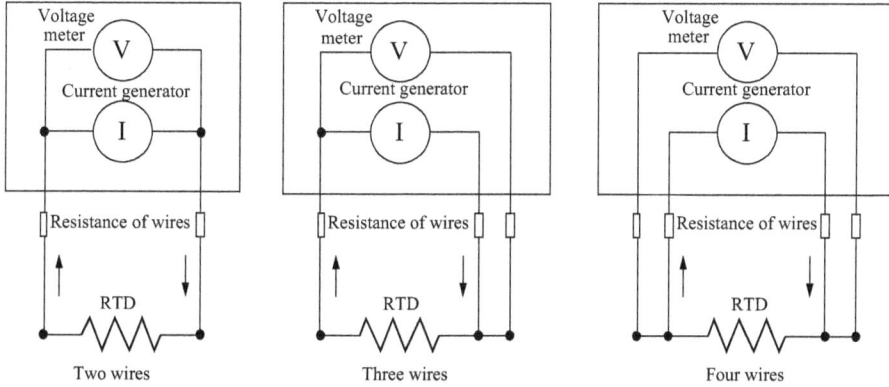

Figure 6.4: Two-wired, three-wired and four-wired methods.

have some resistance and the contacts will always have resistance, too. The two-wired method can cause a huge error in the case of long wires and bad connections. The three-wired method is similar to the two-wired method except that the third wire can compensate for wire resistance. The three-wired method is far better than the two-wired method. Therefore, the three-wired method has become a standard for many industrial applications. In a four-wired approach, the current is applied in a set of the wires and the voltage is sensed on the other set of wires, which fully compensates the lead resistance. There is almost no current in voltage measurement wires, so there is no voltage drop and no error. The four-wired method is the best and most accurate method of measuring RTD.

Typical resistance/temperature characteristic of platinum RTD is shown in Figure 6.5. The resistance value is 100 Ω when the temperature is 0 °C. Platinum RTD has a positive temperature coefficient from −260 to 850 °C. Platinum RTD elements follow a more linear curve than thermocouples and most thermistors.

Advantages of the RTD include the following:
The RTD has a stable output for a long period of time. It is easy to recalibrate and obtain accurate readings within a specified temperature range.

Disadvantages of the RTD include the following:
The RTD has a smaller overall temperature range and higher initial cost. Self-heating is another drawback, which will reduce the accuracy of the measurement.

Typical applications of the RTD include the following:
The RTDs are widely used in food processing, textile production, petrochemical processing, plastics processing and air/gas/liquid temperature measurement.

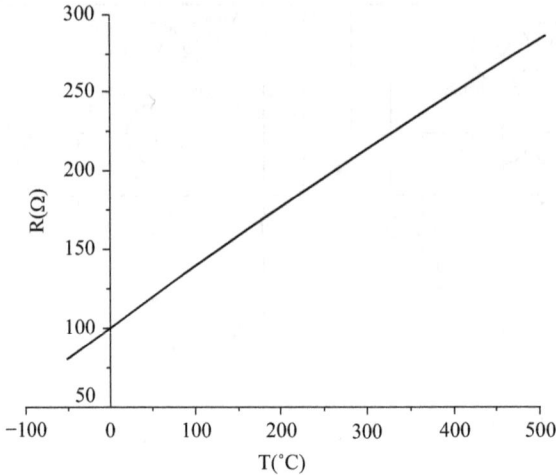

Figure 6.5: Typical resistance/temperature characteristic of platinum RTD.

6.2.3 Thermistors

The name of the thermistor is derived from the combination of the words "thermal" and "resistor." A thermistor is a special type of resistor that its resistance changes rapidly with the small change in temperature, which is relatively inexpensive, adaptable and easy to use. There are two types of thermistors, PTC (positive temperature coefficient) thermistor and NTC (negative temperature coefficient) thermistor. PTC thermistor is an element whose resistance rises with the increase in temperature. PTC thermistors are made of doped polycrystalline ceramic (containing $BaTiO_3$ and other compounds). Elements for PTC thermistors are generally based on ceramics with good reliability and performance. NTC thermistors are elements that reduce the resistance when the temperature rises, which are made into a pressed disc, rod, plate, bead and cast chip of a semiconducting material. NTC thermistors use highly accurate, thermally responsive components. The photographs of the thermistors are shown in Figure 6.6. (The sensors in the photography were made by the U.S. SENSOR Corp., USA.).

The temperature coefficient of most thermistors is negative. Figure 6.7 shows the typical resistance/temperature characteristic of the NTC thermistor. As shown in Figure 6.7, the resistance suddenly drops at a certain critical temperature.

Advantages of the thermistor include the following:
The thermistor is a good temperature sensor with low cost, higher sensitivity, high accuracy (0.05–0.20 °C), fast response, long life, small size and long-term stability.

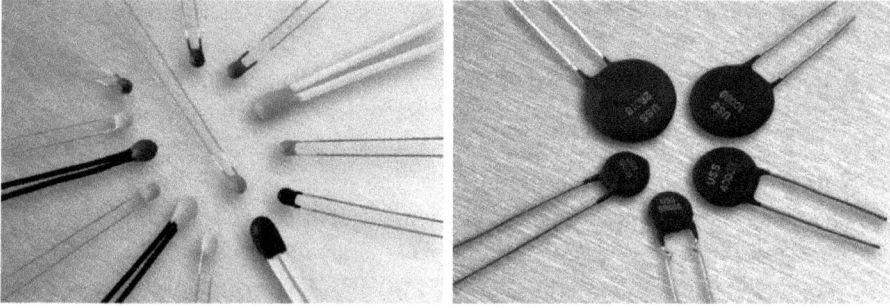

Figure 6.6: Thermistors.

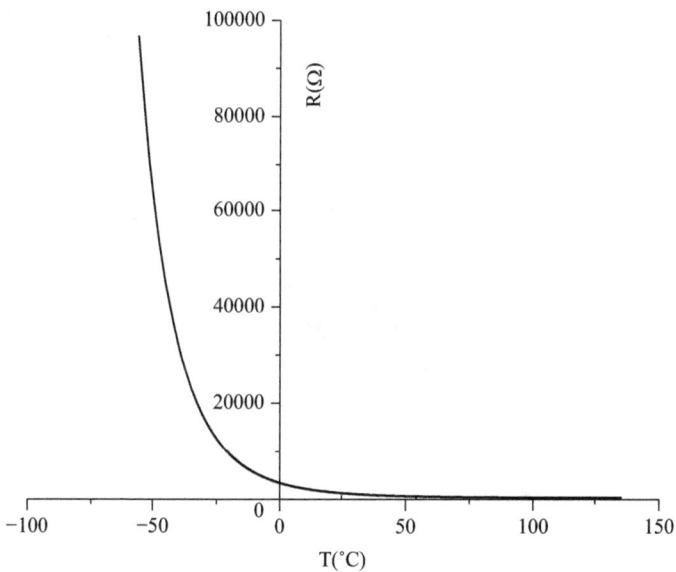

Figure 6.7: Typical resistance/temperature characteristic of NTC thermistor.

Disadvantages of the thermistor include the following:
The thermistor is nonlinear, fragile and self-heating and cannot be used at a very high temperature. In addition, the current source is required.

Typical applications of the thermistor include the following:
The thermistors are widely used in medical equipment, 3d printers, home applian-ces (ovens, hair dryers, toasters, refrigerators, etc.), computers, autoadjust heaters and overcurrent protection.

6.2.4 Integrated circuit temperature sensors

An IC temperature sensor is based on the fact that semiconductor diodes have temperature-sensitive voltage versus current characteristics, which generates an output current proportional to the absolute temperature. The first IC temperature sensor was invented by Texas Instruments in 1970. The sensor package has a small size, low thermal mass and fast response. Compared to thermocouples and RTDs, the measurement range of IC temperature sensors is smaller. Current IC temperature sensors provide an operating range from −55 to +150 °C with high accuracy and linearity. IC temperature sensors are divided into different types: diode temperature sensors, digital output temperature sensor, resistance output silicon temperature sensor, current output temperature sensor and voltage output temperature sensor. The cost of the temperature sensor depends on its accuracy. Figure 6.8 shows an IC temperature probe manufactured by OMEGA Engineering Inc., UK.

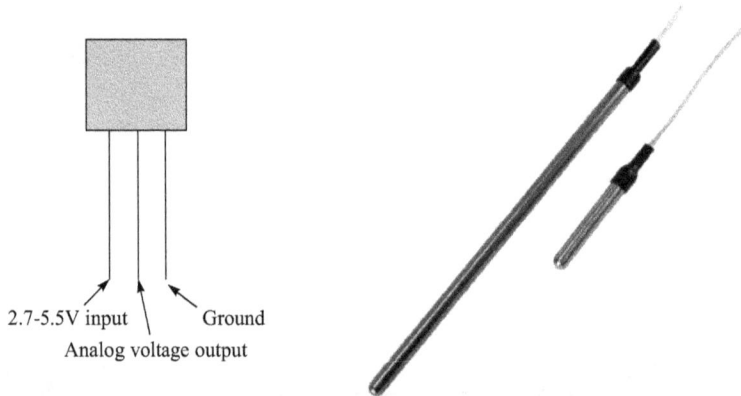

Figure 6.8: Integrated circuit temperature probe.

Advantages of the IC temperature sensor include the following:
The IC temperature sensor is small in size, low in price, easy to interfere with other devices, and has the advantages of high precision, high output, excellent linearity, good repeatability and good stability.

Disadvantages of the IC temperature sensor include the following:
The IC temperature sensor requires a power supply and has a self-heating error, a limited configuration and a narrow temperature range.

Typical applications of the IC temperature sensor include the following:
The IC temperature sensors are used for system monitoring, industrial temperature control, appliance temperature sensing and automotive temperature measurement and control.

6.2.5 Infrared temperature sensors

IR temperature sensor (IR thermometer) is a device for measuring the IR radiation, which is a kind of electromagnetic radiation lower than the visible spectrum emitted by an object. IR temperature sensor works based on black body radiation. IR temperature sensors can detect electromagnetic waves from 700 to 14,000 nm. Any substance above absolute zero moves the molecules in it. The higher the temperature is, the faster the molecules move. When the molecules move, they emit IR radiation, and more moving molecules emit more radiation, including visible light, as they get hotter, which explains why heated metal emits red or white light. IR temperature sensors detect and measure this radiation. The most basic design of an IR temperature sensor includes a lens for focusing the IR thermal radiation onto a detector, which converts the radiant energy into an electric signal. In this way, the temperature of the object can be determined. IR temperature sensors are noncontacting sensors. The photograph of the high-quality handheld IR temperature sensor is shown in Figure 6.9, which was manufactured by OMEGA Engineering Inc. USA.

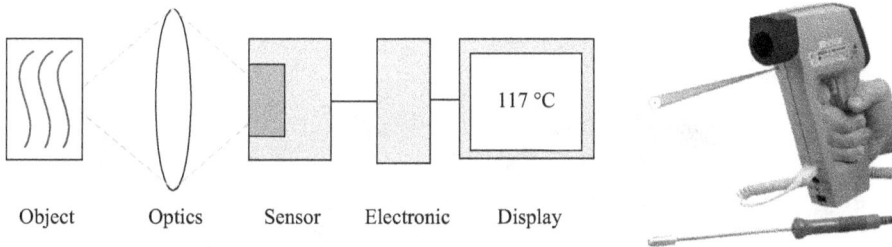

Object   Optics   Sensor   Electronic   Display

Figure 6.9: High-quality handheld infrared temperature sensor.

Advantages of the IR temperature sensor include the following:
The IR temperature sensor doesn't wear out or run longer. More importantly, the IR temperature sensor can read moving objects, so motion can be detected by measuring temperature fluctuations.

Disadvantages of the IR temperature sensor include the following:
The IR temperature sensor can only measure surface temperatures with low accuracy and are susceptible to frost, dust, fog, smoke and other particulate matter (PM) in the air. In addition, the IR temperature sensor does not "see through" glass, liquids or other transparent surfaces.

Typical applications of the IR temperature sensor include the following:
The IR temperature sensors are used for heating and air conditioning, monitoring the performance of motor/engine cooling systems, monitoring plant temperatures and measuring the temperatures of tires, brakes and similar equipment.

6.2.6 Bimetallic temperature devices

The bimetallic temperature device is one of the oldest techniques used to measure temperature. It is made up of a bimetallic system consisting of two strips of dissimilar metals, with different thermal expansion coefficients, inseparably joined together. The strip is helically or spirally wound with one end fixed to the body of the bimetallic temperature device, and the other attached to the axis of a pointer. The bimetallic temperature device uses a helical metal strip that unwinds when heated, and rotates the pointer over a calibrated scale. When it is heated, one metal (like copper) expands more than the other (like Invar), which causes the bimetallic strip unwinding. Invar is an alloy of iron and nickel with a low thermal expansion coefficient. Changes in temperature affect bimetals such as rotating the pointer. When geared correctly to a pointer, the temperature is displayed on a dial. The photograph of the bimetallic temperature device is shown in Figure 6.10 manufactured by OMEGA Engineering Inc. USA.

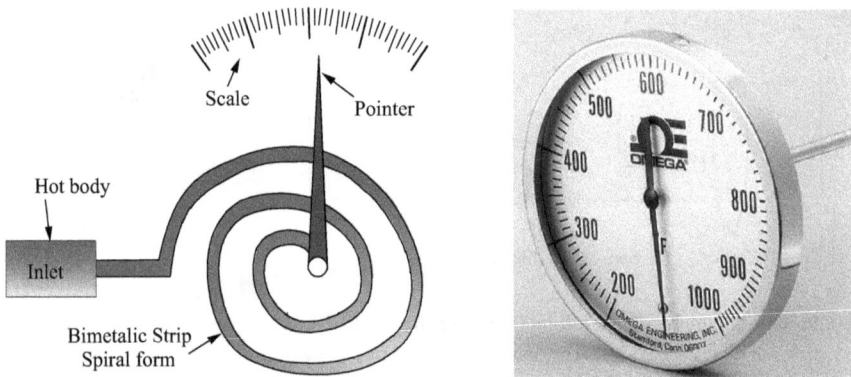

Figure 6.10: The bimetallic temperature device.

Advantages of the bimetallic temperature device include the following:
The bimetallic temperature device has a wide temperature range, low cost and linear response. It is portable, easy to install/maintain and independent of the power supply.

Disadvantages of the bimetallic temperature device include the following:
The bimetallic temperature device has low accuracy, and its temperature value is hard to calibrate, record and display.

Typical applications of the bimetallic temperature device include the following:
The bimetallic temperature devices are widely used in grills, household appliances, thermostat switches, thermometers and circuit breakers for electrical heating devices.

6.3 Pressure sensors

A pressure sensor (or a pressure transducer) converts the physical quantity "pressure" into an electrical signal. It is a device for measuring the pressure of a gas or a liquid. The pressure is an expression of the force required to prevent the expansion of the fluid, usually expressed as a force per unit area. Pressure sensors are used in refrigeration and air-conditioning applications, food industry and petrochemical industry. Different types of pressure sensors discussed in this chapter are strain gauge pressure sensors, piezoelectric pressure sensors, piezoresistive pressure sensors, capacitive pressure sensors, variable reluctance pressure sensors, potentiometric pressure sensors and bourdon tube pressure sensors.

6.3.1 Strain gauge pressure sensors

Strain gauge pressure sensor is used to measure static and dynamic pressures of gas and liquid. The strain gauge pressure sensor employs bonded metallic resistance strain gauge as a sensing element because of its unique operating characteristics. The electrical resistance of the strain gauge changes with the tensile strain. Strain gauges are manufactured using photoetching techniques. The fabrication technique allows the strain gauge manufacturer to produce almost any size and shape of strain gauge to meet the requirements. Any change in pressure will cause an instantaneous deformation of the elastic material, resulting in a change in the resistance of the strain gauge that becomes a useable electrical signal. The photograph of the high-performance thin-film strain gauge pressure sensor manufactured by OMEGA Engineering Inc. USA is shown in Figure 6.11.

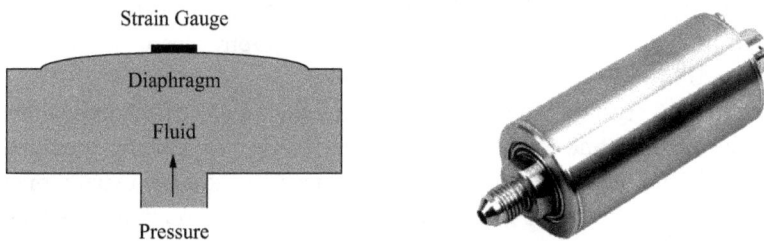

Figure 6.11: High-performance strain gauge pressure sensor.

As shown in Figure 6.11, the sensing element is usually a diaphragm or tube. Its internal volume contains the applied pressure. The element deflects in a predictable manner caused by the fluid pressure, resulting in surface strains and applied force. The strain gauge can be bonded to the nonpressurized surface of the sensing element and respond to the surface strain. The pressure value is obtained by reading the change in output voltage in proportion to the strain gauge resistance change. The strain gauge pressure sensor with millivolt output comes directly from a four-arm Wheatstone bridge strain gauge circuit, with pressures ranging from 10 to 1,000 bar.

Advantages of the strain gauge pressure sensor include the following:
The strain gauge pressure sensor has a fast response time and it is easy to compensate the temperature effects.

Disadvantages of the strain gauge pressure sensor include the following:
The strain gauge pressure sensor has a low level output and does not provide a lower range. Sometimes, there will be creeping because of adhesive agents.

Typical applications of the strain gauge pressure sensor include the following:
The strain gauge pressure sensors are used for measuring pressures of liquids and gases in many fields such as hydraulic pipeline, sealed vessels and gas pressure control.

6.3.2 Piezoelectric pressure sensors

The piezoelectric pressure sensor is a device that uses the piezoelectric effect to convert pressure change into electrical charge for measuring. Piezoelectric effect was discovered by French physicists Pierre Curie and Jacques Curie in 1880. The piezoelectric effect shows the linear electromechanical interaction between the electrical state and the mechanical state in noninverted symmetry crystal materials. This is a reversible process because materials with direct piezoelectric effect also exhibit the reverse piezoelectric effect. The structure of the piezoelectric pressure sensor is shown in Figure 6.12. (The sensor in the photography was made by PCB Piezotronics, Inc., USA.).

There are about 40 crystalline materials that generate an electric charge when strained. Quartz crystal is one of the best choices used in piezoelectric pressure sensors to ensure stable, repeatable operation. The quartz crystals are usually preinstalled in the housings to ensure good linearity. When there is no applied stress on the quartz crystal, the three electric dipole moments are distributed at an angle of $120°$. The positive and negative electron centers coincide, and no charge is generated on the surface of the quartz crystal. As the stress along the x-direction is applied, the vector sum of the three electric dipole moments will be nonzero along the x-direction, resulting in the separation of the positive and the negative electron centers. Therefore, charges are generated on the surface of the unit cell along the x-direction. Similarly, stress along the y-direction and z-direction also

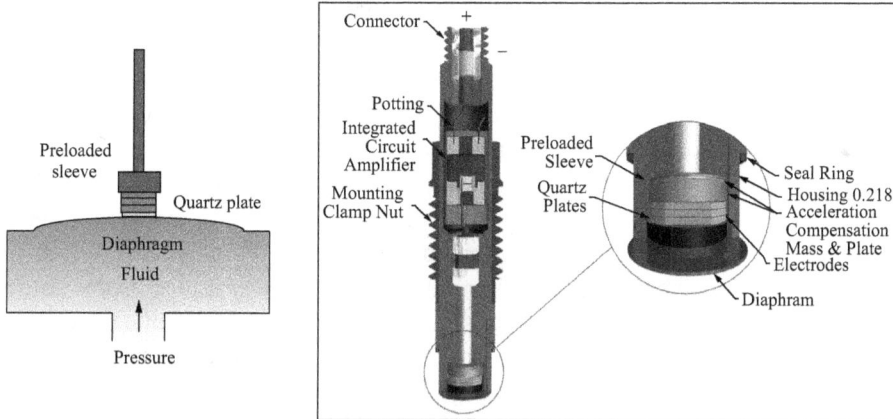

Figure 6.12: Piezoelectric pressure sensor.

results in charges along the x-direction. Therefore, x-direction is called the polarization direction of quartz crystal. In this way, the charge generated on the crystal is proportional to the applied pressure. The resulting electrical signal can be measured as an indication of the pressure applied to the quartz crystal.

Advantages of the piezoelectric pressure sensor include the following:
The piezoelectric pressure sensor is small in size, high in sensitivity, good in frequency response, negligible in phase and structurally sturdy. It can provide the required shape with high mechanical stiffness and a wide frequency range.

Disadvantages of the piezoelectric pressure sensor include the following:
The piezoelectric pressure sensor is only used for dynamic measurement. Some crystals used in the sensor have high-temperature sensitivities and also cannot be used in a high humid environment.

Typical applications of the piezoelectric pressure sensor include the following:
The piezoelectric pressure sensors cannot detect static pressure but can be used to measure rapidly changing pressures, such as turbulent explosions, explosions, pressure pulsations and other sources of vibration.

6.3.3 Piezoresistive pressure sensors

Due to its high sensitivity and linearity, a piezoresistive pressure sensor is considered to be the first MEMS devices on the market. Various industries implement piezoresistive pressure sensors as a good means. Piezoresistive pressure sensors are used for measuring pressure through the strain change on a silicon diaphragm based on piezoresistive principle. The piezoresistive principle can be explained by the fact that a strip of an elastic material tends to move due to some source. If an increase appears in the longitudinal dimension, a decrease will appear in the lateral dimensions and the cross-sectional area. If this is a positive strain, the resistance value will change for the piezoresistive effect. The pressure sensor consists of diffusive piezoresistive silicon diaphragm in a Wheatstone bridge and silicon diaphragm, which is used for converting pressure into mechanical stress. The piezoresistor converts stress into resistance, and eventually the resistance change is converted into an output voltage. In this way, the pressure can be measured. The structure of the piezoresistive pressure sensor structure is shown in Figure 6.13. (The sensor in the photography was made by MICRO SENSOR Co. China).

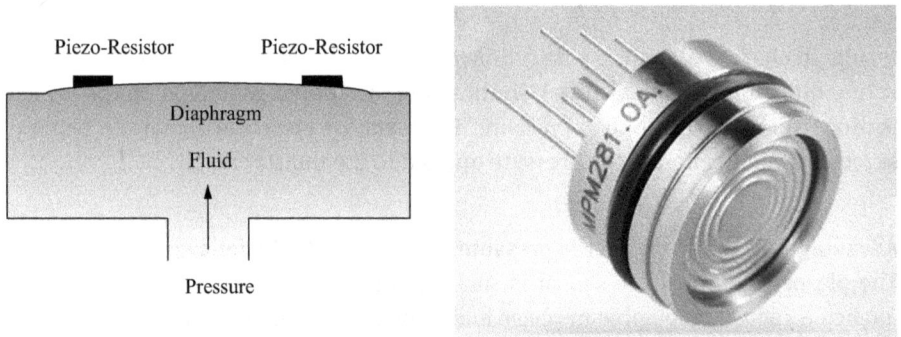

Figure 6.13: Typical piezoresistive pressure sensor structure.

Advantages of the piezoresistive pressure sensor include the following:
The piezoresistive pressure sensor has a high sensitivity (>10 mV/V), good linearity at a constant temperature, low hysteresis and low cost.

Disadvantages of the piezoresistive pressure sensor include the following:
The piezoresistive pressure sensor has a large initial offset, strong nonlinear dependence of the full-scale signal on temperature and large drift.

Typical applications of the piezoresistive pressure sensor include the following:
The piezoresistive pressure sensors are used in household appliances (washing machines and vacuum cleaners), gas/liquid pressure system, oil/gas level detection, blood pressure measurement, tire pressure monitoring systems, industrial process control, aviation and navigation inspection and hydraulic/pneumatic suspension systems.

6.3.4 Capacitive pressure sensors

A capacitance pressure sensor is designed according to the distance change between the plates. The effective area of the plates and the relative permittivity of the dielectric produce a change in the capacitance that reflects the output voltage, where the pressure can be derived. The measurement of the capacitance is accomplished by a resonance circuit. The schematic of the capacitance pressure sensor is shown in Figure 6.14. (The sensor in the photography was made by BCM Sensor Technologies, Belgium.) The capacitance pressure sensor has a capacitive plate (diaphragm), and the other capacitive plate (electrode) is fixed on an unpressurized surface at a distance from the diaphragm. A pressure change will narrow or widen the gap between the two plates, thereby changing the capacitance (ΔC). The change in capacitance that can be mathematically related to the applied pressure can be used for controlling an oscillator's frequency or changing the coupling of an AC signal through the network, and then converted it into a usable signal. In this way, the pressure can be measured.

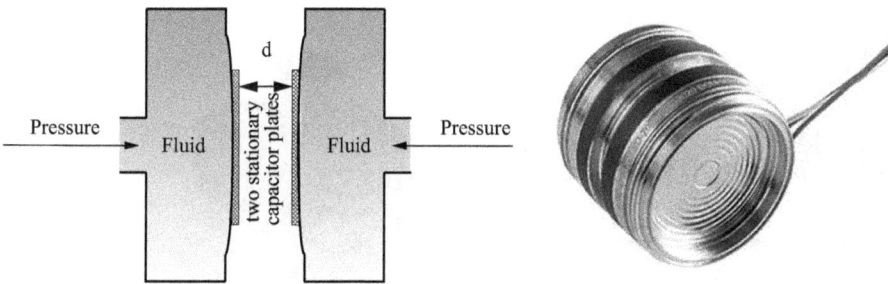

Figure 6.14: Capacitance pressure sensor.

Advantages of the capacitance pressure sensor include the following:
The capacitance pressure sensor has a good frequency response, good linearity, low hysteresis, small loading effect, excellent stability and repeatability.

Disadvantages of the capacitance pressure sensor include the following:
The capacitance pressure sensor has a high sensitivity to temperature and nonlinear relation between the capacitance and pressure. It is difficult to avoid mechanical noise. In addition, the capacitance may change due to the generation of erroneous dust particles and moisture.

Typical applications of the capacitance pressure sensor include the following:
The capacitance pressure sensors are used for measuring positive pressure up to 70 MPa, absolute vacuums, differential pressures, atmospheric pressure and low pressures (0–0.1 MPa).

6.3.5 Variable reluctance pressure sensors

Variable reluctance pressure sensor consists of a pressure sensing diaphragm of a magnetic material. The deflection of the diaphragm controls the air gap in each of two magnetic circuits, as shown in Figure 6.15. (The sensor in the photography was made by Validyne Engineering, USA.) Reluctance is used as a resistance in the magnetic circuit. If the change in pressure changes the gaps between the magnetic flux paths of the two cores, the inductance ratio L_1/L_2 will be related to the change in pressure. The amplitude of the displacement is proportional to the output voltage that is used to indicate the applied pressure. The variable reluctance pressure sensor has a very high output signal and must be excited by the AC voltage. For the high output signals, variable reluctance pressure sensors can be used for applications requiring a relatively small range of high resolution. The range of variable reluctance pressure sensor is from 250 Pa to 70 MPa. Some reluctance pressure units can use replaceable diaphragms to facilitate the replacement of an over-ranged diaphragm.

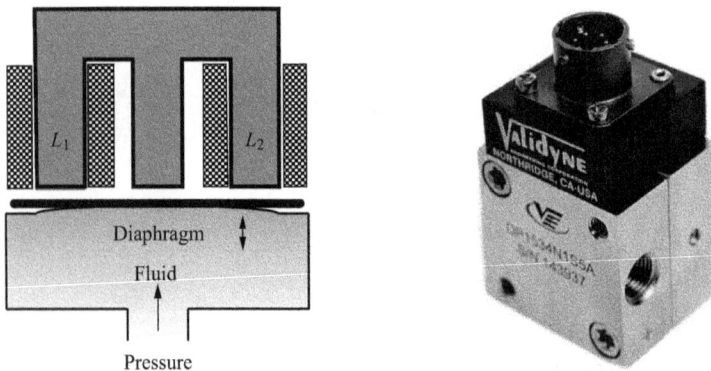

Figure 6.15: Typical structure of variable reluctance pressure sensor.

Advantages of the variable reluctance pressure sensor include the following:
The variable reluctance pressure sensor has a wide temperature range, low sensitivity to vibration and shock, fast response and high output signal. It can be considered as a durable, flexible and rugged sensor.

Disadvantages of the variable reluctance pressure sensor include the following:
Sensitivity of the variable reluctance pressure sensor is affected by the temperature (approximately 2%/100 °C) and the stray magnetic fields. In addition, AC excitation is required in the measuring process.

Typical applications of the variable reluctance pressure sensor include the following:
The variable reluctance pressure sensors are used for measuring the pressure of liquid or gas in a compressed tank or a hydraulic system.

6.3.6 Potentiometric pressure sensors

The potentiometric pressure sensor is one of the basic types of a pressure sensor, which provides a simple way for obtaining an electronic output from a mechanical pressure gauge shown in Figure 6.16. (The sensor in the photography was made by LEEG Instruments Co., Ltd, China.) The device consists of a resistance element having a connection terminal, a sliding track that is connected to the third terminal, the wiper and the housing. The wiper arm is mechanically connected to the bellows, bourdon or a diaphragm element. As the wiper moves, it increases the resistance between the terminals, which is equivalent to the pressure. On the potentiometer, the movement of the wiper arm utilizes a Wheatstone bridge circuit converting the sensor deflection detected mechanically into a resistance measurement. The pressure measurement result is then obtained from the output voltage that reflects the resistance value.

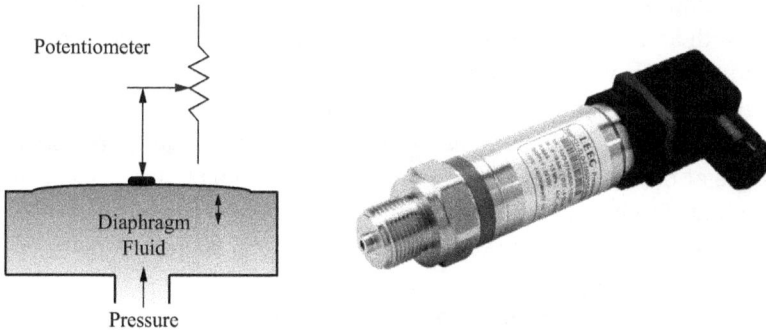

Figure 6.16: Typical potentiometric pressure sensor structure.

Advantages of the potentiometric pressure sensor include the following:
The potentiometric pressure sensor is simpler and less expensive than other types of sensors. Its resistance can be easily converted into a standard voltage or current signal to meet a special requirement.

Disadvantages of the potentiometric pressure sensor include the following:
The potentiometric pressure sensor has a limited resolution (the wiper does not move continuously along the wire). When the wiper moves back and forth on the wire, wear is not inevitable and this creates hysteresis and repeatability errors.

Typical applications of the potentiometric pressure sensor include the following:
The potentiometric pressure sensor is used in industrial and military fields such as measuring the gas flow rate of internal combustion engines, the pressure of fuel gauges and airport ground support equipment.

6.3.7 Bourdon tube pressure sensors

Bourdon tube pressure sensors are the most frequently used mechanical pressure measuring instruments. Bourdon tube was invented by Eugene Bourdon in 1849. The basic idea is that under any form of deformation, the cross-sectional tubing tends to recover its circular form under the action of pressure. The bourdon pressure sensors usually have a slightly elliptical cross section. The tube is usually bent into a C-shape or arc about 270°. The photo of the Bourdon tube pressure sensor is shown in Figure 6.17. (The sensor in the photography was made by SIKA Dr. Siebert & Kühn GmbH & Co. KG, Germany.) A gear mechanism of bourdon tube pressure sensor is shown on the left figure. The pressure is applied to the socket. The other end of the device is sealed by a tip. This prompt connects to a section that is linked by an adjustable length. The segment lever rotates and the spindle fixes the pointer. A mainspring is sometimes used to hold the main shaft of the device frame, which provides the necessary tension to actually engage the gear teeth and free the system to free play. When the fluid pressure enters the bourdon tube, it will attempt to modify it, and due to the available tip, this action will cause the tip moving in free space, and the tube will unwind. The device's main requirement is that the movement of the tip should be the same and the pressure at the tip should return to the initial point as long as the same pressure is applied. The bourdon tube pressure sensor is known for its very high differential pressure measurement range. Its measurement range can reach almost 700 MPa.

In addition to the C shape, the bourdon tube pressure sensor can also be constructed using a helix or spiral, as shown in Figure 6.18. (The sensor in the photography was made by Foxboro, France). They are used for very high pressures, continuously fluctuating pressures and other heavy-duty applications. Helix bourdon tubes are made by winding a few turns of a tube into a helix. There may only be two or three turns or up to 20 layers. The helix shape produces maximum mechanical motion for each applied force, and the increase in sensitivity is changed into an improvement in measurement accuracy. The spiral bourdon tube is made by winding a partially flat metal tube into a spiral shape. The spiral has multiple turns instead of a single C-curved arc. The tip motion of the spiral is equal to the sum of the tip motions of all single C-shaped arcs.

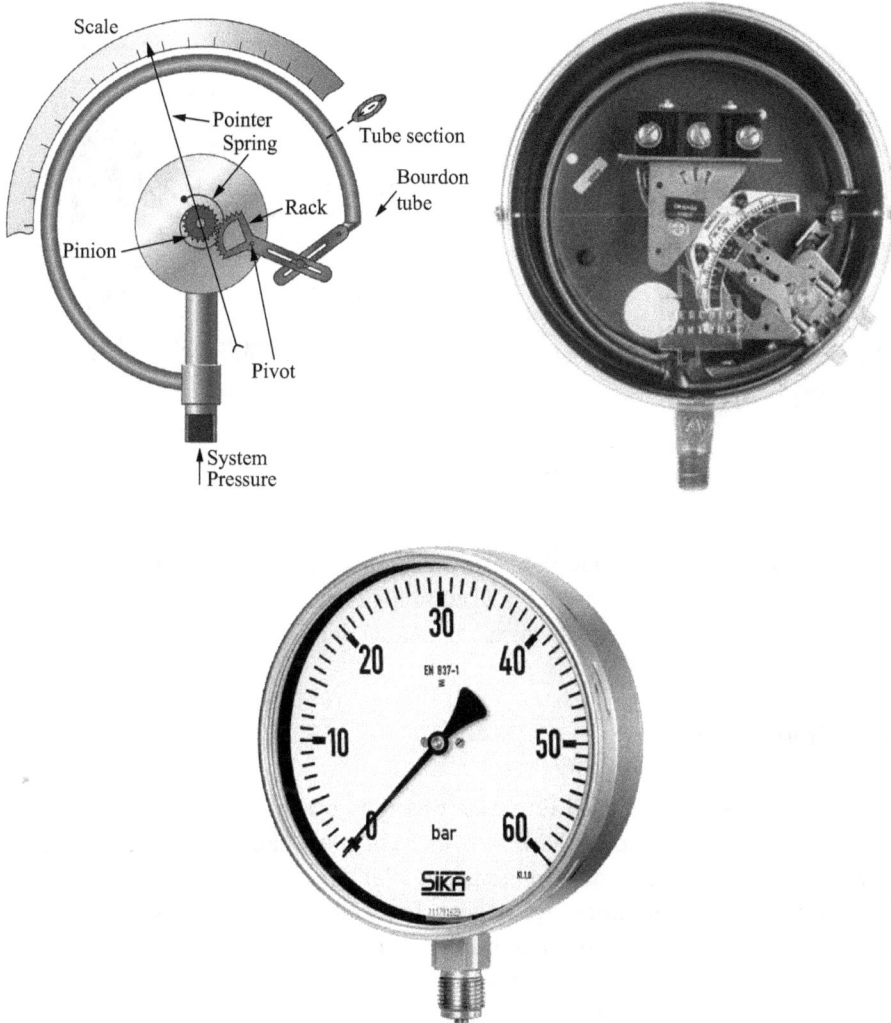

Figure 6.17: The C-shape bourdon tube pressure sensor.

Advantages of the bourdon tube pressure sensor include the following:
The bourdon tube pressure sensor has a simple construction, low cost, high accuracy, excellent sensitivity and so on. It is available for various ranges and can be modified to provide improved electrical output.

Disadvantages of the bourdon tube pressure sensor include the following:
The bourdon tube pressure sensor has hysteresis errors, slow response, low susceptibility to shock and vibration and so on. In addition, it is not suitable for low-pressure applications.

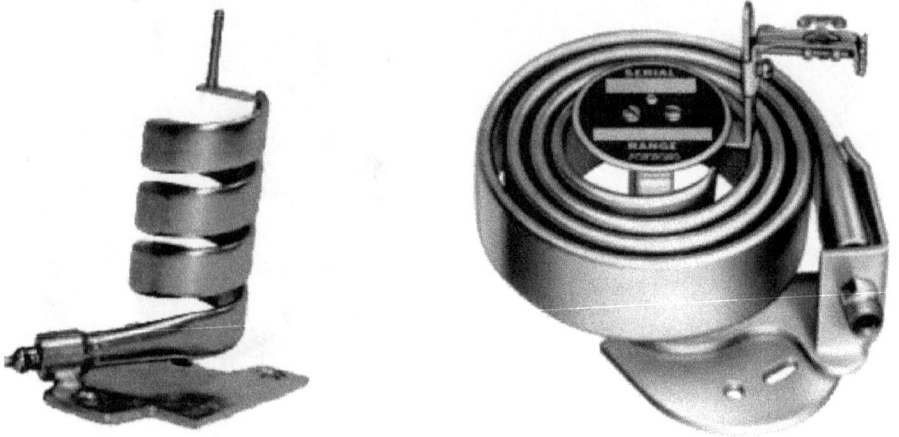

Figure 6.18: The helix and spiral bourdon tube.

Typical applications of the bourdon tube pressure sensor include the following:
The bourdon tube pressure sensor is used in a gas distribution system, hydraulic/
pneumatic installations, petrochemical processing and so on.

6.4 Flow sensors

A flow sensor is a device used for measuring the flow rate, quantity of a gas or a
liquid moving through a pipe. Flow sensors have many names, such as flowmeter,
flow gauge, flow indicator and liquid meter, which are used in the HVAC (Heating
Ventilation and Air Conditioning) system, fuel cells, gas analyzers, low vacuum
control, process control, filter monitoring and extraction hoods or gas measuring
stations. Different types of flow sensors discussed in this chapter are differential
pressure flow element, turbine flowmeter, vortex flowmeter, thermal flowmeter,
Doppler flowmeter, transit time flowmeter and laser Doppler anemometer.

6.4.1 Differential pressure flowmeter

Differential pressure flowmeters measure the flow of a fluid or a gas based on the
Bernoulli's equation by measuring the pressure difference created by placing a structure (E.G., an orifice plate or a Venturi tube or a flow nozzle) in a fluid or a gas flow
and measuring the difference between upstream and downstream. Pressure drop is a
square function of the fluid velocity. As the flow increases, more pressure drops occur.
Fluid flow can be determined by measuring the differential pressure. Differential pressure flowmeter is the most commonly used instrument for measuring fluid flow

through a pipe. The orifice plate, Venturi tube and flow nozzle are three common types, as shown below.

Orifice plate is widely used in flow measurement because it is the simplest and most economical flow detection method. The orifice plate consists of a circular metal disk with a specific hole diameter, which reduces the fluid flow in the tube shown in Figure 6.19. (The sensor in the photography was made by Ztech Control System, India.) The orifice plate is eccentric, and the opening of the orifice is offset from the center of the pipe. On each side of the orifice plate, a pressure tap was added for measuring the differential pressure. According to the loss of energy conservation, the fluid in the pipeline must equal the mass of the fluid, leaving the pipeline at the same time. The velocity of the fluid exiting the orifice is larger than that entering the orifice. Differential pressure flowmeter relies on Bernoulli's equation. The volumetric flow in Figure 6.19 can be written as follows:

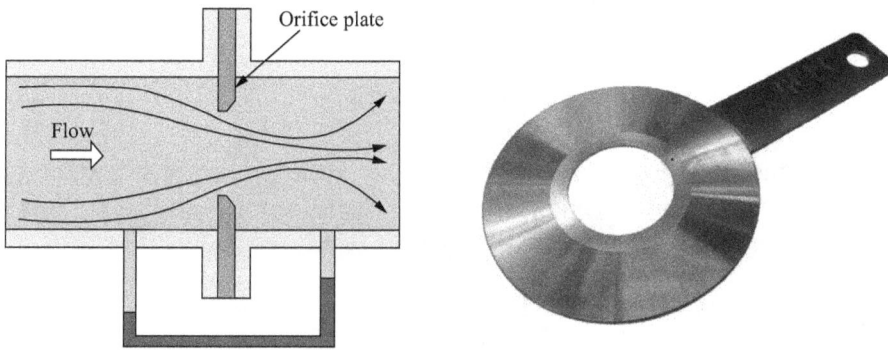

Figure 6.19: The construction of the orifice plate.

$$Q_v = C_d A_o \sqrt{2gh} \qquad (6.1)$$

where Q_v is the volumetric flow rate, C_d is the coefficient of discharge, A_o is the area of orifice, g is the gravitational constant and h is the differential pressure.

Equation (6.1) clearly shows that volume flow rate is proportional to the square root of the pressure drop.

The Venturi tube is an ideal flow measurement device for nonviscous fluids, containing hydrocarbons and other potentially corrosive elements. The device is based on the Venturi effect and is named after the Italian physicist Giovanni Battista Venturi (1746–1822). The Venturi effect occurs when fluid flowing through a pipe is forced through a narrow portion, resulting in a drop in pressure and an increase in speed. The structure of the Venturi tube is shown in Figure 6.20. (The sensor in the photography was made by EMCO Controls, Denmark.). The pressure sensor can be used for measuring the pressure, and then the fluid velocity can be

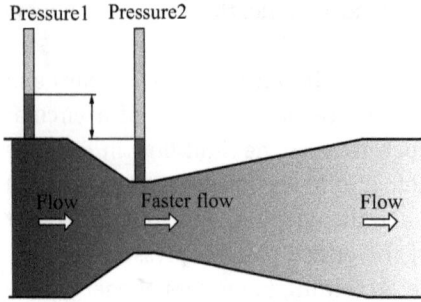

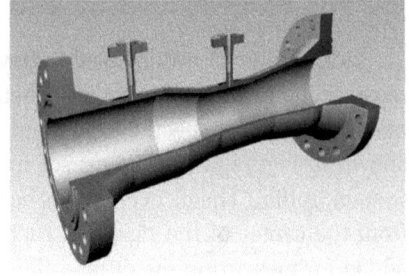

Figure 6.20: The construction of the Venturi tube.

calculated using the Bernoulli equation. The velocity of the fluid is inversely proportional to its pressure. In the Venturi tube, the flow rate is measured by reducing the cross-sectional flow area in the flow path, creating a pressure difference. The pressure difference is given by

$$p_1 - p_2 = \frac{\rho}{2}\left(v_2^2 - v_1^2\right) \tag{6.2}$$

where ρ is the fluid density, v_2 is the fluid velocity where the pipe is narrower; v_1 is the fluid velocity where the pipe is wider.

The volumetric flow rate Q_V can be written as

$$Q_V = v_1 A_1 = \frac{v_1}{4} D_1^2$$
$$= v_2 A_2 = \frac{v_2}{4} D_2^2 \tag{6.3}$$

$$Q_V = \sqrt{\frac{2}{\rho} \frac{(p_1 - p_2)}{(A_1/A_2)^2 - 1}} A_1 \tag{6.4}$$

where Q_V is the volumetric flow rate, D_1 is the diameter of pipe, D_2 is the diameter of throat/constriction, A_1 is the area of the pipe and A_2 is the area of throat/constriction.

In industrial applications, flow nozzles are commonly used as measuring elements for air and gas flow in industrial applications, which cause pressure drops by creating restrictions in the fluid. Flow nozzles are used for high-speed flow measurement, where erosion or cavitation may wear or damage the orifice plates. The flow nozzle is shown in Figure 6.21. (The sensor in the photography was made by EMCO Controls, Denmark.) Flow nozzles do not rely on sharp edges to maintain accuracy, so they have excellent long-term accuracy, less wear and less chance of deformation. The basic formula for the flow nozzle is the same as which for the Venturi tube.

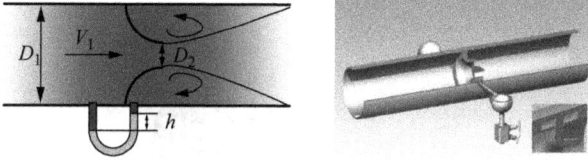

Figure 6.21: The construction of the flow nozzle.

Advantages of the differential pressure flowmeter include the following:
Differential pressure flowmeter is cheap, maintenance free, easy to install or remove, adaptable to different flow rate and pipe sizes, simple and reliable. It has a robust structure, no moving parts and stable performance.

Disadvantages of the differential pressure flowmeter include the following:
Differential pressure flowmeter is easy to leak, difficult to improve the measurement accuracy, difficult for measuring the flow rate of pulsating flow and slurries and inconvenient to install on site. In addition, due to the square root relationship between flow and indenter, its flow range capability is low.

Typical applications of the differential pressure flowmeter include the following:
Differential pressure flowmeters are suitable for measuring the flow rate of both liquid and gas, which are typically utilized in a range of applications, including petroleum, pulp and paper, mining, water and wastewater industries, mineral processing and chemical industries.

6.4.2 Turbine flowmeter

Turbine flowmeter is a volumetric flowmeter that measures the flow of liquids, gases and vapors in a pipeline. The construction of the turbine flowmeter is shown in Figure 6.22. (The sensor in the photography was made by FTI Flow Technology Inc., USA.). The flow sensing element is a freely suspended bladed rotor, which is axially positioned in the flow stream with the flowing fluid that drives the blades. Turbine flowmeters use the mechanical energy of the fluid to rotate the rotor in the flow. When the fluid moves faster, the rotor rotates proportionally faster. The rotational speed of the rotor is proportional to the velocity of the fluid, and the velocity of the fluid can be measured by the electrical pulses in the detector, which are connected to the flowmeter housing close to the rotating rotor. Each pulse represents a discrete fluid volume. The pulse repetition rate or frequency represents the volumetric flow rate. The total number of cumulative pulses represents the total volume of the measurement. In addition, turbine

Figure 6.22: The construction of the turbine flowmeter.

flowmeters should not run at high speeds and applied in dirty liquids to prevent damage to the flowmeter and reduce accuracy.

Advantages of the turbine flowmeter include the following:
Turbine flowmeter is simple, easy to install/operate, suitable for gas or liquid and it has a low cost, good performance and wide operational envelope.

Disadvantages of the turbine flowmeter include the following:
Turbine flowmeter requires clean fluid or gas, and its accuracy is easy to be affected by bearing degradation, viscosity change and cavitation. Calibration work must be done frequently to avoid measuring errors.

Typical applications of the turbine flowmeter include the following:
Turbine flowmeters are typically used in mission-critical applications that require superior accuracy, quality and reliability, such as monitoring clean liquid flows in process chemical (cryogenic liquids, air), petroleum (hydrocarbons and natural gas) and water industries (deionized water, hydraulic fluid).

6.4.3 Vortex flowmeter

Vortex flowmeter is a flow measurement device for measuring the flow based on the vortex shedding principle. The structure of the vortex flowmeter is shown in Figure 6.23. (The sensor in the photography was made by Yokogawa Corp., Japan.) In fluid dynamics, vortex shedding is an oscillating flow, which depends on the

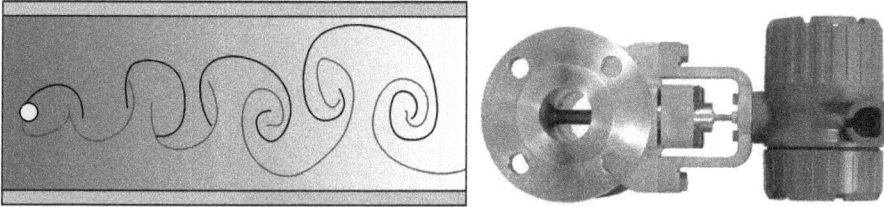

Figure 6.23: The construction of the vortex flowmeter.

size and shape of the body when a fluid such as water or air flows through the bluff body at a certain speed. In this process, vortices are created at the back of the body and are periodically separated from both sides of the body. Theodore von Karman, a Hungarian-American physicist, was the first to describe the effect of placing non-streamline objects in a fast-flow path, which results in the fluid separating alternately from the object on the two downstream sides. This process generates a swirl on one side of the object and vortexes on the other side as the process alternates. Many vortex flowmeters use piezoelectric or capacitive sensors for detecting pressure oscillations around the bluff body. The response of these detectors to pressure oscillations with a low voltage output signal has the same oscillation frequency.

Advantages of the vortex flowmeter include the following:
The vortex flowmeter has a low maintenance cost, low sensitivity to changes in process conditions and low wear rates relative to orifice or turbine flowmeters.

Disadvantages of the vortex flowmeter include the following:
The vortex flowmeter is not good to measure multiphase flow (air bubbles in liquid, solid particles in gas or liquid, droplets in gas), batching flow, intermittent flow or slurries or high viscosity flow. In addition, it has poor accuracy when measuring low pressure (low density) gases.

Typical applications of the vortex flowmeter include the following:
The vortex flowmeter is well suited for low-cost flow measurement of gas and liquid. It is the most popular device for measuring steam in applications.

6.4.4 Thermal flowmeter

The thermal flowmeter is a precision instrument that is used for measuring the flow rate of a fluid flowing in a pipe, which is based on the concept of convective heat transfer. Thermal flowmeters are commonly used for measuring the flow of pure gases. The construction of the thermal flowmeter is shown in Figure 6.24. (The sensor in the

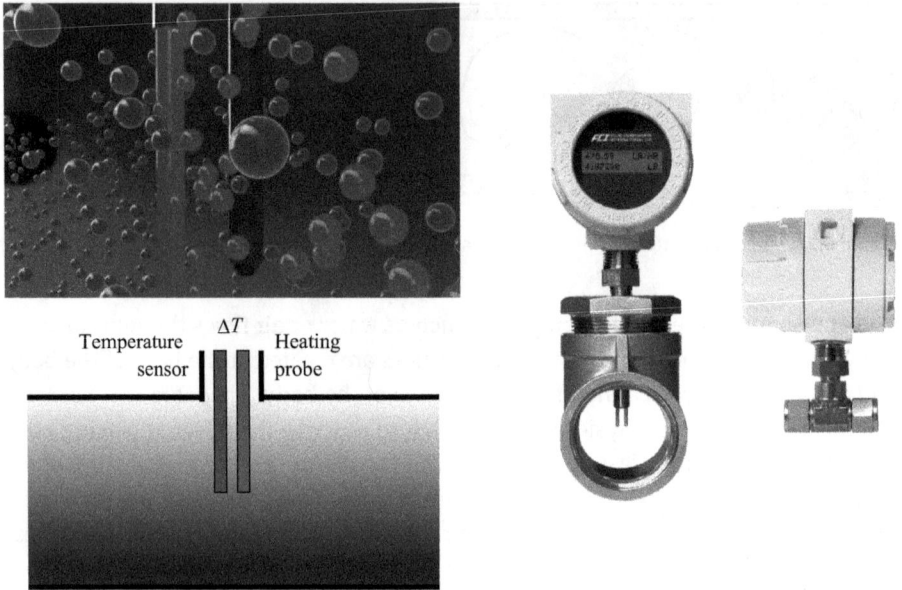

Figure 6.24: The construction of the thermal flowmeter.

photography was made by Fluid Components International, USA.). A typical thermal flowmeter uses two temperature sensors: one sensor is heated by the IC while the other sensor serves as a reference sensor and determines the gas temperature. Special circuits maintain continuous overheating between flow and reference sensors. As the gas flows through the heating sensor, the flowing gas molecules transport heat away from the sensor and, as a result, a portion of the heat is lost to the flowing fluid. The circuit balance is disturbed, and the temperature difference between the heated sensor and the reference sensor has been changed. In 1 s, the circuit regulates the overheating at this temperature by heating the flow sensor to recover the lost energy. The transmitter uses heat input and temperature measurements to determine the fluid flow. As the thermal flow measurement does not depend on the pressure or temperature of the fluid, thermal flowmeters can be used for measuring the mass flow of fluid.

Advantages of the thermal flowmeter include the following:
The thermal flowmeter has excellent accuracy, good repeatability, wide measurement range, short response time, no moving parts, low-cost maintenance and so on. In addition, it can measure flow in large pipes and is insensitive to vibrations.

Disadvantages of the thermal flowmeter include the following:
The thermal flowmeter is not suitable for fluids with varying composition and unknown components (hydrogen-bearing, off-gases) or abrasive fluids (may damage

the sensor). If the fluid composition of actual flowing is different, the accuracy of the thermal flowmeter calibrated for a given fluid mixture will be reduced.

Typical applications of the thermal flowmeter include the following:
The thermal flowmeters are most commonly applied to measure clean, sanitary and corrosive gases with known thermal properties, such as air or compressed air (oxygen, CO_2), ammonia, nitrogen, helium, hydrogen, argon and other industrial gases. It can also be used for laboratory experiment, and in monitoring or controlling quality-related processes such as chemical reactions.

6.4.5 Doppler ultrasonic flowmeter

Doppler ultrasonic flowmeter is a flowmeter used for measuring the flow rate of a liquid or a gas on the principle of ultrasonic waves. It is a very accurate noncontact liquid flow measurement device. The construction of the Doppler ultrasonic flowmeter is shown in Figure 6.25. (The sensor in the photography was made by Badger Meter, Inc., USA.) The basic principle of the sensor is the Doppler effect when the ultrasonic signal is reflected in motion. It was first invented by the Austrian physicist, Christian Doppler in 1842. Doppler effect is the change in the frequency of a wave when the observer moves relative to the source of the wave. All Doppler ultrasonic flowmeters use two piezoelectric crystals in one sensor to transmit ultrasonic energy to the fluid and receive sound from reflectors in the liquid. An ultrasonic wave is transmitted into the pipe with the flowing liquid, and the discontinuity reflects the ultrasonic wave at a slightly different frequency proportional to the flow rate of the liquid. The magnitude of the frequency change is proportional to the speed of the reflector. In this way, the flow rate of a liquid or gas can be measured.

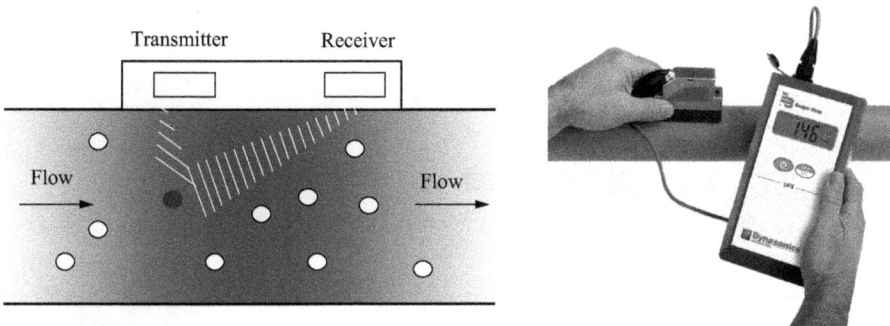

Figure 6.25: The construction of the Doppler ultrasonic flowmeter.

Advantages of the Doppler ultrasonic flowmeter include the following:
Doppler ultrasonic flowmeter is noninvasive, portable and perfect for slurry and aerated liquids. This is the best choice for wastewater applications or any water-based or conductive dirt water and applications requiring low-pressure drop, low maintenance and chemical compatibility.

Disadvantages of the Doppler ultrasonic flowmeter include the following:
Doppler ultrasonic flowmeter is easily affected by the acoustic properties of the fluid and may be affected by density, temperature, viscosity and suspended particulates. It does not apply to clean liquids (distilled water or drinking water).

Typical applications of the Doppler ultrasonic flowmeter include the following:
Doppler ultrasonic flowmeter is used for measuring the flow of slurries, wastewater, liquids with bubbles, acids, gases with sound-reflecting particles, chemicals and viscous liquids in the chemical, mining, petroleum and petrochemical fields. This flowmeter can also be used for measuring the rate of blood flow.

6.4.6 Transit-time ultrasonic flowmeter

Transit-time ultrasonic flowmeter like Doppler ultrasonic flowmeter is a nonintrusive in-line device that uses acoustic vibration to measure the flow rate of the liquid. The construction of the Doppler ultrasonic flowmeter is shown in Figure 6.26. (The sensor in the photography was made by Greyline Instruments Inc., USA.) The transit-time ultrasonic flowmeter must have a pair of sensors, each containing a piezoelectric crystal. It is transmitted by one sensor and received by another sensor. When the fluid in the pipeline is not flowing, upstream and downstream transit time will be the same. When the fluid in the pipeline is not flowing, the sound travels faster in the direction of flow and slower against the flow. The flow rate of the fluid will be proportional to the difference between the upstream and downstream measurements.

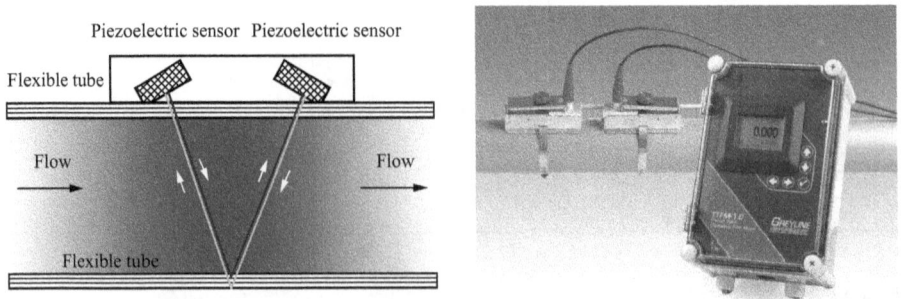

Figure 6.26: The construction of the transit-time ultrasonic flowmeter.

Advantages of the transit-time ultrasonic flowmeter include the following:
The transit-time ultrasonic flowmeter has a wide range of pipe sizes, low installation cost, no maintenance cost and good long-term performance. The transit-time ultrasonic flowmeter is noninvasive (no disruptions to the flow), and it becomes especially cost-effective in large diameter pipelines.

Disadvantages of the transit-time ultrasonic flowmeter include the following:
The transit-time ultrasonic flowmeter requires that the liquid should be acoustically clean enough so that the signal can pass between the two transducers. Therefore, fluids cannot contain significant concentrations of bubbles or solids. Otherwise, the high-frequency sound will be attenuated and too weak to pass through the pipe.

Typical applications of the transit-time ultrasonic flowmeter include the following:
The transit-time ultrasonic flowmeters are widely applied in measuring settled water, backwash, oils, supernatant and chemical additives. The flowmeters have also been widely applied in large-diameter flow transfer.

6.4.7 Laser Doppler anemometer

A laser Doppler anemometer, also known as a laser Doppler velocimeter, is a type of interferometer that used for measuring the instantaneous velocity of a flow field based on Doppler effect. The construction of laser Doppler anemometer is shown in Figure 6.27. (The sensor in the photography was made by Measurement Science Enterprise, Inc., USA.) The main part of the equipment consists of a laser source, a beam splitter, two lenses and a light detector. The laser source sends a beam that is split into two beams by a beam splitter. The two coherent laser beams intersect to create an interference pattern. Flow velocity information is derived from the

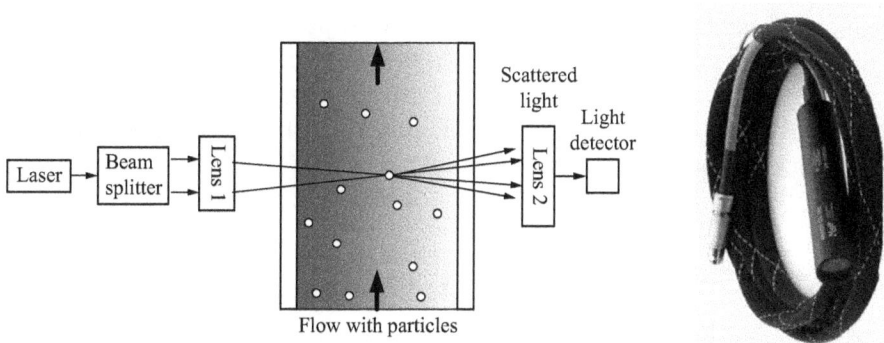

Figure 6.27: The construction of the laser Doppler anemometer.

scattering of the light caused by tiny "seeding" particles carried in the fluid as they pass through the probe volume. The scattered light experiences a Doppler shift in frequency that is proportional to the flow rate. The scattered light is detected by a light detector used for generating a current in proportion to the absorbed photon energy to detect the scattered light. After that, the current will be amplified. The difference between the frequencies of incident and scattered light is called the Doppler shift. The local velocity of the fluid can be determined by analyzing the Doppler-equivalent frequency of the laser scattered by the seeded particles in the flow.

Advantages of the laser Doppler anemometer include the following:
The laser Doppler anemometer has a high accuracy, high-frequency response and small sensing part. It does not disturb the flow during measurement (noncontacting measurement) and can measure flow directly.

Disadvantages of the laser Doppler anemometer include the following:
The laser Doppler anemometer is expensive, and it requires sufficient transparency between the laser source, the target surface and the light detector. In addition, this device cannot be used in clean flows.

Typical applications of the laser Doppler anemometer include the following:
The laser Doppler anemometer can be used for measuring the flow of both gases and liquids (like blood, wind tunnel, microchannels, oil tunnels and high-frequency turbulence fluctuations). It is very suitable for applying where physical sensors are difficult or impossible to use, such as chemical reactions, reverse flow or high-temperature media and rotating machinery.

6.5 Displacement sensors

Displacement sensors can be used for dimensional measurement to determine the range of travel and height, thickness, width and 2D shape of the object. Different types of displacement sensors discussed in this chapter are LVDT (linear variable displacement transformer) displacement sensor, capacitive displacement sensor, inductive displacement sensor, laser displacement sensor, resistive displacement sensor, ultrasonic displacement sensor and grating displacement sensor.

6.5.1 LVDT displacement sensor

LVDT displacement sensor is a device used for position and displacement measurement. It is one of the most rugged and reliable position sensors and converts mechanical motion into variable electrical current or voltage signal. The construction of LVDT displacement sensor is shown in Figure 6.28. (The sensor in the photography was made by OMEGA Engineering Inc. USA.) A typical LVDT displacement sensor has three end-to-end electromagnetic coils. The primary coil is in the center, and the secondary coils are above and below. The purpose of position measurement is to attach to the cylindrical iron core and slides along the axis of the tube. The two secondary voltages of the displacement sensors are equal at the center of the position measurement stroke, but the output of the sensor is zero as they are inversely connected. As the iron core moves away from the center, one of the position sensor secondaries increases and the other decreases. The voltage of the two secondary windings of the alternating current driving primary coil is proportional to the length of the connecting iron core. Therefore, it is possible to obtain a voltage or current output proportional to the measurement position.

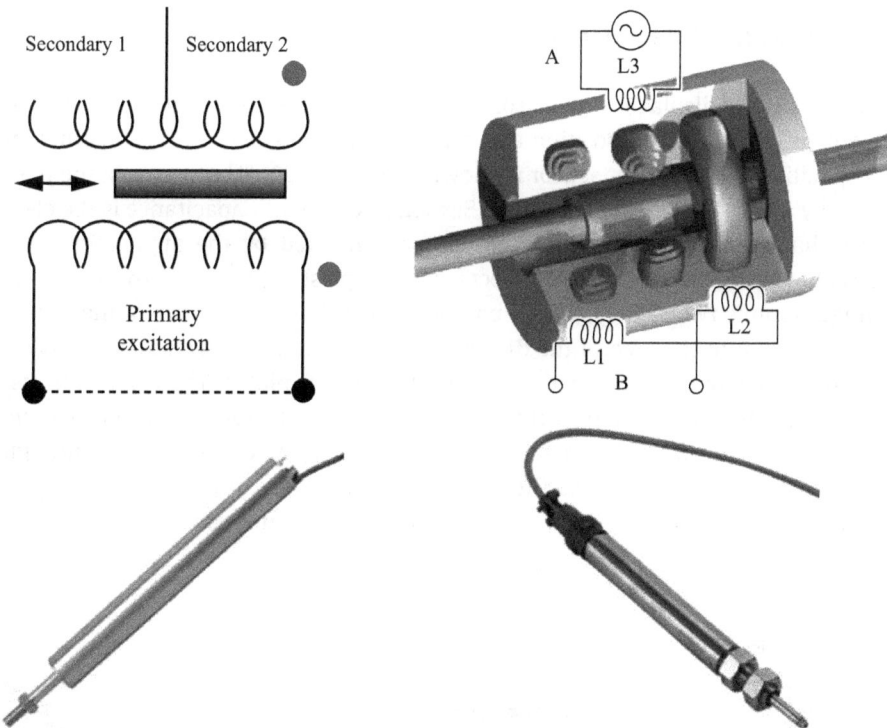

Figure 6.28: The construction of the LVDT displacement sensor.

Advantages of the LVDT displacement sensor include the following:
The LVDT displacement sensor has good linearity, high sensitivity, infinite resolution, low hysteresis, good durability and low power consumption. LVDT

displacement sensor is also very reliable since the sliding core does not touch the inside of the tube.

Disadvantages of the LVDT displacement sensor include the following:
The LVDT displacement sensor is easily affected by the magnetic field, vibrations and temperature changes. In addition, large range (>300 mm) is required for generating high voltages.

Typical applications of the LVDT displacement sensor include the following:
The LVDT displacement sensor is used from precision dimensional gauging to the monitoring of fault-line movement. Many applications can be found in aerospace, flush diaphragm, heavy-duty/industrial, pulp and paper, industrial wastewater neutralization and other process control systems.

6.5.2 Capacitive displacement sensor

The capacitance displacement sensor is a noncontact measurement device for displacement, distance and position and thickness measurements. The construction of capacitive displacement sensor is shown in Figure 6.29. (The sensor in the photography was made by Telemecanique Sensors, Germany.) Capacitance is the electrical characteristic produced by the charge generated by the application of a charge between two conductive objects. The simplest capacitive position sensor configuration is two closely spaced parallel plates. The capacitance of the sensor is proportional to the area of the dielectric constant and the electrodes and is inversely proportional to the space between the parallel plates. The capacitive displacement sensor is inspired by the electrical property of capacitance between the sensor and the target. The constant frequency alternating current passes through the capacitor of the sensor, and the amplitude of the alternating voltage is proportional to the distance between the sensor and the target. In this way, the displacement can be measured.

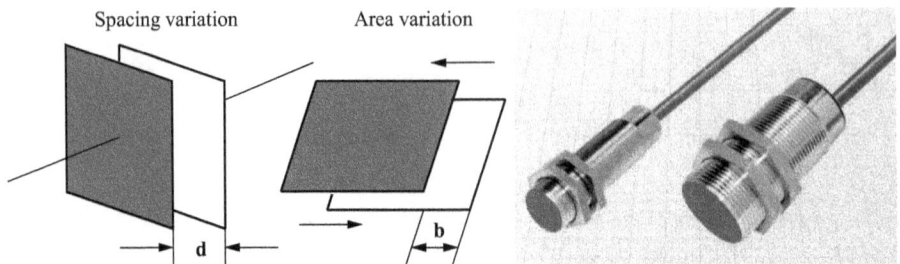

Figure 6.29: The construction of the capacitive displacement sensor.

Advantages of the capacitive displacement sensor include the following:
The capacitive displacement sensor has a superior resolution (down to nanometer level), excellent temperature stability and long-term stability. It is completely non-contact. Capacitive displacement sensor works with conductive, grounded target and offers high measurement bandwidth (up to 100 kHz).

Disadvantages of the capacitive displacement sensor include the following:
The capacitive displacement sensor is sensitive to environmental factors. It is not a good choice for measuring large distances.

Typical applications of the capacitive displacement sensor include the following:
The capacitive displacement sensor is applied in laboratories and industrial measurement tasks. It is commonly applied in position measurement (automation, semiconductor processing, disk drives), dynamic motion measurement (disk drive spindles, tool spindles, high-speed drill spindles and ultrasonic welding machines) and thickness (brake rotor, silicon wafer, disk drive platter and label positioning).

6.5.3 Inductive displacement sensor

The inductive displacement sensor is a noncontact device used for measuring the position of any conductive object with high resolution and is also referred to as eddy-current sensor in this textbook. The construction of the inductive displacement sensor is shown in Figure 6.30. (The sensor in the photography was made by KEYENCE Corp., UK, and Micro-Epsilon, Germany.) An alternating current is generated by the driver in the sensing coil at the end of the probe. An alternating magnetic field is generated in the target material to induce small currents (eddy currents). The eddy current loss increases, and the oscillation amplitude decreases when the target approaches the sensor head. A direct voltage variation proportional to the distance change between the probe and the target is obtained by correcting the amplitude of the oscillation.

Advantages of the inductive displacement sensor include the following:
The inductive displacement sensor is insensitive to material in the gap between the probe and the target. Compared with other noncontact sensing technologies, it is cheaper and smaller. In addition, it can be used in dirty environments.

Disadvantages of the inductive displacement sensor include the following:
The inductive displacement sensor can only detect metallic targets. In these conditions, the inductive displacement sensor is not a good choice for extremely high resolution (capacitive sensor is ideal) and a large gap between the sensor and the target (optical laser is better).

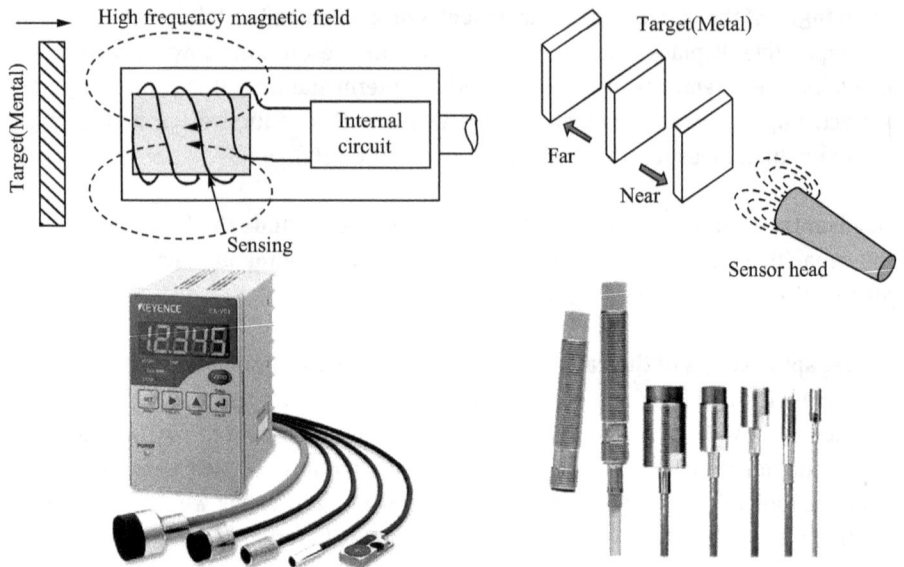

Figure 6.30: The construction of the inductive displacement sensor.

Typical applications of the inductive displacement sensor include the following:
The inductive displacement sensors are widely used for measuring or monitoring the position of conductive targets, especially in dirty environments such as machine tools, heavy industry, robotics, materials handling, aerospace, dimensional inspection, assembly machinery, machine tool monitoring, quality control, medical engineering, hydraulic and pneumatic cylinders, inspection and testing systems, and food and beverage industries.

6.5.4 Laser displacement sensor

The laser displacement sensor is a device to detect the distance between a sensor and an object with laser. Most noncontact laser displacement sensors use the laser triangulation technique to convert the target distance into an output signal. Laser displacement sensors can be divided into two types based on their performance and intended use. High-resolution laser displacement sensors are usually used in displacement and position monitoring that require high accuracy, stability and low-temperature drift. Proximity laser displacement sensors are cheaper and are often used to detect the presence of parts or for counting applications. As shown in Figure 6.31 (the sensor in the photography was made by KEYENCE Corp., UK), the laser irradiates the object. The receiving lens focus the reflected light of the object on the light receiving element. If the distance from the sensor to the target changes, the angle of the reflected light

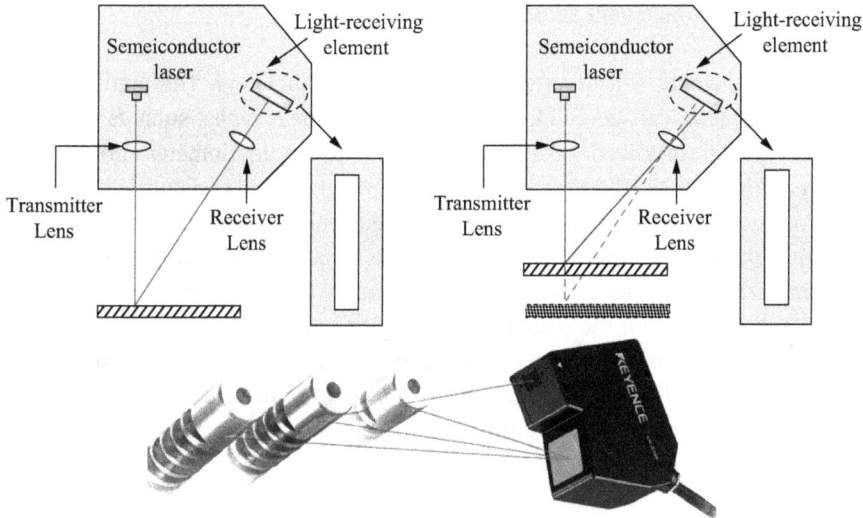

Figure 6.31: The construction of the laser displacement sensor.

will change as the distance between the sensor and the target changes, which results in the position change of the received light on the light receiving element. This change is proportional to the movement amount of the object. In this way, the displacement can be measured. Other principles such as the confocal method and spectral interference method are also used in laser displacement sensors.

Advantages of the laser displacement sensor include the following:
The laser displacement sensor can resolve measurements of less than 1 μm. A larger working distance can provide enough space to reduce possible damage caused by touching the moving target.

Disadvantages of the laser displacement sensor include the following:
The laser displacement sensor is easily affected by temperature, dust, dirt and smoke. Among them, the temperature has the greatest impact. Not only electronic devices exhibit temperature drift but also expansion and contraction of mechanical components can physically change the sensor gap.

Typical applications of the laser displacement sensor include the following:
The laser displacement sensor is widely used in position sensing, robot position, online process monitoring, rail alignment, thickness and dimension measurements, the separation distance between rollers and brake rotor thickness.

6.5.5 Resistive displacement sensor

Resistive displacement sensors are often called potentiometers. The manually adjusted potentiometer can be divided into rotary and linear types, such as a single-turn potentiometer, multiturn potentiometer, dual-gang potentiometer and rheostat as shown in Figure 6.32. The potentiometer is a three-terminal resistor with an electrically conductive wiper. It slides against a fixed resistive element depending on the position or angle of the external shaft. Two terminals are connected to both ends of the resistive element, and the third terminal (wiper) is connected to the sliding contact and moves on the resistive element. The wiper position determines the output voltage of the potentiometer. The potentiometer can basically be used as a variable voltage divider. The output provided by the potentiometer is linearly related to the displacement of the floating wiper.

Single-turn potentiometer

Multi-turn potentiometer

Dual-gang potentiometer

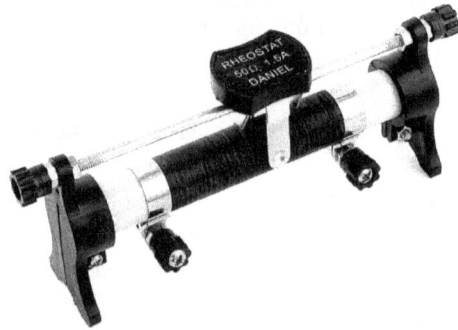

Rheostat

Figure 6.32: Resistive displacement sensors.

Advantages of the resistive displacement sensor include the following:
The resistive displacement sensor has high accuracy, low cost, high sensitivity, simple structure, and lightweight. The resolution of resistive displacement sensor is virtually limitless.

Disadvantages of the resistive displacement sensor include the following:
The resistive displacement sensor has a limited bandwidth, poor linearity and contact noise. The main disadvantage is that wear reduces the life of this sensor.

Typical applications of the resistive displacement sensor include the following:
The resistive displacement sensors are commonly used in a wide range of industries and applications such as control inputs, position measurement and calibration components. For example, potentiometer is often used as a displacement or angle sensor to measure distance or angle. The worm-gear multiturn pot is often used as trimpot on PCB.

6.5.6 Ultrasonic displacement sensor

The ultrasonic displacement sensor is a noncontact device for measuring the distance to a target through ultrasonic waves. The construction of the ultrasonic displacement sensor is shown in Figure 6.33. (The sensor in the photography was made by TAKEX America Inc., USA.) Ultrasonic displacement sensor uses sound waves above 20,000 Hz and exceeds the human hearing range. It sends out a specific frequency of ultrasonic wave and receives the waves reflected by the target to measure the distance. The sensor usually has a piezoelectric sensor that vibrates when electrical energy is applied to the sensor. Vibrations compress and expand the air molecules in waves from the sensor surface to the target object. Another capacitive sensor is used to receive sound (at a sound speed of 343 m/s). By calculating the elapsed time between the generated ultrasonic wave being generated and

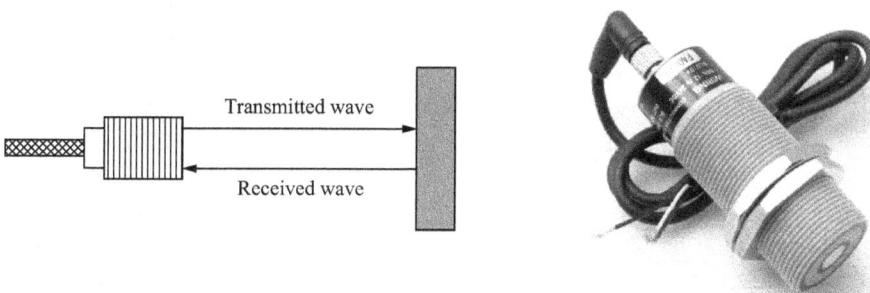

Figure 6.33: The construction of the ultrasonic displacement sensor.

the ultrasonic wave bouncing back, the distance between the ultrasonic displacement sensor and the target can be calculated.

Advantages of the ultrasonic displacement sensor include the following:
The ultrasonic displacement sensor has higher sensing accuracy and longer sensing distance, which is not affected by the color or material of the target object. In addition, it can also work in a humid environment.

Disadvantages of the ultrasonic displacement sensor include the following:
The ultrasonic displacement sensor is affected by the temperature fluctuation, wind and humidity of the air. Temperature fluctuation affects the sound wave speed of the ultrasonic displacement sensor. As temperature increases, the sound wave travels faster to and from the target.

Typical applications of the ultrasonic displacement sensor include the following:
The ultrasonic displacement sensors are used in many automated factories and processing plants to realize liquid level measurement, tire position detection, motion detector systems, micromotion at the bone–implant interface and nondestructive testing. Ultrasonic displacement sensors are widely used as parking sensors in automobiles to help drivers reversing into parking spaces.

6.5.7 Grating displacement sensor

Linear scale and rotary encoder are described as typical grating displacement sensors in this textbook, as shown in Figure 6.34. The linear scale is a sensor associated with the scale of the encoding position. The light source and lens produce parallel light beams (channels A and B) that pass through the four windows of the scanned reticle. The four scanning windows are 90° apart. Then the photoelectric sensor detects the light passing through the glass scale. As the scanning unit moves, the scale converts the detected light beam. The photoelectric sensor detects the light and produces a sinusoidal (square) wave output. Then, the linear encoder system combines the shifted signals to produce two symmetrical sinusoidal (square) outputs with a phase difference of 90°. A reference signal is created when the fifth pattern on the scanning mask is aligned with the same pattern on the scale. The scale is read, and the position is converted into the digital signal output at last. A rotary encoder is a mechanical motion sensor generating digital signals based on motion. As an electromechanical device, a rotary encoder can provide information to the motion control system user regarding position and orientation. A photodetector detects a light-emitting diode passing through the glass disk. This will cause a series of equally spaced pulses when the encoder rotates. The output of

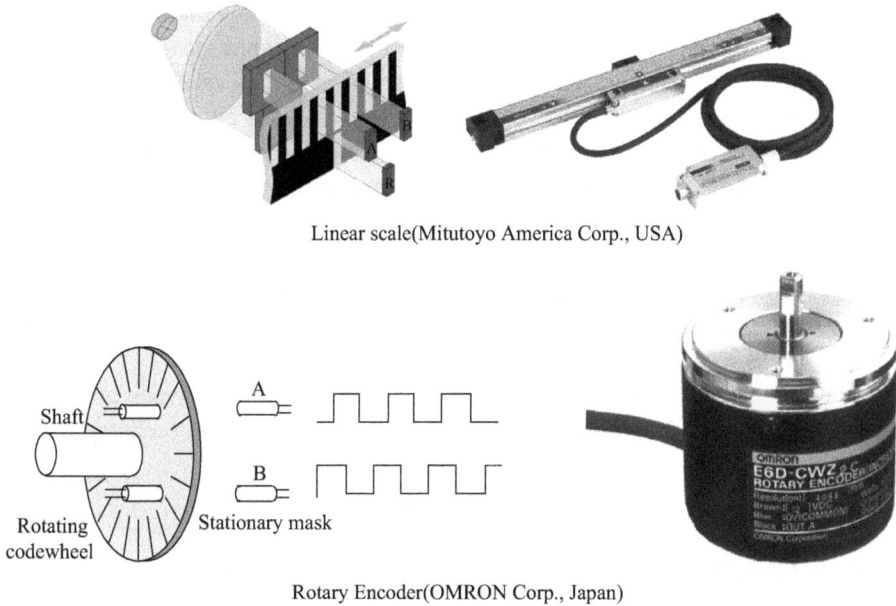

Linear scale(Mitutoyo America Corp., USA)

Shaft

A

B

Rotating
codewheel

Stationary mask

Rotary Encoder(OMRON Corp., Japan)

Figure 6.34: Grating displacement sensors.

the incremental rotary encoder is measured in pulses per revolution for tracking position or determining speed.

Grating displacement sensors are generally divided into incremental sensors and absolute sensors. By providing A and B pulse outputs, the incremental sensors work differently, and the A and B pulse outputs themselves do not provide usable count information. The position of the incremental sensor is completed by the counter accumulating all pulses. Absolute sensors maintain position information when the system is powered off. The position of the sensor is immediately available as soon as the power is on. The relationship between the sensor value and the physical location of the controlled machine is set at the time of assembly.

Advantages of the grating displacement sensor include the following:
Grating displacement sensor has the characteristics of high resolution, small volume, good dynamic performance, high precision, low cost, fast moving speed and good reliability.

Disadvantages of the grating displacement sensor include the following:
Grating displacement sensors are susceptible to contamination, oil and dust contamination, and are affected by direct light source interference.

Typical applications of the grating displacement sensor include the following:
Grating displacement sensors are widely used in CNC machine tools (milling machines, boring machines, grinding machines, small-hole EDM machine tools, wire EDM machines, machining centers), robots, projectors, video measuring machines, railway vehicle lifting jacks, printing presses, elevators, office equipment, labeling machines, medical equipment, assembly machines, printers, testers, robots, packaging machinery and so on.

6.6 Velocity sensor

A velocity sensor is a device that responds to velocity. Different types of velocity sensor discussed in this chapter are piezoelectric velocity sensor, tachometer and electrodynamic velocity sensor (moving coil).

6.6.1 Piezoelectric velocity sensor

The piezoelectric velocity sensor is a device that is used for velocity measurement based on piezoelectric effect. The construction of the inductive displacement sensor is shown in Figure 6.35. (The sensor in the photography was made by Connection Technology Center, INC., USA.) Since it is self-generating, it does not require the help of an external power supply. The piezoelectric effect was discovered by Pierre and Jacques Currie in the late nineteenth century. Detail information about the piezoelectric effect is presented in Section 6.3.2. Piezoelectricity can be defined as the electrical polarization produced by mechanical strain on a certain crystal. Piezoelectric velocity sensor combines piezoelectric accelerometer technology and an integrator to generate an output proportional to velocity.

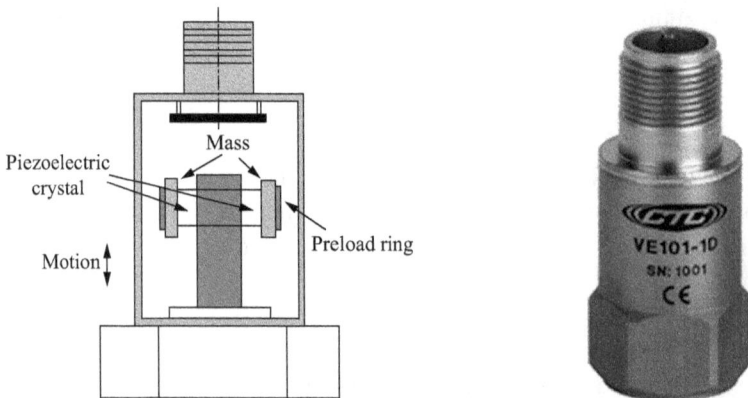

Figure 6.35: The construction of the piezoelectric velocity sensor.

Advantages of the piezoelectric velocity sensor include the following:
The piezoelectric velocity sensors have high sensitivity, small size, wide dynamic range and low noise. It does not require an external power supply.

Disadvantages of the piezoelectric velocity sensor include the following:
The piezoelectric velocity sensor is not suitable for measurement under the static condition and the output may change, due to the temperature variation of the crystal. Relative humidity (above 85% or below 35%) will affect the output.

Typical applications of the piezoelectric velocity sensor include the following:
The piezoelectric velocity sensor is widely used in any vibration measurement application, where velocity units are the desired output including pumps, motors, fans, turbines, compressors, fossil and nuclear energy, oil and gas, petrochemicals.

6.6.2 Tachometer

The tachometer is a device for indicating the angular (rotational) velocity of the rotary shaft. The sensor can be categorized as mechanical tachometer and AC/DC tachometer that indicates instantaneous values of velocity in revolutions per minute. The photograph of the tachometers is shown in Figure 6.36. Mechanical tachometer utilizes that the centrifugal force on the rotating mass is related to the rotation speed and can be used for stretching or compressing the mechanical spring. Inside the sensor, the rotating shaft controls the position of a needle to indicate the speed, which is calibrated to give the speed in the desired units. The AC/DC tachometer provides a voltage value proportional to the speed. The AC tachometer is a small brushless alternator with a rotating multipole permanent magnet. Although the frequency

Hand-Held Mechanical Tachometer
(Electromatic Equipment Co., Inc., USA)

AC/DC Tachometer
(Marsh Bellofram Group of Companies, USA)

Figure 6.36: Tachometers.

of change affects the accuracy of this sensor, the output voltage is still measured by a voltmeter. The DC tachometer is a small DC generator with a permanent magnetic field. The output voltage is proportional to speed and can be measured on a voltmeter calibrated in units of speed.

AC tachometer has primary and secondary stators with fixed windings and a rotor with a permanent magnet. DC tachometer is the most commonly used instrument in robots. It is a DC generator that provides an output voltage that is proportional to the armature angular velocity. The difference between the AC tachometer and the DC tachometer is that the AC tachometer is only used for one-way measurement speed, but the DC tachometer can be used to measure the speed in both directions. Unlike AC tachometer, there is a problem of brush friction and brush at high speed for DC tachometer. The maintenance of the AC tachometer is difficult, and small ripple is inevitable when the DC tachometer outputs.

Advantages of the tachometer include the following:
The tachometer has a simple operation for measuring rotation and surface speed. Only one measurement per revolution is required.

Disadvantages of the tachometer include the following:
The tachometer is not as accurate as the encoder, especially at low speeds. Tachometer has brushes that need to be maintained and magnets that cause erroneous measurements. Tachometer outputs an analog voltage, which is also sensitive to loading effect, especially at low speed.

Typical applications of the tachometer include the following:
The tachometers are used for speed control on various rolling mill drives, machine tools, paper machines, automobiles, aviation, ships, airplanes, trains, light rail vehicles, materials handling, process control and other applications. It is also used for speed adjustment drive scheme and requires high-precision speed measurement.

6.6.3 Electrodynamic velocity sensor

The electrodynamic velocity sensor is a device to measure velocity based on the electrodynamic principle. The construction of the electrodynamic velocity sensor is shown in Figure 6.37. (The sensor in the photography was made by CEMB Iran, Italy.). The electrodynamics principle is used for relative speed sensors. It is based on the induction phenomenon. The magnet is fixed to the vibrating object, and the magnet moves in a noncontact manner. In the coil, a voltage is induced due to the movement of the magnet. A voltage proportional to the vibration velocity is induced in the coil moving in the magnetic field produced by the permanent magnet. By

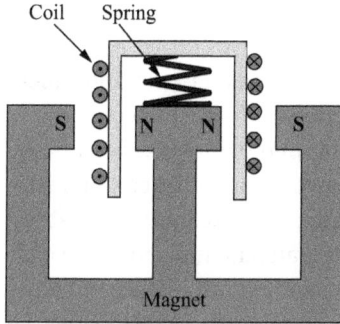

Figure 6.37: The construction of the electrodynamic velocity sensor.

design, the sensor has excellent sensitivity and linearity, which can be reduced to very low vibration levels.

Advantages of the electrodynamic velocity sensor include the following:
The electrodynamic velocity sensor has good linearity and no mechanical contact. It does not require maintenance and is stable under varying temperature conditions. The entire device is included in a magnetically shielded case reducing the effects of stray magnetic fields

Disadvantages of the electrodynamic velocity sensor include the following:
The electrodynamic velocity sensor can be affected by stray magnetic fields. It has limited frequency response and is not very useful for vibration measurement.

Typical applications of the electrodynamic velocity sensor include the following:
The electrodynamic velocity sensors are used to measure velocity in absolute vibration measurements of critical applications such as motors, pumps, fans, steam, gas and water turbines, compressors, pumps and fans.

6.7 Acceleration sensors

An acceleration sensor also named accelerometer is a device that responds to acceleration, which is the change in the rate of the velocity of an object. The type is based on the measurement technique employed within the acceleration sensor. Different types of acceleration sensors discussed in this chapter are piezoelectric acceleration sensor, strain gauge acceleration sensor, variable reluctance acceleration sensor and IC acceleration sensor.

6.7.1 Piezoelectric accelerometer

A piezoelectric accelerometer is a device that is used to make acceleration measurement based on piezoelectric effect. The construction of the piezoelectric accelerometer is shown in Figure 6.38. (The sensor in the photography was made by
PCB Piezotronics, Inc., USA.) Piezoelectric accelerometer utilizes the certain materials' piezoelectric effect for measuring dynamic changes in mechanical variables.
A piezoelectric acceleration sensor consists of a mass attached to a piezoelectric
crystal mounted on the housing. The mass on the crystal is not disturbed by the
space due to inertia when the accelerometer body is vibrated. As a result, the
piezoelectric crystal is compressed and stretched by the mass. According to
Newton's second law, this force is proportional to the acceleration and produces a
charge. With the help of electronics, the charge output is converted into a low-
impedance voltage output.

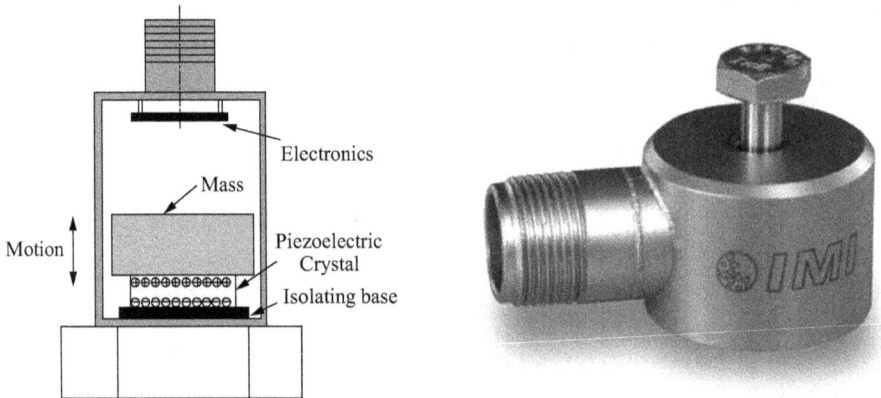

Figure 6.38: The construction of the piezoelectric accelerometer.

Advantages of the piezoelectric accelerometer include the following:
The piezoelectric accelerometers have high-frequency response, high transient response, excellent small linearity, wide frequency range and low output noise. It
does not require an external power supply. Acceleration signals can be integrated
for providing speed and displacement information.

Disadvantages of the piezoelectric accelerometer include the following:
The piezoelectric accelerometer has a low output (lower output from the piezoelectric accelerometer) and high impedance (causing measurement errors). In addition,
it is hard to give a desired shape to the crystals with sufficient strength.

Typical applications of the piezoelectric accelerometer include the following:
The piezoelectric accelerometers are used in medical, chemical and industrial processing equipment. They are more useful for the dynamic measurements (explosions and blast waves) such as impact/vibration testers, engine tests, ballistics, machining systems, metal cutting, machine health monitoring, chassis structure test, structural analysis, reactor, rocket, landing-gear hydraulic system, wind tunnel and modal test.

6.7.2 Strain gauge accelerometer

Strain gauge accelerometer is a device that is used for acceleration measurement based on strain gauge effect. The construction of the strain gauge accelerometer is shown in Figure 6.39. (The sensor in the photography was made by CEC Vibration Products, USA.) The strain gauge accelerometer consists of a cantilever beam, two bounded strain gauges and a mass. When the cantilever beam with a mass attached to the free end is subjected to vibration, vibrational displacement of the mass occurs. According to the displacement of the mass, the beam is deflected so that the beam is pulled tight. The resulting strain is proportional to the vibration displacement of the mass, so the acceleration is measured during calibration. Strain gauge accelerometers use strain gauges as the arm of a full Wheatstone bridge, converting the mechanical strain into a DC output voltage that varies with the applied acceleration. Strain gauge accelerometers can be made more sensitive with the use of stiffer springs, resulting in higher frequency response and output signal amplitude.

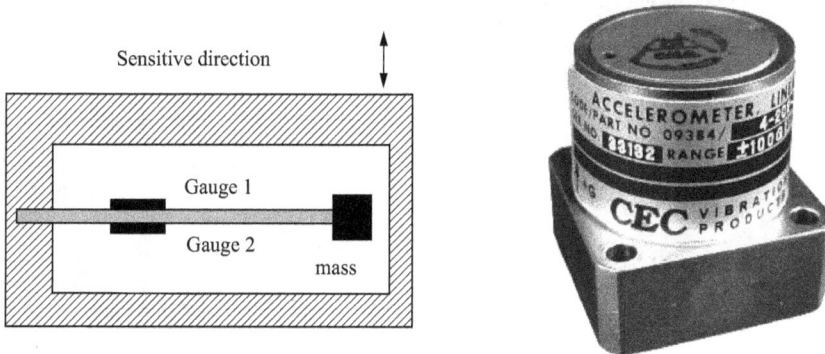

Figure 6.39: The construction of the strain gauge accelerometer.

Advantages of the strain gauge accelerometer include the following:
Strain gauge accelerometer is small, has lightweight and can measure in multiple directions at the same time. Compared with other types of accelerometers, strain gauge accelerometers can measure accelerations as low as 0 Hz.

Disadvantages of the strain gauge accelerometer include the following:
Strain gauge accelerometer is very sensitive to temperature. Compared with piezo-electric acceleration sensors, the response is not very fast.

Typical applications of the strain gauge accelerometer include the following:
Strain gauge accelerometers are widely used in aircraft flight tests, missile/rocket tests, suspension tests, deceleration/braking tests (antilock braking systems and automotive airbags) to measure tilt inertial forces, shocks and vibrations.

6.7.3 Variable reluctance accelerometer

Variable reluctance accelerometer is a device that, when used with very basic electronic circuitry, can detect changes in acceleration. The construction of the variable reluctance accelerometer is shown in Figure 6.40. (The sensor in the photography was made by Brüel & Kjaer, Denmark.) The force is required to accelerate the mass in proportion to the acceleration, the springs supporting the mass deflect in proportion to the acceleration, and thus the displacement measurement achieves an acceleration measurement. The mass is iron and can be used as both an inertial element that transmits acceleration forces and a magnetic circuit element that converts motion into reluctance. If the iron mass is exactly in the middle of two E-frames, the acceleration will be zero and therefore there will be no output voltage. Depending on the motion of the mass, the spring deflects, which corresponds to a changing acceleration. A phase-sensitive demodulator is used to detect the motion of the mass on both sides of zero position. The phase-sensitive output is proportional to the measured acceleration.

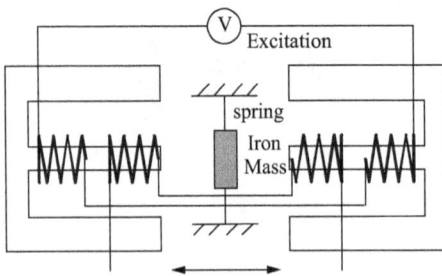

Figure 6.40: The construction of the variable reluctance accelerometer.

Advantages of the variable reluctance accelerometer include the following:
The variable reluctance accelerometer is inexpensive (cost of wire and magnets is low) based on very mature technology. It has advantages in high-precision

measurement of static and slow-change phenomena, with small bias and repeatable thermal errors.

Disadvantages of the variable reluctance accelerometer include the following:
The variable reluctance accelerometers require additional signal processing circuitry to recover useful signals. Measurements for higher frequencies are limited. In addition, the thermal sensitivity of the variable reluctance sensor changes with temperature, which is approximately the same as the way that the LVDT inductance changes with temperature.

Typical applications of the variable reluctance accelerometer include the following:
The variable reluctance accelerometers are used for the control of actuators (E.G., suspension regulating devices), deceleration/brake tests and so on.

6.7.4 MEMS accelerometer

MEMS accelerometer is one of the simplest but most suitable ones. The construction of the MEMS accelerometer is shown in Figure 6.41. (The sensor in the photography was made by Analog Devices, USA.) In the 1990s, MEMS accelerometers were used in the automotive airbag system industry. One of the most commonly used MEMS accelerometers is the capacitive type. The movable plates and fixed outer plates represent capacitors. Capacitors can be arranged as a unilateral or differential pair. MEMS accelerometer is composed of a single planar surface, which is placed along with a mechanical spring between two planar surfaces. The displacement of the movable mass is caused by the acceleration, and it creates a very small capacitance change for correct detection. The sensor depends only on the capacitance that occurs due to the distance between the plates.

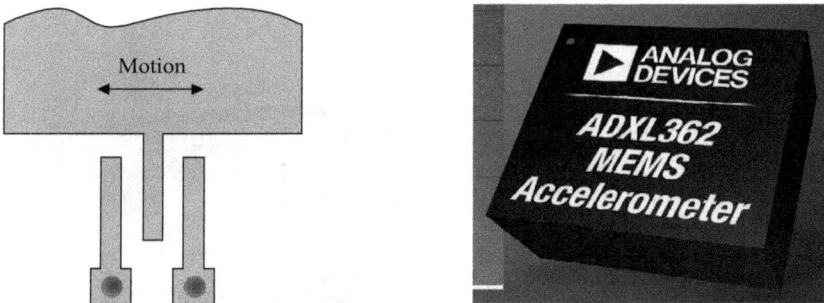

Figure 6.41: The construction of the MEMS accelerometer.

Advantages of the MEMS accelerometer include the following:
The MEMS acceleration sensor has the characteristics of small size, simple structure, low power consumption, high precision, low cost and reliable sensing. Multiple sensors are often combined for providing multiaxis sensing and more accurate data, and they can be easily integrated into MEMS chips with microelectronics.

Disadvantages of the MEMS accelerometer include the following:
The MEMS acceleration sensor is prone to noise and temperature change. It has low bandwidth (up to 5,000 Hz) and brittleness (up to 20,000 g).

Typical applications of the MEMS accelerometer include the following:
The MEMS acceleration sensor is very small and can be applicable for many mechanical purposes, where large measurements are needed. MEMS acceleration sensors have been widely applied in the automotive industry, computers, smart weapon systems (missile launchers, projectiles, etc.) and audio-visual technology. For example, MEMS sensors are being used as step counters in the latest mobile phones and game controllers. Use an accelerometer as the tilt sensor to mark the direction of the picture taken with the built-in camera, detect car crashes and deploy airbags when controlling and monitoring military and aerospace systems.

6.7.5 Angular accelerometer

An angular accelerometer is a device that is used to measure angular acceleration. Angular acceleration is an important parameter for status monitoring and fault diagnosis of rotating machinery. The construction of the angular accelerometer is shown in Figure 6.42. (The sensor in the photography was made by Jewell Instruments LLC, USA.) The sensitivity factor for the transducer is a mechanical-to-electrical transfer function. The electrical output is voltage, and the mechanical input is an angular

Figure 6.42: The construction of the angular accelerometer.

acceleration in units of radians per unit time squared. Indirect acceleration measurement techniques are based on analog or digital postprocessing of available position or velocity signal. Some new accelerometers have been developed for direct sensing of angular acceleration. The simplest measurement method is using a liquid-filled torus having an intercepting diaphragm. As the torus accelerates, the differential pressure is exerted on the diaphragm. A strain gauge is mounted on the diaphragm so that a pressure difference proportional to the angular acceleration can be detected.

Advantages of the angular accelerometer include the following:
The angular accelerometer has low output impedance, low cost, high input range and good reliability. It has a maximum bandwidth of 200 Hz.

Disadvantages of the angular accelerometer include the following:
The angular accelerometer has limited rotation angle and imperfect compensation for the gravitational field. It has strict requirements for the installation location.

Typical applications of the angular accelerometer include the following:
The angle accelerometers can be used for antenna stabilization, acceleration measurement of navigation control systems, servo control, protection monitoring, robotics, vehicle driving analysis, optical frame stabilization, autopilot frame input, vertical and horizontal transport, engine torque estimation and control, flight simulator and dynamic test application.

6.8 Force/torque sensors

A force sensor is a device that continuously monitors the force/torque exerted on it. Different types of force/torque sensors discussed in this chapter are strain gauge force/torque sensor and piezoelectric force/torque sensor.

6.8.1 Strain gauge force/torque sensor

Strain gauge force/torque sensor is a device that converts the material deformation into an electrical signal for strain gauges to measure. The strain gauge is a type of sensor. Its resistance varies with applied force/torque and converts the force/torque into a change in resistance. The strain gauge was invented by Edward E. Simmons and Arthur C. Ruge in 1938. The most common type of strain gauge includes insulating flexible backing, and the most common materials used to fabricate strain gauges are nickel–chromium, copper–nickel, nickel–chromium–molybdenum and platinum–tungsten alloys. Various resistance strain gauges are available for a variety of applications. The foil strain gauge is the most widely used strain gauge. When compared with other type of strain gauges, the foil strain gauge that has significant advantages can be used for force/torque measurement. Figure 6.43 shows the strain gauges used for force, torque and shear. (The sensors in the photography were made by Omega Engineering Inc., USA.) Each strain gauge is designed to measure strain along a well-defined axis so that it can properly align with the strain field.

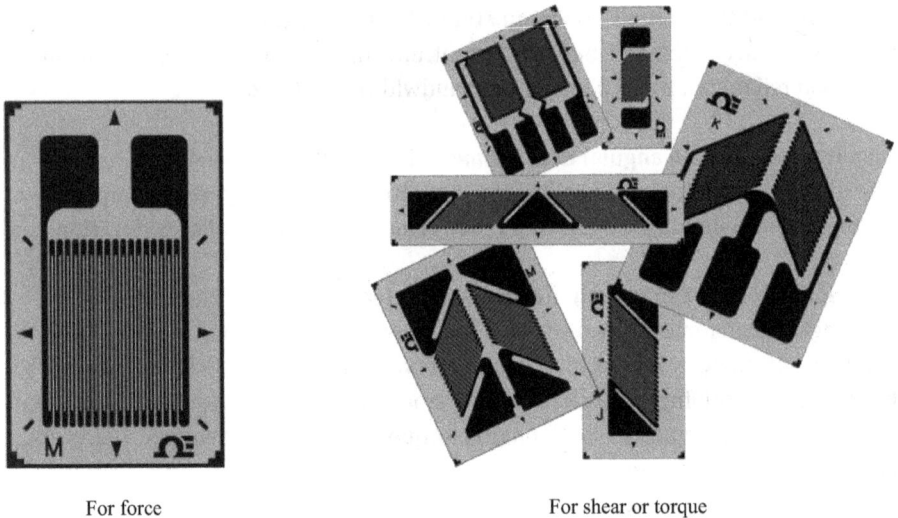

For force   For shear or torque

Figure 6.43: The construction of the strain gauges.

Figure 6.44 shows the strain gauge force/torque sensors including load button, load cell and torque sensor. (The sensors in the photography were made by FUTEK Advanced Sensor Technology, Inc., USA.) In this section, a load cell is used as an example. The load cell is the most common in the industry and uses one strain gauge (quarter bridge) or two strain gauges (half bridge) or four strain gauges (full bridge). The most common load cell uses a curved beam structure. The beam bends slightly as the force acts on it. The bending/strain of the beam material changes the electrical output of the strain gauge mounted on the material. The resistance change of the strain gauge provides an electrical value change. It is calibrated based on the load on the load cell. The Wheatstone bridge is ideal for measuring small changes in resistance and is suitable for measuring strain gauge resistance changes. The electrical signal output is typically on the order of a few millivolts and needs to be amplified by an instrumentation amplifier before it can be used.

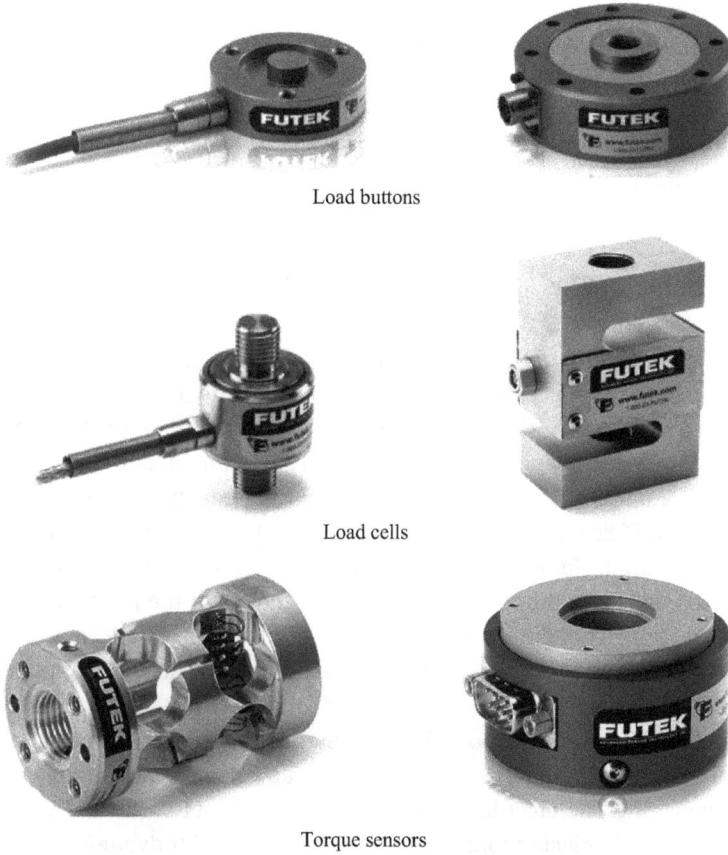

Load buttons

Load cells

Torque sensors

Figure 6.44: The construction of the strain gauge force/torque sensors.

Advantages of the strain gauge force/torque sensor include the following:
The strain gauge force/torque sensor has a small size, high accuracy (<0.1% of full scale), wide range, long life cycle and fast response. In addition, it has no moving parts.

Disadvantages of the strain gauge force/torque sensor include the following:
The strain gauge force/torque sensors are sensitive to temperature and are unstable due to bonding materials. A thermoelastic strain may cause lag. This is not a good solution for OEM applications.

Typical applications of the strain gauge force/torque sensor include the following:
Applications for the strain gauge force/torque sensors range from simple force measurement solutions to new hardware design tests and requirements verification

such as force distribution analysis, hydraulic actuators, sprockets, safety load monitoring systems, conveyor belts, lift measurement, wind turbines, hydrostatic or submersible vehicles used deep in the surface of the ocean, determination of operating force, injection molding force, jacking load, crane measuring system, overload or structural capacity, dam strength, bridge force monitoring, fatigue analysis, preload monitoring and train coupler measurement.

6.8.2 Piezoelectric force sensor

Instead of correlating the strain of the beam with the applied force, the piezoelectric force sensor measures the compressive force directly. The construction of the piezoelectric force sensor is shown in Figure 6.45. (The sensor in the photography was made by Intertechnology Inc., Canada). No power supply is required, and the deformation for generating a signal is very small. It has a high-frequency response for the measuring system. Piezoelectric sensors have a low operating charge and require high impedance cables for electrical interfaces. An electrical charge is formed on the surface of the crystal in proportion to the rate of change of the force when a force is applied to certain crystalline materials. The device needs a charge amplifier for integrating the electric charges to produce a signal proportional to the applied force and large enough to make a measurement. The resulting high stiffness of the piezoelectric force sensor greatly reduces the interference caused by the measurement and provides inherently high natural frequencies and rising time. The piezoelectric effect is dynamic. The electrical output of a gauge is not static and is an impulse function . The piezoelectric force sensor is ideally suited for dynamic measurements, and the electrical output of a sensor is an impulse function. The voltage output is only useful if the strain changes, and no real static measurement is performed. Piezoelectric force sensors can operate over a wide range of temperature and can withstand temperatures up to 350 °C.

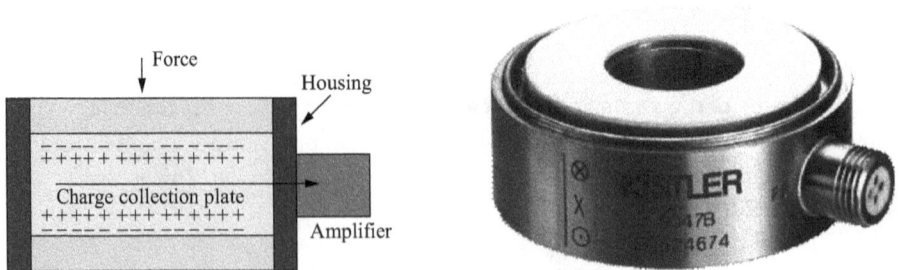

Figure 6.45: The construction of the piezoelectric force sensor.

Advantages of the piezoelectric force sensor include the following:
The piezoelectric force sensor has a thin (0.2 mm) and flexible construction that can be customized and is ideal for innovative product designs – thin, lightweight and low power requirements.

Disadvantages of the piezoelectric force sensor include the following:
The piezoelectric force sensors are less accurate (±5% of full scale) than typical load cells and calibrated by the user.

Typical applications of the piezoelectric force sensor include the following:
The piezoelectric force sensors can measure very fast events (such as shock waves in solids, impact printers, material testing, component insertion, bearing reactions, cutting forces, impact forces, reaction forces in rockets, determination of friction coefficients and press forces).

6.9 Level sensors

A level sensor is used to detect the content of liquids and fluidized solids including slurries, PM and powders that exhibit an upper free surface. Level measurement can be continuous or have point value. The different types of level sensors discussed in this chapter are ultrasonic level sensor, radar level sensor, capacitance level sensor, resistance level sensor, float level switch and rotary paddle level switch.

6.9.1 Ultrasonic level sensor

An ultrasonic level sensor is a device that uses the ultrasonic wave for measuring distance. The ultrasonic level sensor is located at the top of the tank. The construction of the ultrasonic level sensor is shown in Figure 6.46. (The sensor in the photography was made by Microlevel Durko A.Ş, Turkey.) Piezoelectric crystals are used to convert electrical energy into ultrasonic waves. When electrical energy is applied to a piezoelectric crystal, the ultrasonic waves oscillate at high frequencies and vice versa. The sensor head emits an ultrasonic wave and receives the reflected wave from the target. The ultrasonic level sensor measures the distance to the target by measuring the time between the emission and the reception.

Advantages of the ultrasonic level sensor include the following:
The ultrasonic level sensor has no moving parts and can measure the liquid level without being in contact with the object. The detection is not affected by color,

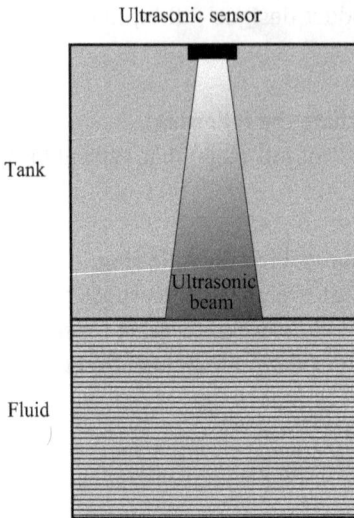

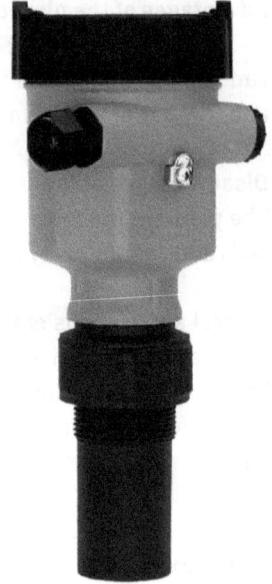

Figure 6.46: The construction of the ultrasonic level sensor.

shape or physical composition of the target material, and even a transparent target can be detected. It does not require maintenance.

Disadvantages of the ultrasonic level sensor include the following:
The ultrasonic level sensors do not work satisfactorily under vacuum or high-pressure conditions. It does not apply to over-smoke, surface mess, foam or high-density moisture. If the depth of the tank is high, it is inconvenient. The object should not be sound-absorbing. The speed of sound waves can sometimes be affected by temperature changes.

Typical applications of the ultrasonic level sensor include the following:
The ultrasonic level sensors can be used to measure liquid levels in tanks with acids, food and beverages, inks and paints, semiconductor process chemicals, slurries, oils, petroleum distillates, waste water and even corrosive, boiling and hazardous chemical tanks.

6.9.2 Radar level sensor

Radar level sensor is a noncontact continuous level measuring device for measuring difficult liquids and solids that are difficult to measure with ultrasonic technology. The construction of the radar level sensor is shown in Figure 6.47. (The sensor in

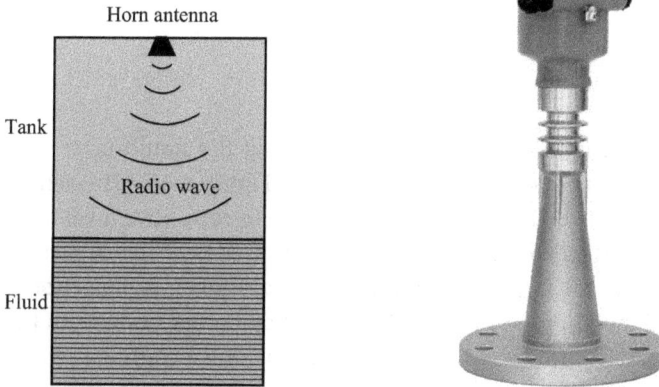

Figure 6.47: The construction of the radar level sensor.

the photography was made by Microlevel Durko A.Ş, Turkey.) Radar level sensor typically uses a horn antenna to send improved microwave beams onto the surface of the liquid being measured. The antenna mounted at the tip of the tank can receive the reflected microwave signal back from the fluid surface, regardless of the tank shape or the environmental conditions. A timing circuit is included in the system that measures the time of flight, which is directly proportional to the distance between the antenna and the fluid level. If the tank geometry is known, the level can be calculated from this variable.

Advantages of the radar level sensor include the following:
The radar level sensors have high sensitivity, long distance, maintenance-free construction, good resistance to pressure, dust, oil, grease, temperature, gases, foam and uneven surfaces.

Disadvantages of the radar level sensor include the following:
The main drawback associated with radar level detectors is that it costs high. In addition, these systems cannot detect the level between interfaces. These devices are only suitable for light dust and dust layers. If the measured fluid is very close to the radar antenna, it will cause an error.

Typical applications of the radar level sensor include the following:
The radar level sensors are used for level measurement and typical applications include but are not limited to sand, aggregate, asphalt, gravel, coal, powder, cement,

solids, minerals, grain, sawdust, rubber products, glass production, building materials, plastics and other solids.

6.9.3 Capacitance level sensor

A capacitance level sensor is a device used for measuring the liquid level. The construction of the capacitance level sensor is shown in Figure 6.48. (The sensor in the photography was made by N.D. Automation, India.) In capacitive level measurement, the sensor and the container form the two electrodes of the capacitor. The capacitance of the sensor is proportional to the liquid level. A capacitor is formed by an insulated metal probe mounted in a metal container forms with the metal wall, and as the dielectric level increases, its capacitance continues to increase. Any change in level results in a capacitance change, which is then converted into a level signal. Phase modulation and the use of higher frequencies make the sensor suitable for applications where dielectric constants are changing.

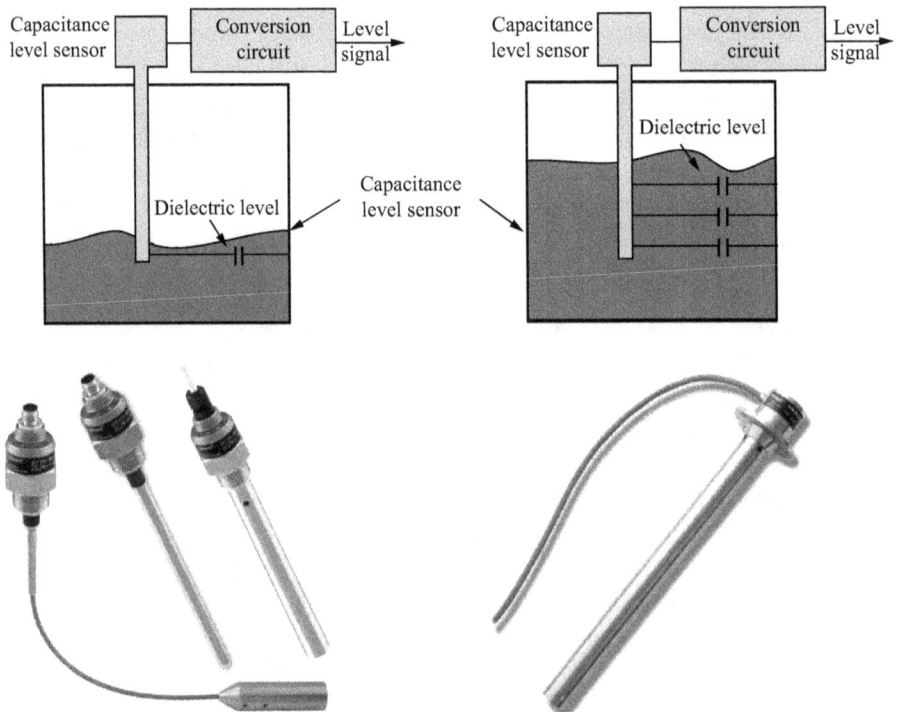

Figure 6.48: The construction of the capacitance level sensor.

Advantages of the capacitance level sensor include the following:
The capacitance level sensor is simple and low cost. It can be measured without blind spots throughout the length of the sensor. In addition, fully insulated capacitive sensors can measure corrosive liquids.

Disadvantages of the capacitance level sensor include the following:
The capacitance level sensor has poor accuracy and is difficult to calibrate. Pressure, density and dielectric constant are required in the container, and installation is required to stop production, clear can and open hole.

Typical applications of the capacitance level sensor include the following:
The capacitive level sensors are good at detecting the presence of various solid, aqueous and organic liquids and slurries. Level measurement applications include home appliances (coffee machines, refrigerators and humidifiers), cars (cleaning fluid, fuel level and coolant level) and medical devices (insulin pumps, drug pens and drop counters).

6.9.4 Resistance level sensor

A resistance level sensor is a device used for measuring the liquid level and it is considered to be very simple in operation. The construction of the resistance level sensor is shown in Figure 6.49. (The sensor in the photography was made by FAFNIR, Germany.) Conductivity sensing or conductivity detection methods have been used for level detection. The liquid in measurement can usually conduct current with a low voltage power supply. The power supply matches the conductivity of the liquid, and the higher voltage version is designed for less conductive media.

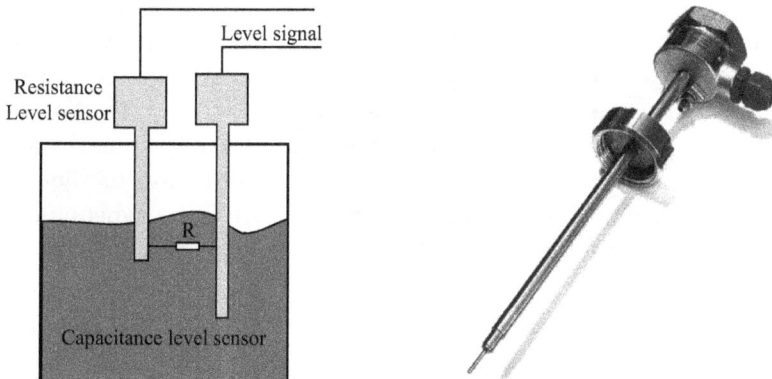

Figure 6.49: The construction of the resistance level sensor.

There are two electrodes: one is inserted into the tank and the other is used as a metal wall of the tank. The conductive liquid contacting the longest probe and the metal tank wall completes a conductive circuit. This causes a change in resistance corresponding to the distance from the top of the sensor to the surface of the fluid. The resistance output of the sensor is inversely proportional to the height of the liquid.

Advantages of the resistance level sensor include the following:
The resistive level sensors are designed for almost all conductive media. It is very simple, low cost, very suitable for two-point or multipoint control and safe to use. The resistive level sensor has a rugged construction and maintenance-free service.

Disadvantages of the resistance level sensor include the following:
Probes of the resistive level sensor must not be contaminated with grease or any other deposits, and the sensor's suitability for different conductivity products is limited.

Typical applications of the resistance level sensor include the following:
The resistive level sensors can be adapted to various applications. The following are the main applications of the sensor: The resistance liquid level sensor is very suitable for point detection of various conductive liquids (such as ink and wastewater). For corrosive conductive liquids, the sensor needs to be preserved.

6.9.5 Float level switch

Float level switch is a level control system that works on the buoyancy principle using a buoyancy sensor. (A float immersed in a liquid is buoyed upward by an equal force to the weight of the displaced liquid.) The construction of the float level switch is shown in Figure 6.50. (The sensor in the photography was made by Tianjin U-ideal Instrument Co., Ltd, China.) Float level switches consist of floats, sensor rods, magnets, reed switches and suspended weights. The float is transmitted through a manipulator or a sliding rod. When the level moves upward, the switch is activated. The float moves up and down with the liquid, which controls the opening and closing of the reed switch. The magnetic float level switches are suitable for level control and alarm for opening tank or pressure containers.

Advantages of the float level switch include the following:
The float level switch is very simple, easy to use, high accuracy, safe and reliable, simple to maintain, best suited for a variety of products. It is faster than normal

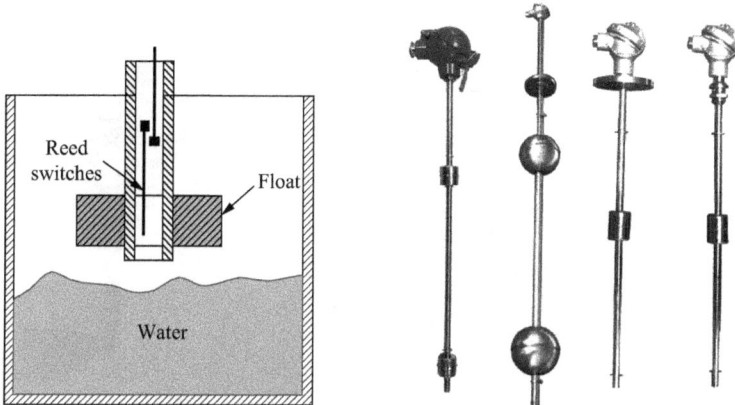

Figure 6.50: The construction of the float level switch.

mechanical switches and has a long working life. Compared with the electronic switch, it has the characteristics of strong resistance to load, and the product can realize multipoint control.

Disadvantages of the float level switch include the following:
The float level switch does not have a self-test function and should be checked and maintained periodically. Its measurement accuracy is poor and requires a variety of mechanical equipment. In addition, floats or buoys are moving parts and can become dirty when using thick or viscous liquids.

Typical applications of the float level switch include the following:
The float level switch is widely used in transportation tanks, storage tanks, water treatment tanks, shipbuilding, printing, papermaking, household appliances (steam engines, washing machines, steam irons, juice presses and automatic coffee machines), petrochemical industry (petrol, diesel and other fuels), food industry, generator equipment, electrician, water treatment, dye industry, hydraulic machinery and so on.

6.9.6 Rotary paddle level switch

A rotary paddle level switch is used for detecting the presence of solid/powdery material in most storage tanks, bins and containers, which is a reliable level switch used for point level detection. The construction of the rotary paddle level switch is shown in Figure 6.51. (The sensor in the photography was made by Endress+Hauser AG, Switzerland.) The device is mounted through the wall of the container so that the paddle protrudes inside the container. A small motor-driven paddle rotates freely without the material. When the paddle is obstructed by the material, the shaft

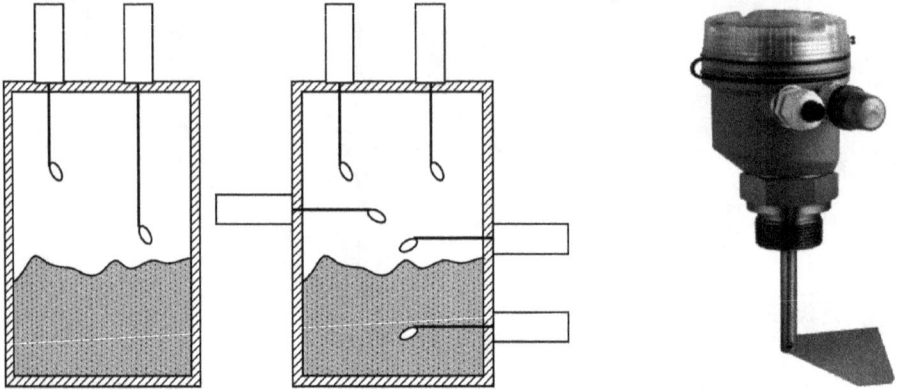

Figure 6.51: The construction of the rotary paddle level switch.

rotation will stop. The motor rotates within the housing and magnetized portion of the motor mounting plate is detected and causes the microswitch to change the state.

Advantages of the rotary paddle level switch include the following:
The rotary paddle level switch is reliable, easy to operate and maintain, compact, easy to install and is suitable for most bulk solids and free from dust.

Disadvantages of the rotary paddle level switch include the following:
Many rotating paddle level switches are needed in a measurement system and will cost more money. In addition, the required connection wires also become complicated.

Typical applications of the rotary paddle level switch include the following:
The rotating paddle level switches are widely used in insulation materials, calcium carbonate, mineral fertilizers, damp feed mixtures, animal feed, fish farms, cereals, broken glass, carbon black storage silo, wood chips, PVC granules, plastic granules, quartz sand, grain, spices and gum.

6.10 Image sensors

An image sensor is an electronic, photosensitive device converting an optical image into an electronic signal. The image sensor is mainly used in camera modules, digital cameras and other imaging devices. It is applied as an image receiver in digital imaging equipment consisting of millions of photodiodes. Different types of image sensors discussed in this chapter are CCD (charge-coupled device) sensor and CMOS (complementary metal-oxide–semiconductor).

6.10.1 CCD sensor

A CCD sensor is a device that converts light into electric charge and processes it into electronic signals based on the photoelectric effect. The device was invented by Dr. Savvas Chamberlain in the late 1960s. The construction of the CCD sensor is shown in Figure 6.52. (The sensor in the photography was made by 1/2.5″ CCD, Canon, Japan.) For modern vision cameras, the CCD sensor is the most usual type sensor. This is because short exposure times and suitability for fast-moving images are allowed by the single step shift to the readout register. When exposed to light, the CCD sensor is a complex electronic component composed of multiple arrays of light-sensitive semiconductor elements. CCD sensor uses millions of tiny photoelectric sensors, and each sensor generates an electrical current. The tiny photoelectric sensor produces only electrical current for 1 min. Before using to create an image, it must be amplified. Some CCDs have an amplifier. This will, in turn, process the current from each sensor. The intensity of the current is proportional to the brightness of the light. The electrical data is captured and converted into an image file. The CCD sensor has extremely low noise, and therefore, can obtain a high signal-to-noise ratio image signal.

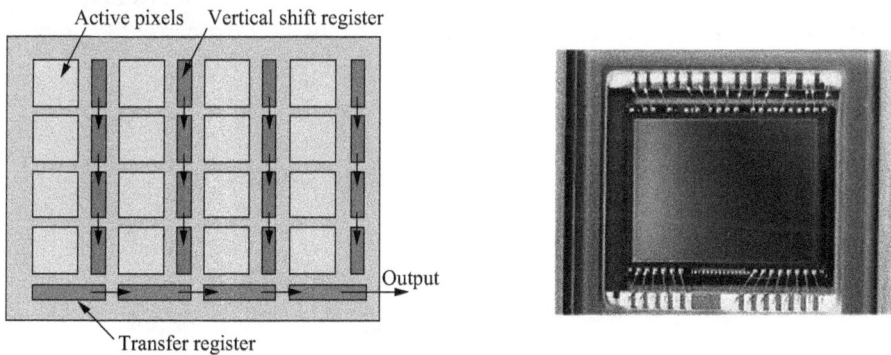

Figure 6.52: The construction of the CCD sensor.

Advantages of the CCD sensor include the following:
The CCD sensors can create high quality, low noise and more pixel images. Traditionally, the CCD sensors are believed to produce high-quality images with a large number of pixels and excellent photosensitivity.

Disadvantages of the CCD sensor include the following:
The CCD sensors consume more power and provide slower data throughput speeds. The CCD amplifier has a higher bandwidth, which leads to higher noise.

Typical applications of the CCD sensor include the following:
The CCD sensors are mainly used for camera modules, digital cameras and other imaging devices that require high-quality images such as applications in the field of dental X-rays including intraoral, panoramic and cephalometric imaging.

6.10.2 CMOS sensor

The CMOS sensor was invented by Dr. Savvas Chamberlain in the late 1970s. A CMOS sensor is a device that converts light into electric charge and converts it into an electronic signal based on the photoelectric effect, and consumes less power while providing high-performance and advanced features (high-speed analog-to-digital conversion and noise reduction). The construction of the CMOS sensor is shown in Figure 6.53. (The sensor in the photography was made by EOS-1Ds Mark II., Canon, Japan.) Nowadays, CMOS sensor is used in cameras increasingly. It allows users capturing 1080p video and applying complex imaging effects easily. CMOS sensor uses different methods to process the charges from millions of photoelectric sensors. In a CMOS sensor, each pixel has its own charge-to-voltage conversion, which means that all the charges can be processed at the same time, clearing the sensors for the next exposure. The sensor also typically includes amplifiers, noise correction and digitizing circuitry, and the strength of the current is proportional to the brightness of the light.

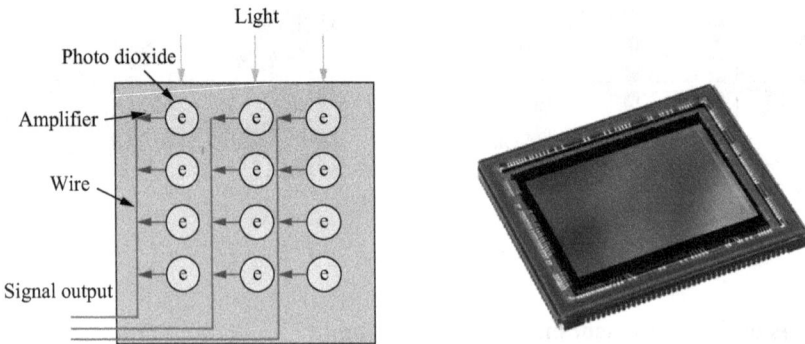

Figure 6.53: The construction of the CMOS sensor.

Advantages of the CMOS sensor include the following:
Compared with the CCD sensors, CMOS consumes very little power and is very cheap. High-speed CMOS units have lower noise than that of high-speed CCD.

Disadvantages of the CMOS sensor include the following:
The traditional CMOS sensors have lower quality, lower resolution, lower sensitivity and are more susceptible to noise. CMOS units have lower light sensitivity than that of CCD.

Typical applications of the CMOS sensor include the following:
The CMOS sensors are designed for mobile phones and high-capacity image sensor applications. In recent years, CMOS image quality has been greatly improved. Therefore, it is worth mentioning that CMOS imagers are superior to CCDs based on almost all imaginable performance parameters in the case of high-volume consumption areas and line scan imagers.

6.11 Proximity or presence sensors

The proximity sensor is a device for detecting the presence of a nearby object without any physical contact based on the changes of electromagnetic field and electromagnetic radiation. Since the physical contact between the sensor and the sensing object does not exist, the proximity sensor can have a longer functional life and higher reliability. Different types of proximity sensors discussed in this chapter are photoelectric proximity sensor, ultrasonic proximity sensor, inductive proximity sensor and capacitive proximity sensor.

6.11.1 Photoelectric proximity sensor

The photoelectric proximity sensor is a device using a light-sensitive element to detect the presence or absence of an object, which is made up of a transmitter and a receiver. The construction of the photoelectric proximity sensor is shown in Figure 6.54. (The sensor in the photography was made by SICK, Germany.) This technology is suitable for long-range sensing or nonmetallic sensing. There are three types of photoelectric proximity sensors: reflective (direct reflection), retroreflective (reflection with reflector) and through-beam mode. Reflective proximity sensor holds the transmitter and receiver together and uses the light reflected directly off the object for detection. Retroreflective proximity sensor requires a reflector to reflect light, which can completely reflect the light. Through-beam proximity sensor fixes the transmitter and receiver, respectively, and detects the object when the beam between the transmitter and receiver is interrupted.

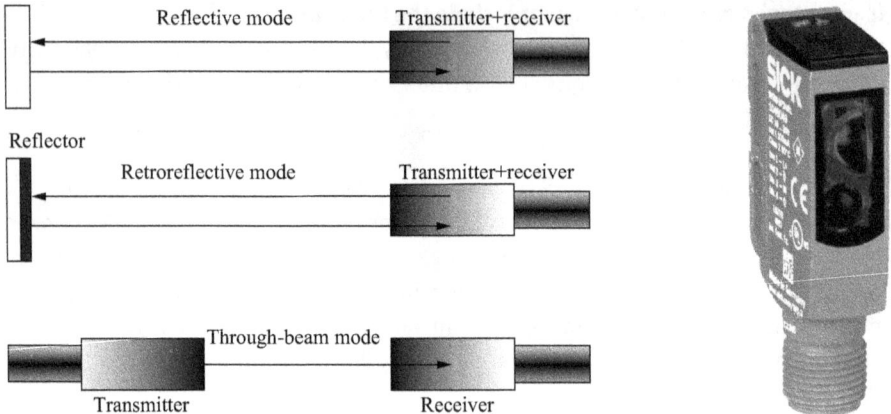

Figure 6.54: The construction of the photoelectric proximity sensor.

Advantages of the photoelectric proximity sensor include the following:
The photoelectric proximity sensor has the longest sensing distance, longer life, lower cost and faster response time. It is accurate and reliable, and can sense a variety of materials.

Disadvantages of the photoelectric proximity sensor include the following:
The sensing range of the photoelectric proximity sensor is affected by the color and reflectance of the target. In through-beam mode, the transmitter and receiver must be installed in the system. This makes the system installation complex.

Typical applications of the photoelectric proximity sensor include the following:
The photoelectric proximity sensors meet applications in many industries such as transportation, food processing, packaging and material handling.

6.11.2 Ultrasonic proximity sensor

An ultrasonic proximity sensor is a device that emits and receives sound waves to detect the absence or presence of an object consisting of a transmitter and a receiver. The construction of the ultrasonic proximity sensor is shown in Figure 6.55. (The sensor in the photography was made by SICK, Germany.) The ultrasonic proximity sensor typically uses piezoelectric sensors to send and detect high-frequency sound waves. It determines the distance to the target by calculating the time interval between sending the signal and receiving the echo. There are two types of ultrasonic proximity sensor: reflective (direct reflection) and through-beam mode. The reflective proximity sensor puts the transmitter and detector together. When a

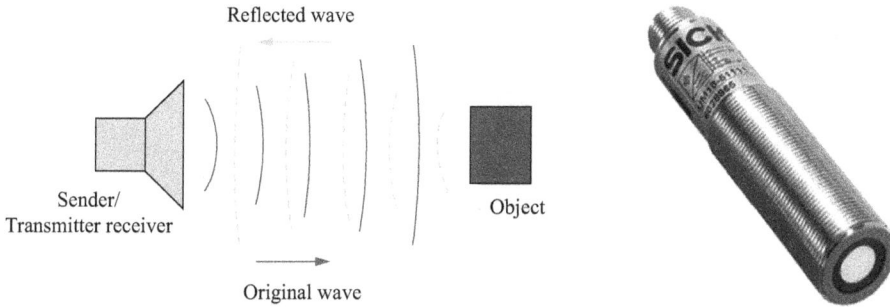

Figure 6.55: The construction of the ultrasonic proximity sensor.

target enters the sensing range of the device, the ultrasonic waves are reflected back to the sensor. Through-beam proximity sensor separates the transmitter from the detector. The detector is mounted opposite the transmitter and blocks the transmitted signal when the object enters the detection range. The output will be switched when the target enters the working range.

Advantages of the ultrasonic proximity sensor include the following:
The ultrasonic proximity sensor can detect various materials/surfaces in adverse climates. It is not affected by color, transparency, gloss or lighting conditions, snow, rain and so on. The target to be tested can be solid, liquid, granular or powdery.

Disadvantages of the ultrasonic proximity sensor include the following:
The ultrasonic proximity sensor has a dead zone at proximity sensing (<5 mm). It may be disturbed by noise.

Typical applications of the ultrasonic proximity sensor include the following:
The ultrasonic proximity sensor can sense most materials such as plastic, wood, metal, liquid and glass.

6.11.3 Inductive proximity sensor

An inductive proximity sensor is a device for generating an electromagnetic field to sense the metal objects passing close to the surface, which is the most commonly used proximity sensors. The construction of the inductive proximity sensor is shown in Figure 6.56. (The sensor in the photography was made by SICK, Germany.) The inductive proximity sensor falls into the category of the noncontact electronic proximity sensor. The basic principle of operation is not difficult to understand. The inductive proximity sensor consists of a coil, a detection circuit, an oscillator and an output

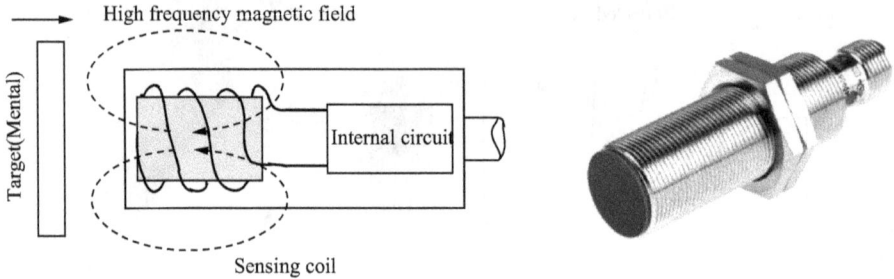

Figure 6.56: The construction of the inductive proximity sensor.

circuit. The inductive coil at the front of the sensor generates a self-sustaining sine wave oscillation at a fixed frequency. When a piece of conductive metal enters the area defined by the boundaries of the electromagnetic field, some of the oscillation energy is transferred into the metal of the target. State changes are detected by an oscillation state detection circuit that triggers the output of the output circuit.

Advantages of the inductive proximity sensor include the following:
The inductive proximity sensors have high repeatability, short response time, long life, high accuracy and high response frequency. They detect objects without any mechanical contact and are not affected by water, dirt, oil and nonmetallic particles. The inductive proximity sensor is insensitive to the target surface finish or target color and can withstand high vibration and impact environments.

Disadvantages of the inductive proximity sensor include the following:
The inductive proximity sensor can only detect metals, and detection is based on heat losses caused by induced currents. It has a short sensing range. In addition, metals such as ferrite that do not allow current to flow cannot be detected.

Typical applications of the inductive proximity sensor include the following:
The inductive proximity sensors are used for locating and detecting metal objects such as steel, aluminum and copper. This is a preferred choice for most applications requiring accurate, noncontact detection of metal objects in mechanical or automated equipment.

6.11.4 Capacitive proximity sensor

A capacitive proximity sensor is a device that generates an electrostatic field to measure the proximity of conductive as well as nonconductive objects with high resolution. The construction of the capacitive proximity sensor is shown in Figure 6.57.

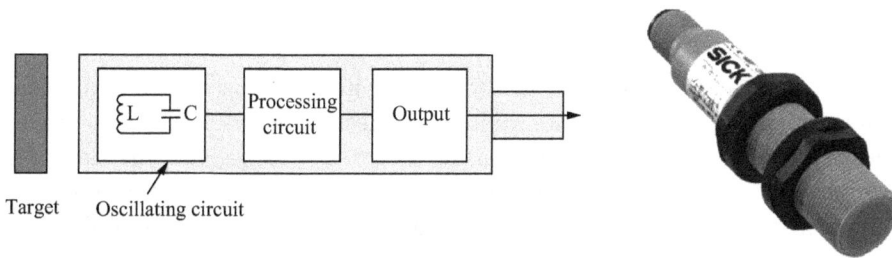

Figure 6.57: The construction of the capacitive proximity sensor.

(The sensor in the photography was made by SICK, Germany.) Capacitive proximity sensors are detecting small objects by the capacitive principle. They can handle both conductive and insulating material, including items moving along the conveyor belt and unprepared mechanical surfaces. Capacitive proximity sensor uses the surface of the sensor as one plate of the capacitor and the surface of the conductive or dielectric target object as the other. In this arrangement, the distance between the capacitor and the capacitor plate varies inversely, and a certain value can be set to trigger the target detection. The capacitive proximity sensor generates an electrostatic field on the surface of the sensor as well as on the target surface area, and the capacitance varies directly with the dielectric medium and inversely with gap distance. The capacitance in the oscillator circuit is changed when an object approaches the sensing surface entering the electrostatic field of the electrodes. As a result, the oscillator starts to oscillate. The trigger circuit reads the amplitude of the oscillator. The sensor's output state changes when a specific level is reached.

Advantages of the capacitive proximity sensor include the following:
The capacitive proximity sensors are not susceptible to dust, contamination, air-jet particles and electromagnetic interference. It can detect metallic and nonmetallic objects and can detect liquid targets.

Disadvantages of the capacitive proximity sensor include the following:
The capacitive proximity sensors have low accuracy and limited range, and they require regular calibration and inspection. In addition, the capacitive proximity sensor has a high sensitivity for the temperature.

Typical applications of the capacitive proximity sensor include the following:
The capacitive proximity sensors can detect both metal and nonmetallic objects such as iron, metal, food, automotive, paper, stone, plastic, wood, powder, granular and liquid. They are ideal for level and feed monitoring.

6.12 Color, contrast and luminescence sensors

Color, contrast and luminescence sensors are suitable for almost all related automation tasks. In the black-and-white world, photoelectric sensors can be very useful. Every photoelectric sensor has two sensing ranges. One range is for the black targets, the other is for the white targets. The photoelectric sensors are also used to sort colored products, identify coded marking, confirm the existence of the date code on the package and others. Color sensor, contrast sensor and luminescence sensor are discussed in this chapter.

6.12.1 Color sensor

A color sensor is a device used to detect the color of the surface based on the RGB principle (principle of wavelength identification of different colors). The construction of the color sensor is shown in Figure 6.58. (The sensor in the photography was made by LEGO Mindstorms, Japan.) Color is the result of interaction between light source, object and observer. The RGB color system is one of the most famous color systems in the world and it combines red, green and blue light to create the colors. Red (wavelength = 580 nm), green (wavelength = 540 nm) and blue (wavelength = 450 nm) are three sets of color light sources. The color sensor usually has a white light transmitter and three independent receivers, and the reflected light is separated from the target into its constituent red, green and blue components. The three colors are emitted in turn and the number of reflected light from the object is registered individually. The high-precision color sensor launches white light, which is first partitioned into the RGB partial spectrum in the receiver. The RGB strength values are determined by comparing with the previous reference values.

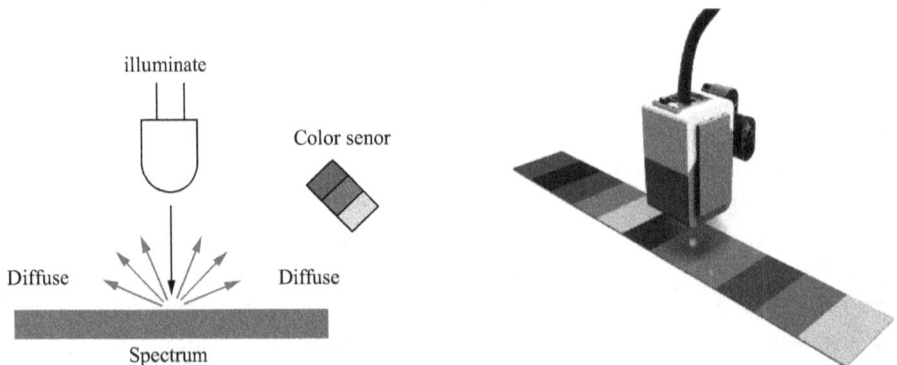

Figure 6.58: The construction of the color sensor.

Advantages of the color sensor include the following:
The color sensor helps to classify objects according to three color methods. It also helps to count objects and help to complete the work in a shorter time. In addition, some powerful large-capacity memory color sensor IC can be obtained at a low cost.

Disadvantages of the color sensor include the following:
The color sensor is costly for small-scale industries. Color matching or recognition is required in applications and the operating distance range needs to be rigorously tested and properly selected in the settings.

Typical applications of the color sensor include the following:
The color sensor is used to check the presence of colored objects, detection of painted products, textile production, inspecting LEDs, pills, plastics, colors and luminescent objects.

6.12.2 Contrast sensor

Contrast sensor is a device that can detect the contrast difference between the presence and absence of an object. The construction of the contrast sensor is shown in Figure 6.59. (The sensor in the photography was made by SICK, Germany.) Contrast sensors operate on the principle of high-energy reflections and detect differences in the gray values of matt, glossy or transparent objects and surfaces. The detection process evaluates the brightness or light level received from the sample presented to the sensor. Most contrast sensors use red or green LED light sources. Different colors absorb different amounts of light, and the wavelength of the light source can be chosen to provide the highest contrast for a given application. The most versatile contrast sensor offers a wide spectrum of white light sources with good resolution

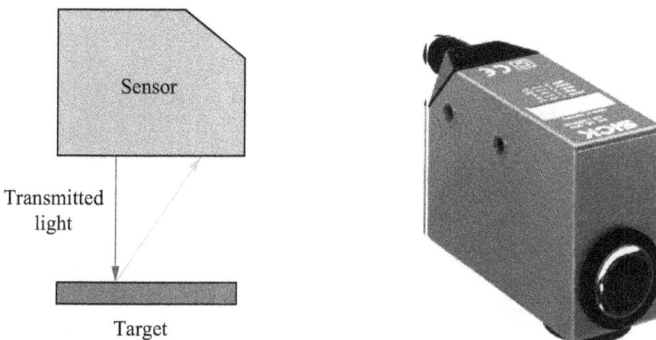

Figure 6.59: The construction of the contrast sensor.

for all color and background combinations. Contrast sensor is mainly used in packaging/printing machines to detect printing or control marks.

Advantages of the contrast sensor include the following:
The contrast sensor has high resolution and high sensitivity. It can detect the best contrast and can be compared with the diffuse reflection sensor that has the white LED.

Disadvantages of the contrast sensor include the following:
The contrast sensors are expensive for small industries and have a limited range of working distances.

Typical applications of the contrast sensor include the following:
The contrast sensors are used in the food, beverage and pharmaceutical industries for precise object positioning in printing presses, packaging plants and labeling machines. Contrast sensors can check the function and integrity of the LED, the printed mark on the package, the glue on the foil, the printed mark, the position of the material or the pipe and others.

6.12.3 Luminescence sensor

A luminescence sensor is a device that responds to materials with luminescent tracers, such as greases, coatings, adhesives and inks. These luminescent tracers emit light in the visible spectrum when excited by an ultraviolet light source. The construction of the contrast sensor is shown in Figure 6.60. (The sensor in the photography was made by SICK, Germany.). The detection process is based on luminophores, the luminescence of certain materials. The luminescence sensor transmits invisible ultraviolet light having a 375 nm wavelength to excite the light emitted by the light emitter in the visible range of the electromagnetic spectrum. The color or wavelength

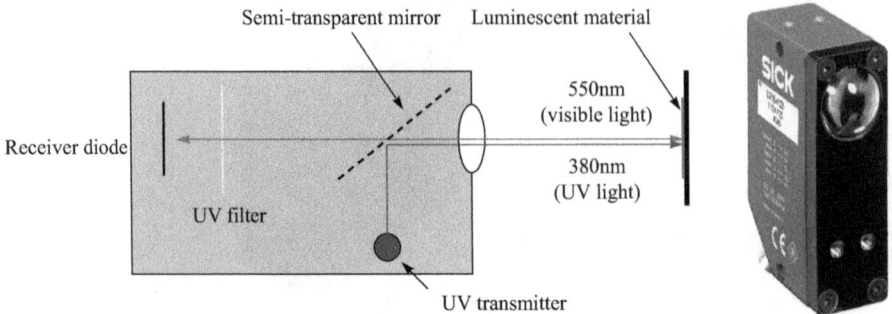

Figure 6.60: The construction of the luminescence sensor.

of the received light depends in part on the type of the luminophore. The luminescence sensor can detect the target when the sensor emits ultraviolet light and receives reflected visible light of a specific wavelength.

Advantages of the luminescence sensor include the following:
The luminescence sensor is particularly useful when detecting objects that are invisible. For example, a marker can be created on the target that appears only when exposed to UV, and this invisible marker does not affect the appearance of the packet. Luminescence sensors are much more sensitive than conventional photoelectrics and are particularly suitable for difficult applications such as detecting the distinction between color markings and similarly colored backgrounds.

Disadvantages of the luminescence sensor include the following:
The luminescent sensors can cause skin cancer if exposed to sufficient UV light. Like wood, the material itself has a luminescent nature with natural light. For example, traces on the wood need to have stronger luminescent properties than the wood for reliable detection.

Typical applications of the luminescence sensor include the following:
The luminescent sensors are used to sense glues, white paper, inks, clear labels, crayon, oils, greases, paints, gum, detergents and chalks with luminescent properties.

6.13 Environmental sensors

The environmental sensor is a device that can measure detailed environmental data with reliable data. This data is very important for well-being, comfort and productivity. Different types of environmental sensors discussed in this chapter are sound level sensor, humidity sensor, CO_2 sensor, particulate matter (PM2.5) sensor and volatile organic compound (VOC) sensor.

6.13.1 Sound level sensor

The sound level sensor, also known as sound pressure level meter, decibel (dB) meter or noise meter, is a device used to measure sound or noise level. The construction of the sound level sensor is shown in Figure 6.61. (The sensor in the photography was made by B & K Sound & Vibration Measurement, Denmark.) The sensor indicates the root-mean-square average of the signal. The simple method is using a microphone to capture sound and the most common unit of acoustic measurement for sound is dB. The sound level sensors can ensure the machine working within the

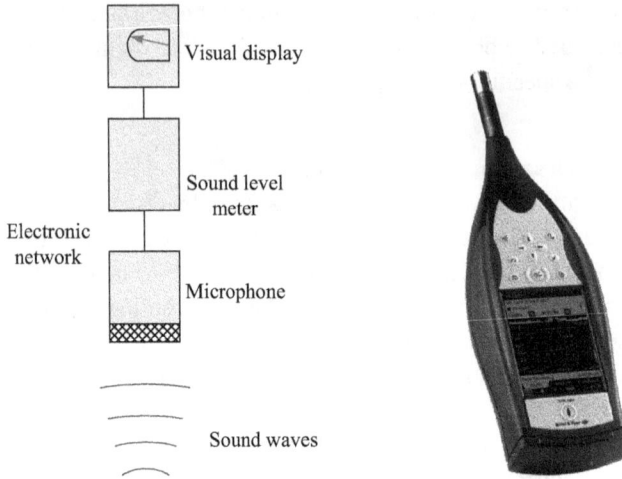

Figure 6.61: The construction of the sound level sensor.

rated noise limits. The IEC 61672-1:2013 is the current international standard for specifying the functionality and the performances of sound level meter.

Advantages of the sound level sensor include the following:
The sound level sensors are easy to use and operate in real time. It is used in security systems and can help staff make informed decisions about the potential hazards of hearing.

Disadvantages of the sound level sensor include the following:
The sound level sensor is expensive and not very accurate. When it is close to the radio signal, it is necessary to eliminate the interference.

Typical applications of the sound level sensor include the following:
The sound level sensors are used to measure construction noise, traffic noise in classrooms, architecture, industry, construction services, trade and so on.

6.13.2 Humidity sensor

A humidity sensor is a device for detecting and measuring water vapor. Based on the substance used for the measurement, there are various humidity sensors, including a resistive humidity sensor and a capacitive humidity sensor. The construction of the humidity sensor is shown in Figure 6.62. (The sensor in the photography was made by Aosong Electronics Co., Ltd, China.) Resistive humidity sensors measure the change in

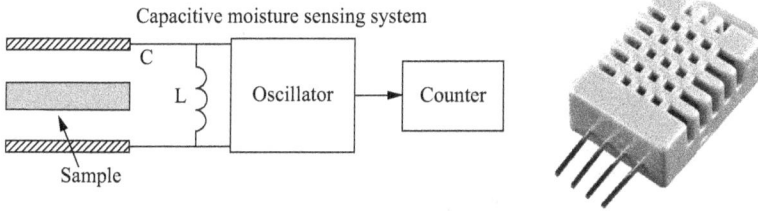

Figure 6.62: The construction of the humidity sensor.

electrical impedance of a hygroscopic medium (conductive polymer, salt or treated substrate). Capacitive humidity sensors are widely used in industrial, commercial and weather telemetry applications, and their operating principles are similar to those of plate capacitors. The lower electrode is deposited on a carrier substrate (glass, ceramic or silicon). It consists of a substrate where a thin film of metal oxide or polymer is deposited between two conductive electrodes. The water vapor molecules will enter or leave the hygroscopic polymer until the water vapor content is balanced with the ambient air or gas. The dielectric strength of the polymer is directly proportional to the water vapor content. The incremental change in the dielectric constant of a capacitive humidity sensor is almost proportional to the relative humidity of the surrounding environment.

Advantages of the humidity sensor include the following:
The humidity sensor has a low-temperature coefficient and can operate at high temperatures (up to 200 °C) and has reasonable chemical vapor resistance. It is flexible and does not require long maintenance.

Disadvantages of the humidity sensor include the following:
The humidity sensors have limited long-term stability. It is sensitive to condensation and certain corrosive substances.

Typical applications of the humidity sensor include the following:
The humidity sensors are widely used in the industrial, automotive, heavy truck, agriculture, aerospace, home appliances, commercial and weather telemetry applications.

6.13.3 CO_2 sensor

A carbon dioxide (CO_2) sensor is a device used to monitor air quality. The construction of the CO_2 sensor is shown in Figure 6.63. (The sensor in the photography was made by Parallax Inc., USA.) The most common sensor for measuring CO_2 is a

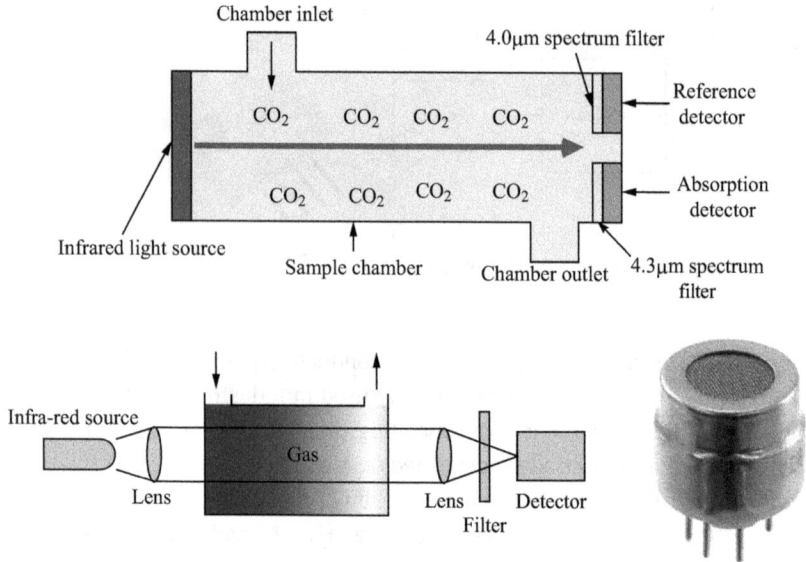

Figure 6.63: The construction of the CO_2 sensor.

nondispersive IR CO_2 sensor. Carbon dioxide, which is composed of two different atoms, absorbs IR radiation in a characteristic unique manner. The gas absorbs energy from a specific wavelength light when the light flows through a gas stream containing carbon dioxide. The remaining light is filtered to a wavelength specific to carbon dioxide. The difference between the amount of light radiated by the IR lamp and the amount of IR light received by the detector is measured, which is proportional to the number of CO_2 molecules.

Advantages of the CO_2 sensor include the following:
The CO_2 sensor is stable and highly selective to the measured gas. It is very important to improve soda water or air quality.

Disadvantages of the CO_2 sensor include the following:
The CO_2 sensor has a huge cross-sensitivity for other gases, and requires operating temperature requirements and limitations.

Typical applications of the CO_2 sensor include the following:
The CO_2 sensors are used in an atmosphere, indoor air quality, cellar and gas storage, marine vessels, greenhouses, landfill gas, confined spaces, ventilation management, and mining to monitor and improve air quality.

6.13.4 Particulate matter sensor

Atmospheric PM are tiny air pollutants that float in the air. They come from building materials, smoking, dust, car exhaust, cooking, charcoal power plant and so on. PM less than 10 μm in diameter is called PM10, and PM less than 2.5 μm in diameter is called PM2.5. PM2.5 particles can penetrate deeper into the lungs, and the body does not have a mechanism to repel them. PM2.5 is particularly harmful to the human respiratory system. Long-term exposure to the high concentration of PM2.5 can lead to lung cancer, mesothelioma, heart attacks, bronchitis and many other diseases.

PM sensor is a device used for air quality dust monitoring based on laser scattering principle. The construction of the PM2.5 sensor is shown in Figure 6.64. (The sensor in the photography was made by Xiaomi, China.) PM sensor utilizes laser scattering technology to scatter suspended particles in the air, and then collects scattered light to obtain a curve of scattered light over time. Light scattering can be caused when particles pass through the detection area. The scattered light is converted into electrical signals that are amplified and processed. Since the signal waveform has a certain relationship with the particle size, the equivalent particle diameter and the number of particles of different diameter per unit volume can be obtained through analysis.

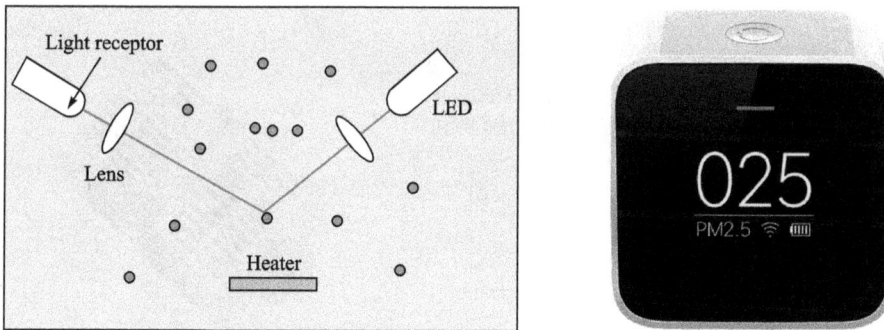

Figure 6.64: The construction of the PM2.5 sensor.

Advantages of the PM sensor include the following:
The PM sensor has high sensitivity, small size, lightweight, simple maintenance and long-term stability. It is a low-cost way to measure the particle diameter and the number of particles in the air.

Disadvantages of the PM sensor include the following:
The PM sensor can provide size information, but may encounter difficulties at high particle concentrations.

Typical applications of the PM sensor include the following:
The PM sensor is mainly used to measure PM in the air. The PM comes from building materials, dust, smoking, cooking, car exhaust and so on.

6.13.5 Volatile organic compound sensor

A VOC sensor is a device used for monitoring VOCs, which are organic chemicals with a high vapor pressure at room temperature. The construction of the PM2.5 sensor is shown in Figure 6.65. (The sensor in the photography was made by LOGOELE, China.) Typical sources of VOCs include paint, glue, furniture, permanent marking and emissions from the oil or gas industry. Some VOCs are harmful to human health and may cause environmental damage. The MOS-type gas sensor is commonly used to measure VOCs. In the clear air, the oxygen concentration is 21%. When the sensor is exposed to VOC gas, the oxidation reaction of the gas with adsorbed oxygen takes place on the surface of tin dioxide. The density of oxygen adsorbed on the surface of tin dioxide decreases. By measuring the resistance change of the MOS-type gas sensor, the VOC gas concentration in the air can be detected.

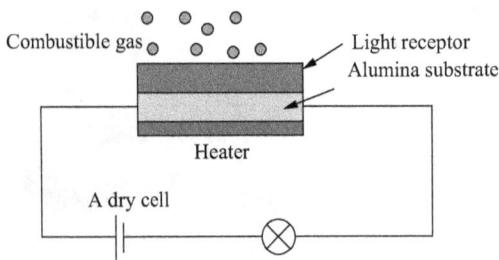

Figure 6.65: The construction of the VOC sensor.

Advantages of the VOC sensor include the following:
The VOC sensor can measure air pollutants other than breathing, such as building materials, detergents, perfumes and furniture and carpet degassers. The further benefit is that it does not require extra work.

Disadvantages of the VOC sensor include the following:
The VOC sensor is complicated to operate. This is because VOCs do not retain well and must be trapped in stronger adsorbent materials.

Typical applications of the VOC sensor include the following:
The VOC sensors are widely used to monitor indoor air quality, urban air, roadside air, industrial processes, vehicle emissions, food, perfume, evaporation losses from gasoline storage and forest fires.

References

Ahmed, S., Huang, B., Shah, S. L. Novel identification method from step response. Control Engineering Practice. 2007, 15(5): 545–556.

Ahmed, Salim, Huang, Biao, Shah, Sirish L. Identification from step responses with transient initial conditions. Journal of Process Control. 2008, 18 (2): 121–130.

Alciatore, David G., Histand, Michael B. Introduction to Mechatronics and Measurement Systems (Fourth Edition). New York: McGraw-Hill Science/Engineering/Math. 2002, 496 pages.

Alonso-Martín, Fernando, Castro-González, Aívaro, de Gorostiza Luengo, Francisco Javier Fernandez, Ángel Salichs, Miguel. Augmented robotics dialog system for enhancing human–robot interaction. Sensors 2015, 15(7): 15799–15829.

Atul Kr. Dewangan, Meenu Gu pta, Pratibha Patel. Automation of railway gate control using frequency modulation techniques. VSRD International Journal of Electrical, Electronics and Communication Engineering. 2012, 2(6): 288–298.

Azmaiparashvili, Z. A. Method for measuring the resonant frequency of an oscillating system. Measurement Technologies. 2004, 47(9): 920–925.

Bartali, A El, Aubin, V., Degallaix, S. Fatigue damage analysis in a duplex stainless steel by digital image correlation technique. Fatigue and Fracture of Engineering Materials and Structures. 2008, 31(2): 137–151.

Becker, T. H, Mostafavi, M, Tait, R. B, Marrow TJ. An approach to calculate the J-integral by digital image correlation displacement field measurement. Fatigue and Fracture of Engineering Materials and Structures. 2012, 35(10): 971–984.

Beckwith, Thomas G., Marangoni, Roy D., Lienhard V., John H. Mechanical Measurements (6th Edition). London: Pearson. 2006, 784 pages.

Berfield, T A., Patel, J K., Shimmin, R G., Braun, P V., Lambros, J., Sottos, N R. Micro- and nanoscale deformation measurement of surface and internal planes via digital image correlation. Exp. 2007, 47(1): 51–62.

Bing, P., Hui-min, X., Tao, H., Asundi, A. Measurement of coefficient of thermal expansion of films using digital image correlation method. Polymer Testing. 2009, 28(1): 75–83.

Bocarnea, Mihai C., Reynolds, Rodney A., Baker, Jason D. Online Instruments, Data Collection, and Electronic Measurements: Organizational Advancements 1st Edition. Hershey: IGI Global. 2012, 397 pages.

Brachman, R W I., McLeod, H A., Moore, I D., Take, W A. Three-dimensional ground displacements from static pipe bursting in stiff clay. Canadian Geotechnical Journal. 2010, 47(4): 439–450.

Brunelli, Alessandro. Calibration Handbook of Measuring Instruments 1st Edition. Research Triangle Park: International Society of Automation. 2017, 352 pages.

Chang, S., Wang, C. S, Xiong, C. Y, Fang, J. Nanoscale in-plane displacement evaluation by AFM scanning and digital image correlation processing. Nanotechnology. 2005, 16(4): 344–349.

Cheatle, Keith R. Fundamentals of Test Measurement Instrumentation. Research Triangle Park: Instrumentation, Systems, and Automation Society. 2006, 329 pages.

Chu, T. C., Ranson, W. F., Sutton, M. A. Applications of digital-image-correlation techniques to experimental mechanics. Experimental Mechanics. 1985, 25(3): 232–244.

Clausen, Jesper, Knudsen, Asger. Nondestructive testing of bridge decks and tunnel linings using impulse-response. American Concrete Institute, ACI Special Publication. 2009, 263–275

Dautriat, J., Bornert M., Gland, N., Dimanov A., Raphanel, J. Localized deformation induced by heterogeneities in porous carbonate analysed by multi-scale digital image correlation. Tectonophysics. 2011, 503(1): 100–116.

https://doi.org/10.1515/9783110624397-007

Dickinson, AS, Taylor, AC, Ozturk, H, Browne, M. Experimental validation of a finite element model of the proximal femur using digital image correlation and a composite bone model. Journal of Biomechanical Engineering. 2011, 133(1): 014504.

Dominguez-Lopez, Alejandro, Soto, Marcelo A., Martin-Lopez, Sonia, etc. Resolving 1 million sensing points in an optimized differential time-domain Brillouin sensor. Optics Letters. 2017, 42(10): 1903–1906.

Fercher, A. F., Briers, J. D. Flow visualization by means of single-exposure speckle photography. Optics Communications. 1981, 37(5): 326–330.

Figliola, Richard S., Beasley, Donald E. Theory and Design for Mechanical Measurements (Sixth Edition). Hoboken: John Wiley & Sons, Inc. 2014, 614 pages.

Giacomo Lionello, Camille Sirieix, Massimiliano Baleani. An effective procedure to create a speckle pattern on biological soft tissue for digital image correlation measurements. Journal of the Mechanical Behavior of Biomedical Materials. 2014, 39(39): 1–8.

Gilchrist, C. L., Xia, J. Q., Setton, L. A., Hsu, E. W. High-resolution determination of soft tissue deformations using MRI and first-order texture correlation. IEEE transactions on medical imaging. 2004, 23(5): 546–553.

Godara, A., Raabe D., Bergmann I., Putz R., Müller U. Influence of additives on the global mechanical behavior and the microscopic strain localization in wood reinforced polypropylene composites during tensile deformation investigated using digital image correlation. Composites Science and Technology. 2009, 69(2): 139–146.

Grant, B. M, Stone, H. J, Withers, P. J, Preuss, M. High-temperature strain field measurement using digital image correlation. The Journal of Strain Analysis for Engineering Design. 2009, 44(4): 263–271.

Guo, Xiang, Liang, Jin, Tang, Zhengzong, Cao, Binggang, Yu, Miao High-temperature digital image correlation method for full-field deformation measurement captured with filters at 2,600°C using spraying to form speckle patterns. Optical Engineering 2014, 53(6): 063101.

Gustavsson, I. Survey of applications of identification in chemical and physical processes. Proc. 3rd IFAC Symposium, the Hague/ Delft, the Netherlands. 1973, 67–85.

Hand, David J. Measurement: A Very Short Introduction (Very Short Introductions) 1st Edition. Oxford: Oxford University Press. 2016, 144 pages.

Heinz, S. R, Wiggins, J. S. Uniaxial compression analysis of glassy polymer networks using digital image correlation. Polymer Testing. 2010, 29(8): 925–932.

Hughes, Ifan, Hase, Thomas. Measurements and their Uncertainties: A practical guide to modern error analysis 1st Edition. Oxford: Oxford University Press. 2010, 160 pages.

Hung, P. C, Voloshin, A. S. In-plane strain measurement by digital image correlation. Journal of the Brazilian Society of Mechanical Sciences and Engineering. 2003, 25(3): 215–221.

Hughes, Thomas A. Measurement and Control Basics, 4rd Edition. Research Triangle Park: Instrumentation, Systems, and Automation Society. 2006, 375 pages.

Jin, T. L., H. a, N. S., Goo, N. S. A study of the thermal buckling behavior of a circular aluminum plate using the digital image correlation technique and finite element analysis. Thin-Walled Structures. 2014, 77(77): 187–197.

Kamakshaiah, S., Amarnath, J., Krishna Murthy., Pannala. Electrical Measurements and Measuring Instruments. New Delhi: IK International Publishing House. 2011, 480 pages.

Kammers, A D., Daly, S. Small-scale patterning methods for digital image correlation under scanning electron microscopy. Measurement Scienceand Technology. 2011, 2(12): 125501.

Y. Kong, P. X. Du, Z. H. Tan, "Simulation System of Railway Signal Transmission and On-Board Cab Signal Receiving," 2009 IEEE 70th Vehicular Technology Conference Fall, Anchorage. AK. 2009, 1–5.

Krehbiel, J. D., Lambros, J., Viator, J. A., Sottos, N. R. Digital image correlation for improved detection of basal cell carcinoma. Experimental Mechanics. 2010, 50(6): 813–824.

Kreis, T. Holographic interferometry: principles and methods. Simulation and Experiment in Laser Metrology: Proceedings of the International Symposium on Laser Applications in Precision Measurements Held in Balatonfüred/Hungary. 1996, 2: 323.

Kulakowski, Bohdan T., Gardner, John F., Lowen Shearer., J. Dynamic Modeling and Control of Engineering Systems. New York: Cambridge University Press, 2014. 502 pages.

Kularatna, Nihal. Digital and Analogue Instrumentation: Testing and measurement. London: The Institution of Engineering and Technology. 2003, 675 pages.

YanJie, Li, HuiMin, Xie, Qiang, Luo, ChangZhi, Gu, ZhenXing, Hu, PengWan, Chen, QingMing, Zhang. Fabrication technique of micro/nano-scale speckle patterns with focused ion beam. Science China Physics, Mechanics and Astronomy. 2012, 55(6): 1037–1044.

Li, Ning, Guo, Siming, Sutton, Michael A. Recent Progress in E-Beam Lithography for SEM Patterning. MEMS and Nanotechnology. 2011, 2: 163–166.

Li, Y., Xie, H., Luo, Q., Gu, C., Hu, Z., Chen, P., Zhang, Q. Fabrication technique of speckle patterns with focused ion beam. Science China Physics Mechanics 2012, 55(6): 1037–1044.

Li,Y. J, Xie, H. M, Luo, Q, Gu, C. Z, Hu, Z. X, Chen, P. W. Fabrication and optimization of micro-scale speckle patterns for digital image correlation. Measurement Science and Technology. 2016, 27(1): 015203.

Libertiaux, V., Pascon, F., S. Cescotto. Experimental verification of brain tissue incompressibility using digital image correlation. Journal of the Mechanical Behavior of Biomedical Materials, 2011, 4: 1177–1185.

Lin, Chuhong, Kätelhön, Enno, Sepunaru, Lior, Compton, Richard G. Understanding single enzyme activity via the nano-impact technique. Chemical Science. 2017, 8(9): 6423–6432.

Lin, Liwei, Pisano, A. P., Howe, R. T. A micro strain gauge with mechanical amplifier. Journal of Microelectromechanical Systems. 1997, 6(4): 313–321.

Lionello, Giacomo, Cristofolini, Luca. A practical approach to optimizing the preparation of speckle patterns for digital-image correlation. Measurement Science and Technology. 2014, 25(10): 107001.

Liptak, Bela G. Instrument Engineers' Handbook: Process Control 3rd Edition. Boca Raton: CRC Press. 1995, 1584 pages.

Lockhart, Paul. Measurement Reprint Edition. Cambridge: Belknap Press. 2014, 416 pages

Malaric, Roman. Instrumentation and Measurement in Electrical Engineering. Boca Raton: BrownWalker press. 2011, 219 pages.

Matsubara, Atsushi, Ibaraki, Soichi. Monitoring and control of cutting forces in machining processes: A review. International Journal of Automation Technology. 2009, 3(4): 445–457.

Mazzoleni, Paolo, Zappa, Emanuele, Matta, Fabio, Sutton, Michael A. Thermo-mechanical toner transfer for high-quality digital image correlation speckle patterns. Optics and Lasers in Engineering. 2015, 75: 72–80.

Morris, Alan S., Langari, Reza. Measurement and Instrumentation: Theory and Application. Oxford: Butterworth-Heinemann. 2011, 640 pages.

Nakadate, S., Yatagai, T., Saito, H. Electronic speckle pattern interferometry using digital image processing techniques. Applied Optics. 1980, 19(11): 1879–1883.

Nakra, B. C., Chaudhry, K. K. Instrumentation, Measurement and Analysis. New Delhi: Tata McGraw-Hill Education. 2003, 632 pages.

Nguyen, T. L, Hall, S. A, Vacher, P., Viggiani, G. Fracture mechanisms in soft rock: Identification and quantification of evolving displacement discontinuities by extended digital image correlation. Tectonophysics. 2011, 503(1): 117–128.

Pan, B., Wang, B., Lubineau, G. Moussawi. Comparison of subset-based local and finite element-based global digital image correlation. Experimental Mechanics. 2015, 55(5): 887–901.

Pan, B., Wang, B., Lubineau, G. Comparison of subset-based local and FE-based global digital image correlation: Theoretical error analysis and validation. Optics and Lasers in Engineering. 2016, 82: 148–158.

Pan, B., Wu, D., Wang, Z., Xia, Y. High-temperature digital image correlation method for full-field deformation measurement at 1200 C. Measurement science and technology. 2010a, 22(22): 015701.

Pan, B., Wu, D., Xia, Y. High-temperature deformation field measurement by combining transient aerodynamic heating simulation system and reliability-guided digital image correlation. Optics and Lasers in Engineering. 2010b, 48(9): 841–848.

Pan, B., Xie, H., Guo, Z., Hua, T. Full-field strain measurement using a two-dimensional savitzky-golay digital differentiator in digital image correlation. Optical Engineering. 2007, 46(3): 033601.

Passieux, J.-C., Bugarin, F., David, C., Périé, J.-N., Robert, L. Multiscale displacement field measurement using digital image correlation: Application to the identification of elastic properties. Experimental Mechanics. 2015, 55(1): 121–137.

Peters, W. H., Ranson, W. F. Digital imaging techniques in experimental stress analysis. Optical Engineering. 1982, 21(3): 427–432.

Pitre, Krishna, Tiwari., Sweety. Microfaradaic electrochemical biosensors for the study of anticancer action of DNA intercalating drug: Epirubicin. Biosensors for Health, Environment and Biosecurity. London: InTech. 2011, 550 pages.

Pittari, III J, Subhash, G. Fracture toughness testing of advanced silicon carbide ceramics using digital image correlation. Dynamic Behavior of Materials. 2015, 1: 207–212.

Pritchard, R. H, Lava, P., Debruyne, D., Terentjev, E. M. Precise determination of the Poisson ratio in soft materials with 2D digital image correlation. Soft Matter. 2013, 9: 6037–6045.

Rajput, R. K. Electrical Measurements and Measuring Instruments. New Delhi: S Chand & Co Ltd. 2007, 741 pages.

Réthoré, J., Roux, S., Hild, F. An extended and integrated digital image correlation technique applied to the analysis of fractured samples: The equilibrium gap method as a mechanical filter. European Journal of Computational Mechanics/Revue Européenne de MécaniqueNumérique. 2009, 18(3–4): 285–306.

Risbet, M., Feissel, P., Roland, T., Brancherie, D., Roelandt, J. M. Digital Image Correlation technique: Application to early fatigue damage detection in stainless steel. Procedia Engineering. 2010, 2(1): 2219–2227.

Salmanpour, A., Mojsilovic, N. Application of Digital Image Correlation for strain measurements of large masonry walls. Proceedings of the 5th Asia Pacific Congress on Computational Mechanics. Queens Town, Singapore. 2013: 11–14.

Sánchez-Arévalo F. M, Pulos G. Use of digital image correlation to determine the mechanical behavior of materials. Materials Characterization. 2008, 59(11): 1572–1579.

Tung, S. H, Shih, M. H, Kuo, J. C. Application of digital image correlation for anisotropic plastic deformation during tension testing. Optics and Lasers in Engineering. 2010, 48(5): 636–641.

Shibo, Xiong, Changyi, Huang. Measurement Technology of Mechanical Engineering (3rd Edition). Beijing: China Machine Press, 2007, 283 pages. (only available in Chinese)

Stinville, J. C., Echlin, M. P., Texier, D., Bridier, F., Bocher, P., Pollock, T. M. Sub-grain scale digital image correlation by electron microscopy for polycrystalline materials during elastic and plastic deformation. Experimental Mechanics. 2016, 56(20): 1–20.

Stoica, Petre, Moses, Randolph. Spectral Analysis of Signals. Upper Saddle River, N.J.: Pearson/ Prentice Hall. 2005, 452 pages.

Sun Y; Pang J. H; Fan W. Nanoscale deformation measurement of microscale interconnection assemblies by a digital image correlation technique. Nanotechnology. 2007, 18(39): 395504.

Sutton, M. A., Wolters, W. J., Peters, W. H., Ranson, W. F., McNeill, S. F. Determination of displacements using an improved digital correlation method. Image and Vision Computing. 1983, 1(3): 133–139.

Sztefek, P., Vanleene, M., Olsson, R. Using digital image correlation to determine bone surface strains during loading and after adaptation of the mouse tibia. Journal of Biomechanics. 2010, 43(4): 599–605.

Tang, Z., Liang, J., Xiao, Z., Guo, C. Large deformation measurement scheme for 3D digital image correlation method. Optics and Lasers in Engineering. 2012, 50(2): 122–130.

Tarigopula, V., Hopperstad, O. S, Langseth, M., Clausen A. H, Hild F. A study of localisation in dual-phase high-strength steels under dynamic loading using digital image correlation and FE analysis. International Journal of Solids and Structures. 2008, 45(2): 601–619.

Thompson, M. S, Schell, H., Lienau, J., Duda, G. N. Digital image correlation: a technique for determining local mechanical conditions within early bone callus. Medical Engineering and Physics. 2007, 29(7): 820–823.

Van Etten, Wim C. Introduction to Random Signals and Noise. Hoboken: John Wiley & Sons, Inc. 2005, 270 pages.

Vendroux, G., Schmidt, N., Knauss, W. Submicron deformation field measurements: Part 3. Demonstration of deformation determinations. Experimental Mechanics. 1998, 38 (3): 154–160.

Venkateshan, S.P. Mechanical Measurements (2nd Edition). Hoboken: John Wiley & Sons, Inc. 2015, 550 pages

Vijayachitra, S. Transducers Engineering. Delhi: PHI Learning Limited, Rimjhim House, 2016

Wang, P., Pierron, F., Thomsen, O. T. Identification of material parameters of PVC foams using digital image correlation and the virtual fields method. Experimental Mechanics. 2013, 53(6): 1001–1015.

Webster, John G. Electrical Measurement, Signal Processing, and Displays (Principles and Applications in Engineering) 1st Edition. Boca Raton: CRC Press. 2003, 768 pages

Winiarski, B., Schajer, G. S., Withers, P. J. Surface decoration for improving the accuracy of displacement measurements by digital image correlation in SEM. Experimental Mechanics. 2012, 52(7): 793–804.

Witte, Robert A. Electronic Test Instruments: Analog and Digital Measurements (2nd Edition). Upper Saddle River: Prentice Hall. 2002, 400 pages.

Wojciechowski, S., Twardowski, P. Cutting forces and vibrations analysis in milling of tungsten carbide with CBN cutters. Proceedings of 4th CIRP International Conference on High Performance Cutting, 24–26 october, 2010, Nagaragawa Convention Center, Gifu, Japan, 2010.

Xu, Z. H., Sutton, M. A., Li, X. D. Mapping nanoscale wear field by combined atomic force microscopy and digital image correlation techniques. Acta Materialia. 2008, 56(20): 6304–6309.

Yamaguchi, I. A laser-speckle strain gauge. Journal of Physics E: Scientific Instruments. 2000, 14 (11): 1270–1273

Yin, Juan, Cao, Yuan, Li, Yu-Huai, Liao, Sheng-Kai, Zhang, Liang, Ren, Ji-Gang, etc. Satellite-based entanglement distribution over 1200 kilometers. Science. 2017, 356(6343): 1140–1144.

Yoneyama, S., Morimoto, Y., Takashi, M. Automatic Evaluation of Mixed-mode Stress Intensity Factors Utilizing Digital Image Correlation. Strain. 2006, 42(1): 21–29.

Yu, L., Pan, B. Single-camera stereo-digital image correlation with a four-mirror adapter: optimized design and validation. Optics and Lasers in Engineering. 2016, 87: 120–128.

Websites

http://encyclopedia2.thefreedictionary.com/Measurement+Technology

https://en.wikipedia.org/wiki/Measurement

https://en.wikipedia.org/wiki/Filter_(signal_processing)

https://en.wikipedia.org/wiki/Integrator

https://encyclopedia.thefreedictionary.com/Sound+level+meter

http://www.robotplatform.com/knowledge/sensors/types_of_robot_sensors.html

http://www.laas.fr/robots/jido/data/en/robot.php

http://augmenting.me/cte/programs/small_grants/2005/mobile_robot.htm

https://www.tradeindia.com/fp346247/Automated-Guided-Vehicle.html

https://en.wikipedia.org/wiki/Automated_guided_vehicle

http://www.mfg.mtu.edu/cyberman/quality/metrology/force.html

http://mfgnewsweb.com/archives/4/47918/Automation-Equip-and-Systems-dec16/Digital-Factory-Solution.aspx

http://www.hitachi.com/businesses/infrastructure/product_site/car/

https://www.labnews.co.uk/features/measurement-in-medicine-a-delicate-balance-20-07-2010/

https://en.wikipedia.org/wiki/Medical_device

https://www.campbellsci.com.au/agriculture-plant-physiology

https://atlasofscience.org/nano-impacts-a-new-perspective-on-enzymes/

http://www.swansea.ac.uk/engineering/nanohealth/researchareas/newsensorsdevices/

https://phys.org/news/2017-05-fiber-based-sensor-quickly-problems-bridges.html

https://www.techopedia.com/definition/31462/intelligent-sensor

https://en.wikipedia.org/wiki/Intelligent_sensor

https://www.oilandgasmiddleeast.com/article-10129-top-10-instruments-providers

https://www.plantautomation-technology.com/articles/top-industrial-automation-companies-in-the-world

http://www.ni.com/en-us.html

http://nptel.ac.in/courses/117106090/

http://nptel.ac.in/courses/117106090/Pdfs/1_6.pdf

https://en.wikipedia.org/wiki/Dirac_delta_function

http://web.eecs.utk.edu/~roberts/ECE342/RandomSignalsAndNoise.pdf

https://en.wikipedia.org/wiki/Almost_periodic_function

http://signalsandsystems.wikidot.com/notes-signals-energy

http://mathworld.wolfram.com/FourierSeries.html

http://mathworld.wolfram.com/FourierSeriesSawtoothWave.html

https://users.wpi.edu/~goulet/Matlab/overlap/efs.html

https://meettechniek.info/compendium/average-effective.html

http://circuitglobe.com/what-is-peak-value-average-value-and-rms-value.html

https://en.wikipedia.org/wiki/Window_function#Processing_gain_and_losses

http://rfteststation.com/null-type-and-deflection-type-instruments/

https://www.onosokki.co.jp/English/hp_e/products/application/FFT/fft_v_1_2.htm

http://www.olsoninstruments.com/slab-impulse-response-ndt.php

http://www.olsoninstruments.com/sonic-echo-impulse-response-ndt.php

https://tarabah.me/roomresponse1/

http://aurora-plugins.forumfree.it/?t=53443032

https://www.slideshare.net/asadwarraichc/tacoma-narrows-suspension-bridge

https://en.wikipedia.org/wiki/Nyquist_plot

http://lpsa.swarthmore.edu/Transient/TransInputs/TransStep.html

https://en.wikipedia.org/wiki/Capacitive_coupling
https://www.primuscable.com/store/p/
http://www.circuitstoday.com/strain-gauge
https://store.chipkin.com/articles/strain-gauge
https://fujihita.wordpress.com/2017/11/06/memo-strain-gauge-bridge-circuits/
https://electronicsproject.org/wheatstone-ac-bridge/
https://en.wikipedia.org/wiki/Maxwell_bridge
https://en.wikipedia.org/wiki/Wien_bridge
https://circuitglobe.com/hays-bridge.html
https://en.wikipedia.org/wiki/Schering_Bridge
https://circuitglobe.com/ac-bridge.html
https://www.allaboutcircuits.com/textbook/alternating-current/chpt-12/
ac-bridge-circuits/
http://portal.unimap.edu.my/portal/page/portal30/Lecturer%20Notes/
https://electronicspost.com/ring-modulator-for-the-double-sideband-suppressed-carrier-
generation/
https://en.wikipedia.org/wiki/Ring_modulation
http://www.tml.jp/e/product/instrument/instrument_sub/dc004p.html
http://www.yourdictionary.com/discriminator#1MRr7LagFltArWhX.99
https://www.edgefx.in/types-of-modulation-techniques-with-applications/
https://en.wikipedia.org/wiki/Filter_(signal_processing)
https://en.wikipedia.org/wiki/Low-pass_filter
https://en.wikipedia.org/wiki/Filter_bank
http://www.linear.com/product/LTC1564
http://www.microwavefilter.com/pdffiles/pg14.pdf
https://www.honeywellprocess.com/en-US/explore/products/instrumentation/recorders-and-data
-acquisition/circular-chart-recorders/
https://en.wikipedia.org/wiki/Chart_recorder
http://www.eeeguide.com/galvanometer-type-recorder/
http://www.industrial-electronics.com/DAQ/mi_8.html
https://ocw.mit.edu/courses/mechanical-engineering/2-161-signal-processing-continuous-and-
discrete-fall-2008/lecture-notes/
https://en.wikipedia.org/wiki/Autocorrelation
https://en.wikipedia.org/wiki/Cross-correlation
https://uk.rs-online.com/web/p/thermocouples/3971264/
http://www.ussensor.com/precision-interchangeable-thermistors-05%C2%
B0c-and-10%C2%B0c-accuracy
https://www.omega.com/pptst/OM-2628.html
http://www.a-n-instruments.com/bi_metal.htm
http://www.pcb.com/Resources/Technical-Information/Tech_Pres
http://validyne.com/category/Pressure-Sensors
https://www.flowcontrolnetwork.com/selecting-pressure-gauges-new-advances-old-technology/
http://www.spiraxsarco.com/Resources/Pages/Steam-Engineering-Tutorials/flowmetering/princi
ples-of-flowmetering.aspx
http://www.emcocontrols.com/353/venturi-tube-type-kvr
http://www.alicat.com/alicat-blog/mass-flow-measurement-techniques-radar/
http://www.emcocontrols.com/358/emco-asme-flow-nozzle
http://www.odealsplus.com/webpage/FTI%20TB%20Flowmeter.html
http://www.ftimeters.com/products/turbine_flowmeters.htm

https://en.wikipedia.org/wiki/Vortex_shedding
http://www.sierrainstruments.com/products/heavyindustry.html
http://www.fluidcomponents.com/
https://www.badgermeter.com/business-lines/flow-instrumentation/
ufx-handheld-doppler-ultrasonic-flow-meters/
http://greyline.com/index.php/products/water-wastewater/ultrasonic-flow-meters/ttfm-1-0-
transit-time-flowmeter-detail
https://www.omega.com/technical-learning/dif-between-doppler-transit-time-ultrasonic-flow-
meters.html
http://measurementsci.com/products/minildv/
http://instrumentationandcontrollers.blogspot.in/2012/11/laser-doppler-anemometer.html
https://www.omega.com/technical-learning/linear-variable-displacement-transducers.html
http://www.rdpe.com/ex/hiw-lvdt.htm
https://en.wikipedia.org/wiki/Capacitive_displacement_sensor
http://www.chenyang-ism.com/CapaSensorPosi.htm
https://www.keyence.com/ss/products/sensor/sensorbasics/eddy_current/
https://www.keyence.com/ss/products/measure/measurement_library/type/laser_1d/
http://www.anaheimautomation.com/manuals/forms/encoder-guide.php#sthash.H4HzemJW.dpbs
https://commons.wikimedia.org/wiki/File:Linear_Scale_Scheme.svg
http://hades.mech.northwestern.edu/index.php/File:Encoder_diagram.png
https://www.bestech.com.au/isotron-iepe/
https://www.checkline.com/product/HTM
https://www.bestech.com.au/piezoelectric/
https://www.motavera.com/how-does-an-accelerometer-work.html
http://www.analog.com/media/en/technical-documentation/data-sheets/ADXL362.pdf
https://www.tekscan.com/resources/ebook/load-cell-vs-force-sensor
http://www.npl.co.uk/reference/faqs/how-many-different-types-of-force-transducer-are-there
-(faq-force)#pressure
http://www.msdkr.com/news/articleView.html?idxno=44
http://www.microlevel.com.tr/EN/Urunler/Detay.aspx?CID=4&KID=1&ID=27
http://www.pvl.co.uk/capacitive-level-sensors.html
http://www.pvl.co.uk/conductive-level-sensors.html
https://www.babbittinternational.com/mls-4ex-multi-point-float-level-switches.html
http://www.you-ideal.com/ProductView.asp?ID=158
https://www.omega.com/pptst/LVD-803_LVD-804.html
https://www.uwt.de/en/products/point-level-measurement/rotonivor-rotary-paddle/rn-3000-
series.html
https://www.stemmer-imaging.co.uk/en/knowledge-base/ccd/
https://www.smartinfoblog.com/cmos-vs-ccd-sensor/
http://www.fargocontrols.com/sensors/photo_op.html
http://www.minipedia.org.ua/exo-otskakivayushhij-zvuk/
https://www.keyence.com/ss/products/sensor/sensorbasics/proximity/info/

Index

2D-DIC technique 209

A/D converters 6
ABB 21
absolute sensor 249
AC bridge circuit 136, 145
AC count signal 167
AC/DC tachometer 251
acceleration sensor 9, 253
Accuracy 79
accuracy as percentage of scale span 79
accuracy as percentage of true value 79
Active differentiator 7
active instruments 71
alcohol-in-glass thermometer 4
aliasing 185
almost periodic signals 31
AM wave 154
amplification 6
amplitude modulation 153
analog filters 170
Analog instrument 73
analog output 6
Angular accelerometer 258
anti-aliasing filter 186
aperiodic signal (Non-periodic signal) 31
Aperiodic transient signal 50
aser-guided vehicles (LGV) 9
Atmospheric particulate matters (PM) 285
Autocorrelation 197
automated guided vehicle 7
automatic diagnosis 11
Automatic type instrument 75
Average Absolute Value 48
Average Power 48, 50
Average Value 48
axial strain 137

balanced bridge circuit 136
balanced bridge circuits 144
Band-pass filter 169, 177
Band-stop filter 169
Bandwidth 172
Basic AC bridge circuit 151
Bias 85
bimetallic temperature device 220

Bode plot 105
Bourdon tube 228
Bourdon tube pressure sensors 228
bridge circuit 135
Building monitoring and security systems 13

Cable shield 134
capacitance bridge circuit 146
Capacitance displacement sensor 242
Capacitance level sensor 266
capacitance pressure sensor 225
Capacitive coupling 132
capacitive high-pass (CR) filter 175
capacitive humidity sensor 282
capacitive proximity sensor 276
Carbon dioxide (CO_2) sensor 283
Cauchy–Schwarz inequality 197
CCD sensor 271
charge amplifier 262
circular chart recorder 189
CMOS sensor 272
color sensor 278
combined drift 87
Complex Fourier Coefficient 44
composite periodic signal 30
computer monitor 188
confocal method 245
contact sensor 8
Contacting type instrument 75
continuous random sequences 69
continuous stochastic process 69
continuous swept-sine method 93
Continuous-time signals 27
Contrast sensor 279
Convolution 60, 89
correlation coefficient 195
Critically damped case 115
Cross-correlation function 200
current 27
Cutoff frequency 171
Cutting force measurements 10

damped oscillating signals 50
damping ratio 99
Danaher 24
DC (Wheatstone) bridge circuit 138

DC bridge circuit 136
dead zone 85
Deflection-type instrument 73
Delay time 120
Demodulation 153
demodulation/ detection 158
deterministic signal 28
differential operation 27
Differential pressure flowmeter 230
Differential property 77
Differentiation and Integration 60
differentiators 6
Digital filters 170
Digital Image Correlation method (DIC) 205
Digital instrument 73
Digital recorders 191
Dirac comb 65
Dirac Delta function 62
discrete random sequence 69
discrete stochastic process 69
Discrete-time signals 27
displacement amplitude 27
displacement frequency 27
displacement phase information 27
Displacement sensor 240
distance sensor 9
Distributed fiber-optic sensors 18
door sensor 13
Doppler Effect 239
Doppler ultrasonic flowmeter 237
Drift 87
dynamic characteristics 75
dynamic error 88
dynamic strain 162
dynamic strain gauge 162

electrical isolation 6
electrical signal 27
electrical signals 6
Electrodynamic velocity sensor 252
electro-hydraulic force excitation
       instrument 94
Emerson 22
end-point straight-line method 82
Endress+Hauser 22
energy signal 33
environmental sensor 281
equal resolution filterbank 179
Equipment Monitoring 11

Error 88
even signal 31
exponential decay signals 50
Exponential Fourier Series Representation 37

fidelity 88
filter 169
filtering 6
fire alarm and home security monitoring
       system 13
First order measuring system 123
first order systems 104
Float level switch 268
Flow nozzle 232
Flow sensor 230
FM ratio detector 165
foil strain gauge 260
force sensor 259
Fourier Decomposition of A Sawtooth
       Wave 35
Fourier transform pair 51
Frequency discrimination 165
frequency domain analysis 34
Frequency Modulation (FM) 162
frequency modulation 153
Frequency property 78
frequency response function 90
full-bridge strain gauge circuit 143
full-wave non-phase-sensitive detection 159

GE Measurement & Control 23
global quantum communication network 19
Grating displacement sensor 249
ground loops 7

half-bridge strain gauge circuit 142
half-wave non-phase-sensitive detection 159
half-wave phase-sensitive detection 159
Hay bridge circuit 150
heat sensor 13
High-pass filter 169, 175
High-resolution laser displacement
       sensor 244
home and office automation system (AVCS,
       Kenya) 14
Honeywell 23
Hooke's Law 137
Humidity sensor 282
Hysteresis 86

IC (Integrated circuit) temperature sensor 218
Identification of Dynamic Characteristics 123
Image filtering 182
image sensor 270
impulse response function (IRF) method 95
impulse response of the room 98
incremental sensor 249
Indicating instrument 73
inductance bridge circuit 147
Inductive coupling 133
Inductive displacement sensor 243
inductive high-pass (RL) filter 175
inductive proximity sensor 275
Infrared (IR) temperature sensor (Infrared
    thermometer) 219
Instrument with a signal output 73
Integral property 77
integrators 6
Intelligent Sensors 19

Laplace Transforms 76
Laser displacement sensor 244
Laser Doppler Anemometer 239
laser radars 12
laser triangulation technique 244
LED digital display 188
Level sensor 263
Light-beam oscillograph 109
Linear scale 248
linear time invariant system (LTI system) 77
linear time-invariant system 76
Linearity 55, 82
Loading effect 127
Low-pass filter 169, 173
Luminescence sensor 280
LVDT (Linear Variable Displacement
    Transformer) displacement sensor 240

Magnetrol 23
Manually operated type instrument 75
manufacturing process monitoring 7, 10
Maxwell bridge circuit 149
Measurement Technology 1
measuring lag 88
mechanical tachometer 251
MEMS (Micro-electro-mechanical Systems)
    accelerometer 257
mercury-in-glass thermometer 4
modulation 153

motion sensor 13
multiple signal outputs 6
multipoint strip chart recorder 189

nanotechnology 18
National Instruments 25
natural vibration frequency 99
navigation/positioning sensor 9
Nichols plot 105
noise 67
noise-free measuring system 120
non-contact measurement system 12
Non-contacting type instrument 75
non-deterministic signal 29
non-electricity 27
Non-smart instrument 75
Novel printed/flexible (P/F) sensors 19
NTC (Negative Temperature Coefficient) 216
Null-type Instrument 71
Nyquist frequency 185
Nyquist plot 100, 105
Nyquist-Shannon sampling theorem 184

octave filter bank 180
odd signal 32
Okazaki 23
Omron automation 25
Orifice plate 231
over damped response 116
Overdamped case 115
Owen bridge circuit 149

paperless chart recorder 190
passive differential circuit 7
passive instruments 71
Peak time 118
Peak Value 48
Peak-to-Peak Value 48
pen strip chart recorder 189
Percent Overshoot 119
periodic inputs 88
periodic signal 29
phase modulation 153
phase response 91
photoelectric proximity sensor 273
photo-etching techniques 221
physical measurand 6
physical quantity 6
Piezoelectric accelerometer 254

Piezoelectric effect 222, 250
piezoelectric force sensor 262
Piezoelectric pressure sensor 222
Piezoelectric velocity sensor 250
piezo-resistive pressure sensor 223
piezo-resistive principle 224
PM sensor 285
PM10 285
PM2.5 285
potentiometer 246
potentiometric pressure sensor 227
power signal 33
Precision 81
Pressure 221
pressure sensor 9, 221
Programmable filter 181
proportional resolution filterbank 179, 180
Proximity laser displacement sensor 244
proximity sensor 8, 273
PTC (Positive Temperature Coefficient) 216

Q (Quality) factor 172
Quality Performance Inspection 12
quarter-bridge strain gauge circuit 140
Quartz crystal 222

R.M.S (Root- Mean-Square) Value 48
Radar 202
Radar level sensor 264
railway signal 167
random signal 29
Random signals 66
range or span 81
RC circuit 104
reaction torque sensor 167
rectangular pulse signals 50
Rectangular Window Function 61
regular periodic function 40
Relevance 193
remote process monitoring 11
Repeatability 87
Resistance level sensor 267
Resistance Temperature Detector (RTD) 213
resistance thermometer 213
Resistive displacement sensor 246
resistive humidity sensor 282
Resolution 84
resonance 116
response speed 88

reverse piezoelectric effect 222
ring demodulator 161
ring modulation 160
Ripple 172
Rise time 117
Rockwell automation 24
rotary encoder 248
Rotary paddle level switch 269
rotary torque sensor 167

sampling 183
sampling interval 183
sampling period 183
Scaling property 77
Schering bridge circuit 150
Schneider Electric 25
Second order measuring system 124
second order system 109
Seebeck effect 212
sensitivity 82
sensitivity drift 87
Settling time 118
shock waves 50
Siemens 24
sifting property of the delta function 63
Signal 27
signal conditioner 6
Signal conditioning 135
Signal noise filtering 181
Signal Synthesis 33
sine/cosine function 65
single pulse signals 50
Smart instrument 73
smoke sensor 13
Sonic echo-impulse response system 97
sound level meter 5
Sound level sensor 281
sound sensor 8
Spectral analysis 204
spectral interference method 245
spectrum of a signal 41
standard deviation 194
static characteristics 75
Steady-state error 120
stepped swept-sine method 93
STM32 19
Strain 137
Strain gauge accelerometer 255
Strain gauge force/torque sensor 259

strain gauge measurement 7
Strain gauge pressure senso 221
Strain gauges 136
Superposition principle 77
Swept-sine methods 93
switch bounce 50
Symmetry/Duality 56
synchronous demodulation techniques 156
system identification 76

Tachometer 251
TC 212
temperature sensor 8, 211
the critical damping of the response 116
The Exponential Fourier Series 44
The impulse response method 95
The impulse response of the first order
    system 108
the inverse Fourier Transform 51
the least squares method 82
the principle of ultrasonic waves 237
the spectrum analyzer 179
Thermal flowmeter 235
thermal noise 29
thermistor 216
Thermocouple 212
thermo-electric effect 212
thermoelectric enzyme sensor 17
third-octave filter bank 180
Threshold 85
time constant 108
time domain analysis 34
Time scaling 57
Time shift and Frequency shift 58
Tolerance 88
torque sensor 166
transfer function 90

transient inputs 88
transient signals 31
triangle wave 41
Trigonometric Fourier Series Representation 37
Turbine flowmeter 233

ull-wave phase-sensitive detection 159
ultrasonic displacement sensor 247
Ultrasonic level sensor 263
ultrasonic proximity sensor 274
unbalanced bridge circuit 136
unbalanced bridge circuits 144
Undamped case 115
under damped response 116
Underdamped system 115

Variable reluctance accelerometer 256
Variable reluctance pressure sensor 226
VEGA 25
velocity sensor 250
Venturi tube 231
Virtual instruments 20
VOCs sensor 286
voltage 27
Vortex flowmeter 234
vortex shedding 234

water sensor 13
Wein bridge circuit 148
Wheatstone bridge 135

Yokogawa 24

zero drift 87
zeroth order systems 103

δ sampling property/sifting property 62